DATE DUE

DE 2 0'00			

DEMCO 38-296

Levels of Selection in Evolution

MONOGRAPHS IN
BEHAVIOR AND ECOLOGY

Edited by John R. Krebs and
Tim Clutton-Brock

Levels of Selection in Evolution

LAURENT KELLER, EDITOR

Princeton University Press
Princeton, New Jersey

Copyright © 1999 by Princeton University Press
Published by Princeton University Press,
41 William Street, Princeton, New Jersey 08540
In the United Kingdom: Princeton University Press,
Chichester, West Sussex

Library of Congress Cataloging-in-Publication Data

Levels of selection in evolution / Laurent Keller, editor.
p.   cm. — (Monographs in behavior and ecology)
Includes bibliographical references (p. 253).
ISBN 0-691-00703-9 (cl : alk. paper).
— ISBN 0-691-00704-7 (pb : alk. paper)
1. Natural selection. 2. Evolution
(Biology) I. Keller, Laurent. II. Series.
QH375.L48   1999
576.8′2—dc21   99-22314
                                                             CIP

This book has been composed in Times Roman

The paper used in this publication meets the minimum
requirements of ANSI/NISO Z39.48-1992 (R1997)
(*Permanence of Paper*)

http://pup.princeton.edu

Printed in the United States of America

10 9 8 7 6 5 4 3 2 1

10 9 8 7 6 5 4 3 2 1
(Pbk.)

I am pretending to a unity that, deep inside myself, I know does not exist. I am fundamentally mixed, male with female, parent with offspring, warring segment of chromosomes that interlocked in strife millions of years before the River Severn ever saw the Celts and Saxons of Housman's poem.

—W. D. Hamilton,
Narrow Roads of Gene Land

Contents

Acknowledgments

I thank all the contributors to the book for having made great efforts to structure their chapters according to the underlying theme of the book and for having complied with my numerous requests. I would also like to thank them, as well as Andrew Bourke, John Brookfield, Austin Burt, Tim Clutton-Brock, Robin Dunbar, Dieter Ebert, Marc Feldman, Scott Forbes, Steven Frank, David Haig, Laurence Hurst, Rufus Johnstone, Finn Kjelberg, Walt Koenig, Doug Mock, Linda Partridge, Anne Pusey, John Thompson, Jack Werren, George Williams, and D. S. Wilson for reviewing one or more chapters. The book also benefited from the comments of several post-docs and graduate students in our Institute, including William Brown, Else Fjerdingstad, Michael Krieger, Cathy Liautard, Drude Molbo, and Max Reuter.

Sam Elworthy has been supportive throughout this project and very helpful in finding two excellent reviewers. Teresa Carson has copyedited the manuscript with great care and on occasion made helpful suggestions on matters of substance. Finally, I thank Cathy Liautard and Drude Molbo for their help in modifying several figures and for compiling all the references. I also acknowledge the continuous support of the Swiss National Science Foundation as well as a grant of "Commission de Publications" from the University of Lausanne.

Contributors

H. Charles J. Godfray
Department of Biology
Imperial College at Silwood Park
Ascot, Berks, U.K.

Edward Allen Herre
Smithsonian Tropical Research Institute
Balboa, Panama

Laurent Keller
Institute of Ecology
University of Lausanne
Lausanne, Switzerland

Dawn M. Kitchen
Department of Ecology, Evolution & Behavior
University of Minnesota
St. Paul, Minnesota

Egbert Giles Leigh, Jr.
Smithsonian Tropical Research Institute
Balboa, Panama

Catherine M. Lessells
NIOO
Heteren, Netherlands

John Maynard Smith
School of Biological Sciences
University of Sussex
Flamer, Brighton, U.K.

Richard E. Michod
Department of Ecology and Evolutionary Biology
University of Arizona
Tucson, Arizona

Leonard Nunney
Department of Biology
University of California
Riverside, California

Craig Packer
Department of Ecology, Evolution & Behavior
University of Minnesota
St. Paul, Minnesota

Andrew Pomiankowski
Galton Laboratory
Department of Biology
University College London
London, U.K.

H. Kern Reeve
Section of Neurobiology and Behavior
Cornell University
Ithaca, New York

Eörs Szathmáry
Collegium Budapest
Budapest, Hungary

Levels of Selection in Evolution

1

Levels of Selection: Burying the Units-of-Selection Debate and Unearthing the Crucial New Issues

H. Kern Reeve and Laurent Keller

The purpose of this volume is to sample current theoretical and empirical research on (1) how natural selection among lower-level biological units (e.g., organisms) creates higher-level units (e.g., societies), and (2) given that multiple levels exist, how natural selection at one biological level affects selection at lower or higher levels. These two problems together constitute what Leigh (chap. 2) calls the "fundamental problem of ethology." Indeed, as Leigh further suggests, they could be viewed jointly as the "fundamental problem of biology," when genes and organisms are also included as adjacent levels in the biological hierarchy. This generalization has the desirable property of immediately removing the long-standing conceptual chasm between organismal and molecular biologists.

These two problems are just beginning to be addressed, but their study promises in the decades ahead to generate crucial insights (perhaps *the* crucial insight) into biological evolution both on our planet and on imagined planets. To appreciate just how intriguingly intricate these problems are, we use an analogy from particle physics to generate a heuristically useful picture of the myriad interlocking and concatenated selective forces acting simultaneously at different levels of biological organization (fig. 1.1). This picture can be thought of as a visual guide to the kinds of multilevel selection issues addressed in the chapters in this volume.

First, however, we wish to make yet one more attempt to bury the issue that usually usurps discussions of the levels of selection at the expense of the truly interesting issues raised by these two problems; that is, the question of what unit is the "true" fundamental unit of selection. This issue emerges in cyclic debates about (a) whether genes or individuals are best seen as the true units of selection, and (b) whether groups of individuals *can* be units of selection. In our opinion, these questions have been satisfactorily answered repeatedly, only to reappear subsequently with naive ferocity in new biological subdisciplines (e.g., the group-selection controversy is currently generating copious amounts of smoke within the human sciences; see, e.g., Wilson and

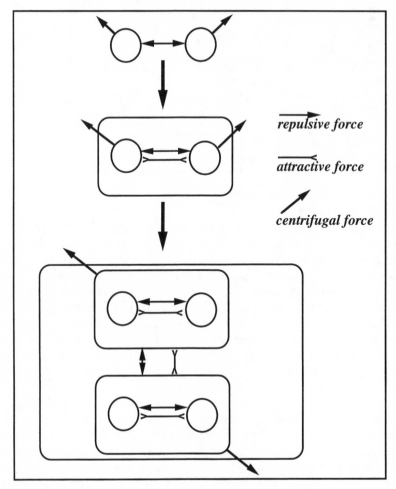

FIG. 1.1. The formation of higher-level vehicles from lower-level vehicles. A higher-level vehicle forms when the attractive inclusive fitness force (see box 1.1) for lower-level vehicles overcomes the repulsive and centrifugal inclusive fitness forces.

Sober 1994 and responses; Sober and Wilson 1998). The particularly frustrating aspect of these constantly renewed debates is that, even though they seemed to be sparked by rival theories about how evolution works, in fact, they often involve only rival metaphors for the very same evolutionary logic and are thus empirically empty.

Thus, we first pause to heap one more shovelful of dirt on the units-of-selection debates (a) and (b) above by very briefly reviewing what we believe to be their well-established, correct (if not universally known) resolutions.

Burying the Debate over Whether Genes or Individuals Are the Units of Selection

Organisms themselves are not replicated in the process of reproduction. They die, and only their genes are passed on. This led Dawkins (1976, p. 12) to propose that "the fundamental unit of selection, and therefore of self-interest, is not the species, nor the group, nor even strictly, the individual. It is the gene, the unit of heredity." Dawkins referred to this unit of self-interest as the *replicator*, or enduring unit of replication. Dawkins's view, which builds on previous ideas by Hamilton (1964a) and Williams (1966a), has been criticized as too reductionist by those who argue that genes are not directly visible to natural selection (e.g., Gould 1984; Sober and Lewontin 1984). That is, selection simply cannot pick among genes directly, but must select among packages created by and containing these and other genes (e.g., organisms). Dawkins (1982), however, recognized organisms and perhaps higher-level or laterally extended units as being *vehicles*, that is, the units directly confronting selection. We expect that, as the result of natural selection, vehicles will possess properties that maximize the replication success of the set of genes that cocreated them. This picture is slightly modified by the possibility of intragenomic selection (e.g., meiotic drive) favoring certain genes. In this case, selection may seem to choose among genes directly. However, a useful distinction still can be made between the replicator as a piece of genetic information and the "vehicle" as the physical stretch of DNA containing this genetic information. Thus, even in this case, it is only a vehicle (albeit a *replicator-level* vehicle) that directly confronts selection.

Thus, one internally consistent logical picture is that the unit of replication is the gene (or, more precisely, the information contained in a gene), and the organism is one kind of vehicle for such genes, a vehicle being the entity on which selection acts directly. The debate is resolved: Dawkins (1976) emphasized that genes (i.e., bits of genetic information) are the enduring units of replication, whereas Sober (1984) and Sober and Lewontin (1984) emphasized that individuals and possibly higher-level units, and not genes (as bits of genetic information), are vehicles. Case closed.

Burying the Old Group-Selection Debate

It is still embarassingly common to read inaccurate statements in newspapers and even in professional biological literature that frogs have to produce many eggs to ensure the survival of the species because tadpoles suffer extremely high rates of predation, or that wolves have evolved ritualized displays to establish dominance hierarchies because physical combats would be too disadvantageous for the species. These naive statements betray a widespread and persistent misunderstanding of the level at which natural selection most commonly operates.

Wynne-Edwards (1962, 1993) has been the leading modern proponent of the idea that animals behave for the good of the group. He suggested that a population would become extinct if it overexploited its food resources; such between-population selection has fixed population-level adaptations to prevent extinction (such as animal displays to signal population density and thereby limit the risk of resource overexploitation). The most important criticism of this idea was formulated by Williams (1966a). Although selection at the population level is theoretically possible, in practice, such selection will be weak because of the high speed of within-population (between-individual) selection relative to that of between-population selection. Moreover, virtually all examples of group selection given by Wynne-Edwards (1962) have been shown to be better understood with the individual-selection paradigm (e.g., Alcock 1998; Kitchen and Packer, chap. 9). In Dawkins's terms, overwhelmingly strong theoretical arguments and empirical evidence tell us that individuals, far more commonly than populations, are the vehicles.

More recently, formal models of within-population group selection—that is, selection that occurs when a single breeding population is temporarily broken up into subgroups within which both cooperative and competitive actions can occur—have been developed under the rubric of "new group selection," "intrademic group selection," or "trait-group selection." These models simply partition ordinary individual fitness into within- and between-group components—often using the clever covariance approach of Price (1972)—and allow detailed predictions of the circumstances favoring the evolution of traits affecting both within- and between-group fitness in various ways (e.g., Wilson 1975; Wilson and Sober 1989). These models are mathematically equivalent to individual-selection (i.e., inclusive fitness) models, however, and therefore do not point to a fundamentally different kind of evolution (e.g., Dugatkin and Reeve 1994; Bourke and Franks 1995). Thus, acknowledging the utility of these models should not be taken as a tiptoed retreat to Wynne-Edwardsian interpopulation selection, as is often mistakenly feared because of the shared label of "group selection." Acceptance of these models also does *not* commit one to a particular view about the relative balance of cooperation and conflict in nature, because either can have any degree of strength in these models (Dugatkin and Reeve 1994). Furthermore, these models fit comfortably into Dawkins's (1982) conceptual scheme because the "groups" in these models (e.g., animal societies) can be viewed simply as vehicles above the level of the individual (Seeley 1997). A distinct virtue of intrademic group-selection models is that they provide a simple, standardized means of unveiling the structure of selection working simultaneously at different hierarchical levels (e.g., Dugatkin and Reeve 1994; Reeve and Keller 1997; Keller and Reeve, chap. 8).

Unearthing the New Issues

Most class lectures on levels of selection begin and end with a discussion of the two (now stale) debates above. However, the current theoretical excitement in theoretical and empirical research in multilevel selection centers on the two problems set forth at the beginning of this introduction, namely, (1) how natural selection among lower-level biological vehicles creates higher-level vehicles, and (2) given that multiple levels of vehicles exist, how natural selection at one level affects selection at lower or higher levels. The richness of these two questions can be conveyed with the help of figure 1.1, which pictures interactions within and between lower- and higher-level vehicles (e.g., for vehicles ranging from single-celled organisms, to multicellular individuals, to social groups of individuals).

An analogy from particle physics is useful here. Higher-level vehicles can be seen as composites of lower-level vehicles, each of which experiences both evolutionary repulsive ($\leftrightarrow$) and attractive ($>\!\!-\!\!<$) bipolar forces with other units at the same level (fig. 1.1). The separated unipolar forces can be viewed as having magnitudes equal to the absolute inclusive fitnesses for peaceful cooperation with a same-level partner unit ($-\!\!<$) or for competitive suppression (e.g., killing) of the same partner unit ($\longrightarrow$). The outgoing arrows ($\nearrow$) refer to the absolute inclusive fitness of a vehicle that leaves the group; thus, this represents a second evolutionary force, which we call "centrifugal force," that tends to break apart the group. In this scheme, a cooperative group of lower-level units will be stable only if, for every unit, the attractive force exceeds the maximum of the repulsive and centrifugal forces also acting on that unit. (See box 1.1 for elaboration of the exact nature of these forces.)

Figure 1.1 makes explicit several key features of the evolution of higher-level vehicles from lower-level ones. First, a higher-level vehicle is created from a lower-level vehicle whenever an attractive force arises that exceeds both the maximal repulsive and centrifugal forces. Interestingly, repulsive forces among unbound lower-level units can *create* binding forces between other such units, for example, as when ancestral multicellularity increased the fused cells' ability to outcompete single-celled organisms for resources, or when social grouping increased the ability of individuals to defend resources from intruding robbers.

Second, because the magnitude of each of the forces depends on inclusive fitness, which in turn depends on both genetic relatedness and multiple ecologically determined costs and benefits of cooperation and noncooperation, it follows that understanding higher-level vehicle formation requires knowing both genetic and ecological factors that generate attractive, repulsive, and centrifugal forces. Ecology will be crucially important in determining the magnitude of the centrifugal force, by strongly affecting the expected repro-

BOX 1.1. THE ABSOLUTE INCLUSIVE FITNESS "FORCE"

By absolute inclusive fitness, we mean a focal vehicle's direct reproductive output plus the sum, over all related vehicles, of the product of its relatedness to the vehicle times the reproductive output of that vehicle. Note that we use *absolute* outputs, rather than *changes* in outputs caused by the focal vehicle (the latter is used in most verbal formulations of inclusive fitness). The outputs of vehicles unaffected by the focal vehicle's actions will appropriately vanish when the absolute inclusive fitnesses associated with the two actions by the focal vehicle are compared by subtraction, because such outputs will have exactly the same value in the compared inclusive fitnesses. For example, suppose a phenotype A causes the focal animal to have x offspring and a relative (of relatedness r) to have y offspring. The corresponding offspring numbers for phenotype B are z and w. The magnitudes of the corresponding absolute inclusive fitnesses (i.e., of the "forces") are

$$x + ry$$

for A and

$$z + rw$$

for B. The "net force" is then obtained by subtraction and is readily seen to equal

$$(x - z) + r(y - w),$$

which is the same as Hamilton's rule when set greater than zero (Grafen 1982, 1984, 1985). If action B did not change the reproductive output of the focal individual's relative, then $y = w$ and the term $r(y - w)$ would simply vanish. If the net force is greater than zero, phenotype A is favored.

The use of absolute offspring number in the above inclusive fitness calculations may sound wrong to some because of well-known theoretical admonitions against (1) including personal offspring added because of help received from others, and (2) giving inclusive fitness credit for the reproductive outputs of relatives that are unaffected by the phenotype, both of which can cause gross overestimation of the kin-selective value of a cooperative strategy (Grafen 1982, 1984). However, this is an error only when a strategy's absolute inclusive fitness is compared with zero, not when the absolute inclusive fitness for one strategy is compared (by subtraction) with the absolute inclusive fitness of another strategy. The latter procedure automatically yields the appropriate description of net selective force by generating Hamilton's rule. It should be mentioned, however, that the "absolute inclusive fitness force" approach is precisely true only if there are additive costs and benefits and weak selection

BOX 1.1. CONT.

(Grafen 1984, 1985). Conditionality of phenotypic expression can make these assumptions more likely to hold (Parker 1989).

Our characterization of the magnitudes of the attractive, repulsive, and centrifugal forces properly ties vehicle behavior to the interests of the ultimate replicators, the genes that create the vehicles. Why? This scheme correctly specifies when kin selection favors cooperation as summarized in Hamilton's rule. That is, the sign of net inclusive fitness force determines whether kin selection favors cooperation over killing or ejecting the partner and also over leaving the group to reproduce independently. (The physical analogy breaks down a bit here, because in the physical case, the net attractive force would simply be the vector sum of all three forces, not the difference between the attractive force and the maximum of the repulsive and centrifugal forces. Despite this, the physical picture is useful.)

ductive output of a dispersing, solitary vehicle. The most complete theories of vehicle formation will thus be those that specify both the ecological and the genetic contexts for vehicle creation.

Third, even if the creation of higher-level vehicles requires that attractive forces exceed repulsive and centrifugal forces, this does not imply that the latter two forces will disappear once the higher-level vehicles are formed. They may continue to operate and shape the features of the higher-level vehicle (just as the conformation of a stable molecule will depend on the internal electrical repulsive forces). Indeed, repulsive forces may sometimes strengthen sufficiently to cause subsequent vehicle breakdown. For example, the attractive forces will often be sufficiently weak and variable that a composite vehicle lasts only a short time, just as an unstable, heavy particle created in an particle accelerator may leave only a short track on a photographic plate before disintegrating into component particles. Analogously, in many if not most animal species, the only cooperative groups are fleeting associations of mates during courtship, copulation, and mate defense; that is, the inclusive fitness for cooperation (attractive force) exceeds that for noncooperation (repulsive and centrifugal forces) only until mating is completed. A complete theory of social evolution will tell us not only the contexts in which higher-level vehicles form, but also the contexts in which they break down.

Finally, this model, represented in figure 1.1, predicts that larger cooperative groups are inherently less likely to be stable. Suppose there are n lower-level vehicles within the cooperative group (i.e., higher-level vehicle). If the group is to be completely stable, the attractive forces must exceed the repu-

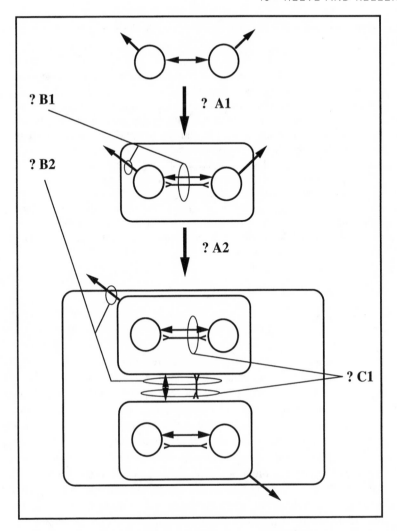

Fɪɢ. 1.2. Major problems in understanding the evolution of higher-level vehicles. How do attractive forces come to exceed repulsive and centrifugal forces (A1 and A2; see text)? How does the balance of these forces affect properties of the higher-level vehicle (B1 and B2; see text)? How does the interaction among forces within lower-level vehicles affect the properties of the higher-level vehicle (C1; see text)?

sive forces for all $n(n-1) = n^2 - n$ polar interactions, and, in addition, the attractive forces must exceed the centrifugal forces for all n cases, for a total of $n^2 - n + n = n^2$ requirements. Thus, the number of Hamilton's rule requirements for group stability increases as the square of the number of group members! This immediately suggests that larger groups will be progressively less stable, unless high genetic relatedness, positive correlation among subunits in the values of their inclusive fitness parameters, or some kind of between-subunit interaction somehow forces the multiple Hamilton's rule requirements to be satisfied en masse. Furthermore, as lower-level vehicles are nested to form higher-level vehicles, say from lower-level vehicles consisting of n_l subunits each to a higher-level vehicle of n_h lower-level vehicles, the total number of Hamilton's rule requirements rapidly becomes compounded to $(n_l n_h)^2$. This immediately suggests that vehicles created by the nesting of successively higher-level vehicles will become progressively less stable, again unless some condition or process causes these requirements to be satisfied at once.

Now we can represent the questions that together form the "fundamental problem of biology," and thus the conceptual structure of this book, in terms of the picture in figure 1.1. Questions A–C below refer to processes A–C in figure 1.2.

A1. What attractive evolutionary forces bind low-level vehicles (i.e., vehicles nearly at the same level as the replicators themselves), like physical stretches of DNA (replicators being the genetic *information* encoded in such stretches), chromosomes, and cells, into intermediate-level vehicles, like multicellular organisms? Under what conditions do these attractive forces exceed the repulsive and centrifugal forces and under what conditions do they not?

This topic is addressed in chapters 3 and 4. A central question in the study of the origin of life is how cooperating groups of small replicator-level vehicles could have arisen and how they could have protected themselves against invasion by molecular parasites. Szathmáry (chap. 3) argues that synergism (i.e., division of labor and complementation of functions) provided the most important attractive force leading the first replicator-level vehicles to associate. Cooperation was also facilitated by genetic compartmentalization that resulted from limited dispersal and bonding of different replicator-level vehicles (which were therefore obliged to "sit in the same boat"; Szathmáry, chap. 3). Compartmentalization represented an important step in the overriding of repulsive and centrifugal forces and also probably led to their subsequent weakening. Finally, the benefits of division of labor, together with the many advantages of larger size, were probably the two important attractive forces that favored the transition from unicellular to multicellular life (Michod, chap. 4).

A2. Similarly (as we move up the hierarchy of nested vehicles), what attractive evolutionary forces bind intermediate-level vehicles, such as organisms, into higher-level vehicles, such as social groups of individuals? Under what conditions do these attractive forces exceed the repulsive and centrifugal forces?

This topic is addressed in chapters 5, 6, and 8–11. In sexual species, the necessity of finding a mate provides an inescapable attractive force. In most species, however, this attractive force is transient because males and females typically have low genetic interest in each other's future (Lessells, chap. 5). Another attractive force may keep parents together: their common genetic interest in rearing their offspring. The magnitude of this attractive force directly depends on the degree to which greater parental investment increases offspring reproductive success (Godfray, chap. 6). This positive force is opposed by the centrifugal force created by mating opportunities outside of the pair bond. Thus, the dynamics of the attractive and centrifugal forces set the stage for a variety of conflicts (new, subtle repulsive forces) between mates over their relative investment in parental care (Lessells, chap. 5; Godfray, chap. 6).

Attractive forces may also lead individuals other than mates to cooperate when this increases either their survival or number of offspring produced or the survival and fecundity of relatives. Higher relatedness increases the magnitude of the attractive forces and decreases the magnitude of the repulsive forces (because increased relatedness between interacting individuals enhances the inclusive fitness payoffs for cooperation and reduces the inclusive fitness payoffs for group-destructive selfishness). Increased relatedness thus increases the scope both for reproductive altruism (whereby individuals forgo direct reproduction to help others) and possibly for group stability, although models of optimal reproductive skew (Keller and Reeve, chap. 8) predict that dominant members of animal societies may actually increase the attractive force for potential subordinate helpers when the latter are less related, erasing any net effect of relatedness on group stability (Reeve and Ratnieks 1993; Reeve 1998a). Not surprisingly, unreciprocated altruism occurs nearly exclusively in groups formed by closely related individuals (Keller and Reeve, chap. 8; Kitchen and Packer, chap. 9; Maynard Smith, chap. 10). Groups of unrelated individuals are generally stable only when group living provides direct reproductive benefits to all group members, when it requires no reproductive altruism, and opportunities for cheating are limited (i.e., repulsive forces are weakened) (Kitchen and Packer, chap. 9). The other important factor shaping social life is the ability for individuals to disperse successully and reproduce outside the group. Groups will be inherently more stable when such opportunities are limited (weak centrifugal forces).

Interspecific mutualism provides another interesting case of attractive forces being stronger than repulsive and centrifugal forces, the two forces that generally predominate in interspecific interactions. Interestingly, the same positive force (the benefits of division of labor) that facilitated the evolution of early life is also probably important in shaping the nature of interspecific cooperation (Herre, chap. 11). Moreover, Herre provides examples showing that stable interspecific cooperation (or reduced virulence) is facilitated by the long-term association of interspecific individuals and parallel vertical transmission of the symbionts (from parents to offspring). The consequence of symbionts being only vertically transmitted is similar to the effect of compartmentalization of replicator-level vehicles during the early evolution of life because, in both cases, the interests of replicators are aligned, increasing the magnitude of the net attractive force.

B1. How do attractive, repulsive, and centrifugal forces among lower-level vehicles interact to shape the properties of intermediate-level vehicles like individuals? Can different repulsive forces sometimes nullify each other within intermediate-level vehicles (as when there is some mechanism of policing against intragenomic selfishness) and thus leave an imprint on the characteristics of the intermediate-level vehicle (such as increased reproductive efficiency resulting from greater internal cooperation)? Are there attractive forces (perhaps arising only *after* the creation of the intermediate-level vehicle) that would overcome all or some of the original repulsive forces and thus leave an imprint on the characteristics of the intermediate-level vehicle?

These topics are addressed in chapters 4, 7, and 12. The two main forces shaping the integrity of the organism are the attractive and repulsive forces because lower-level vehicles (genes and cells) have little or no opportunity to leave the organism and embark on independent and solitary lives (except in some primitive multicellular organisms). The repulsive forces stem from the benefits that lower-level vehicles (gene-level vehicles and cells) may gain by increasing their reproductive rates at the expense of the other vehicles forming the organism. Thus, genes may increase their reproduction by subverting meiosis in diploid organisms. Similarly, cells may reap a short-term reproductive benefit at the long-term expense of the organism through uncontrolled cell proliferation (cancer). Michod (chap. 4) and Pomiankowksi (chap. 7) provide examples of how repulsive forces can nullify each other to decrease conflicts between genes and enforce fair meiosis. For example, it is the mutual interest of genes in multicellular organisms in decreasing repulsive forces that probably led to the sequestration of a cell lineage set early in development for the production of gametes (Michod, chap. 4). Mutual competition (repulsion) between cell lineages might result in no net advantage for either; moreover, such competition might greatly limit the efficiency of the vehicle formed by their cooperation. The separation of the germ line reduced the opportunity for conflict (greatly reducing repulsive forces) and thus was a first step toward the evolution of individuality (i.e., a higher-level vehicle with stronger attractive than repulsive forces). Similarly, because most genes in the genome suffer from the detrimental effects of meiotic-drive genes (unless linked with them), they are selectively favored to suppress the selfish actions of such genes (Pomiankowski, chap. 7). Nunney (chap. 12) suggests that between-lineage species selection may cause the long-run predominance of genetic architectures that decrease the risk of cancer (detrimental to the organism) and also that decrease the probability of a shift from from sexual to asexual reproduction (the latter being detrimental to the species). (Note that this is *not* Wynne-Edwardsian group selection, because Nunney is only speaking of differential extinction among lineages that have different biological characteristics, the latter characteristics all having been fixed by within-population selection.) Under this intriguing view, lineages with relatively high repulsive and low attractive forces (i.e., those in which lower-level vehicles are less likely to form higher-level vehicles) are more likely to become extinct, leading to a long-term lineage selection for clades that exhibit well-elaborated, high-level vehicles.

B2. Similarly, how do attractive, repulsive, and centrifugal forces interact to shape the properties of high-level vehicles like animal societies? Can repulsive forces sometimes nullify each other within high-level vehicles (as in policing against selfishness) and thus leave an imprint on the characteristics of the high-level vehicle (such as increased efficiency resulting from greater internal cooperation)? Are there attractive forces (perhaps originating only *after* the initial creation of the high-level vehicle) that would nullify all or some of the original repulsive forces and thus leave an imprint on the characteristics of the high-level vehicle?

These topics are addressed in chapters 8–11. For example, Keller and Reeve (chap. 8) discuss how policing and bribing can promote intragroup cooperation within animal societies by in effect weakening repulsive forces or strengthening attractive forces. Similarly, Maynard Smith (chap. 10) investigates the conditions favoring the emergence and enforcement of social contract strategies to punish selfish behaviors in human societies. Finally, the evolution of co-adapted traits in obligately mutualistic species (e.g., figs and their associates; Herre, chap. 11) provide yet another example of attractive forces that arose or strengthened after the initial creation of a higher-level vehicle from the mutualistic pair of organisms, that is, following the evolution of complete reproductive interdependence.

C1. Perhaps the most unexplored question concerns how interactions between lower-level vehicles might affect the interactions between intermediate-level vehicles and thus affect the properties of the highest-level vehicle. For example, Keller and Reeve (chap. 8) describe one of Reeve's (1998b) hypotheses for the absence of nepotism within insect societies. Intragenomic selection on parentally imprinted alleles involved in kin recognition (lowest-level vehicles) might favor sabotaging of the potential nepotism-dispensing machinery of individuals (intermediate-level vehicles), leading to the lack of nepotism and thus increased cooperation within hymenopteran societies (highest-level vehicles).

Acknowledgments

We thank E. Leigh for useful comments on the manuscript. We were supported by grants from the Swiss and U.S. NSF.

2

Levels of Selection, Potential Conflicts, and Their Resolution: The Role of the "Common Good"

Egbert Giles Leigh, Jr.

Adaptation is shaped by the competitive process of natural selection (Darwin 1859). Genes are the units whose "self-interest" drives natural selection. In other words, nothing nonrandom happens in the evolution of a species unless it "serves the self-interest"—causes the differential reproduction—of some gene (Dawkins 1976; Bourke and Franks 1995). Yet no one of these genes, ultimate unit of self-interest though it might be, can do a thing outside the context provided by the rest of its genome and the organism for which that genome is appropriate. Alone, a gene is as useless as a fragment of a computer program without the rest of the program, a computer suited to run the program, and an operator capable of using the program and the machine. In short, the units of competitive self-interest that make up a genome are utterly interdependent. How did the competitive process of natural selection shape so intricate a mutualism?

Ecological communities are structured to a large extent by competition: competition among individuals for food or space, and competition between consumers and their potential prey over who uses the resources in these prey's bodies (Hutchinson 1959; Paine 1966). Competition among plants for light, water, and nutrients, and between consumers and their intended prey, is particularly intense in tropical forest (Robinson 1985; Richards 1996). Yet tropical forest is not only a climax of competition but an apex of mutualism. Plants depend on fungi for the uptake of nutrients (Allen 1991) and on animals for pollination of their flowers, dispersal of their seeds, sometimes for burial of their seeds out of the reach of insect pests (Corner 1964; Smythe 1989; Forget 1991). These mutualisms make possible the diversity and luxuriance of tropical forest (Corner 1964; Regal 1977; Crepet 1984). They constitute an extraordinary web of interdependence. A tree species that needs agoutis to bury its seeds needs other tree species to keep the agoutis fed when it itself is not fruiting (Forget 1994). The durian whose flowers are pollinated by bats needs mangroves to keep these bats in nectar when the durian's forest has few plants in flower (Lee 1980). Although ecological communities are theaters of competition, their species depend on each other in many ways. No species can survive outside an ecosystem that provides it food, shelter, even the air that it breathes. How has competition among species brought forth the interdependence that characterizes ecological communities?

Such questions are avatars of a fundamental problem in ethology (Moynihan 1998). All vertebrates, and many other animals, depend on some conspecific for help of some sort, at least in conceiving offspring. Yet mates are potential competitors (Lessells, chap. 5). Who can forget Fabre's (1989, vol. 1, pp. 1105–1106) account of the headless trunk of a male mantis, still continuing to impregnate the female that has already eaten its head and is now chowing down on its thorax? In view of such possibilities as this, what keeps competition from destroying the common good that could be created by cooperating?

The problem of which social mechanisms maintain cooperation among potential competitors (Moynihan 1998), and how this cooperation evolved to begin with (Hamilton 1964a), is most acute for animals that live in groups (Keller and Reeve, chap. 8; Kitchen and Packer, chap. 9). Fellow members of a group depend on each other for the advantages they derive from group life: increased safety from predators or competitors, benefits of mutual assistance such as grooming for ectoparasites, and the like. Yet fellow group members are also each other's closest competitors for food, mates, and other resources. What keeps competition among a group's members from overwhelming their common interest in their group's effectiveness and annihilating the common good of their cooperation? Ethology is now a marginal subject: The more fashionable of its practitioners have hastened to label themselves sociobiologists or behavioral ecologists. Yet this "fundamental problem of ethology" is the unifying theme of this book. Moreover, we wrote this book because we think that the recurrence of this problem at various levels of biological organization is one of the grand unifying themes of biology and anthropology.

Avatars of the Ethologist's Problem

This fundamental problem of ethology is parallelled at other levels of biological and social organization, not least in human societies (Maynard Smith, chap. 10). Perhaps it is no accident that a clear formulation and discussion of this type of problem is already given in Aristotle's *Politics* (Barnes 1984, pp. 1986–2129). Here, Aristotle is concerned with how best to achieve harmony between the good of a city-state and the enlightened self-interest of its inhabitants. Aristotle observed, "In all arts and sciences the end is a good, and the greatest good and in the highest degree a good in the most authoritative [science] of all—this is the political science of which the good is justice, in other words, the common interest" (*Politics* 1282b, 14–16; p. 2035 in Barnes 1984). Aristotle considered that a city-state's organization (constitution) was more likely to persist if it clearly served the common good of its inhabitants; otherwise, it would be more liable to overthrow by conspiracy or revolution. Indeed, the common good turns out to play a crucial role in all avatars of our problem, for a mutualism will evolve only if it serves the common interest of all participants (Leigh 1991).

An obvious, yet mysterious, avatar of the ethologist's problem concerns species in ecosystems. A crucial feature of ecosystems is the variety of interdependence among species whose members are all competing for resources needed to survive and reproduce. Ecosystems can be viewed as functional entities, with producers, transformers, decomposers, etc. We speak of ecosystems being injured by human disturbance, as if ecosystems are organized to fulfill functions. Aristotle (*Physics* 199b4, p. 340 in Barnes 1984) remarked that in systems organized to function, the abnormal and the disrupted are usually less functional. Indeed, only if a system is organized to fulfill a function can we speak of it as being impaired by change (Fisher 1958, pp. 41–44). Such talk leaves much unanswered. What is the "function" of an ecosystem? Ecosystems are not units of selection, organized to reproduce themselves. It makes more sense to view ecosystems, like human societies, as commonwealths in whose integrity member species share a common stake. The mystery lies in the nature of this common interest, and in whether (and if so, how) this common interest affects the evolution of the ecosystem's species (cf. D. S. Wilson 1980).

Unlike nation-states, ecosystems, and many animal societies, the harmony of cellular organization and developmental process seems so absolute that it was long taken for granted. Yet, just as dysfunctions induced by gene mutations provide essential clues to the mechanisms of gene action and developmental process, so other dysfunctions—segregation-distorters, cytoplasmic sex-ratio mutants, cancer, and the like—reveal that organismic harmonies we once took for granted originated from cooperation among relatively independent entities (Hurst et al. 1996; Michod, chap. 4; Pomiankowski, chap. 7). Indeed, each of these conflicts reveals a past evolutionary breakthrough, a major evolutionary transition (Maynard Smith and Szathmáry 1995), whereby groups of relatively independent, cooperating entities cohered into more integral wholes that then became the central units of selection. The ethologist's problem thus has fundamental parallels in developmental and cellular biology. Indeed, it now appears that even molecular biologists would do well to ask where and how the fundamental problem of ethology relates to their work, as the following examples show.

Some alleles spread by biasing meiotic segregation-ratios in their own favor. Nevertheless, at most chromosomal loci, meiosis is one of the fairest

lotteries known to art or nature, in that a gamete of an individual heterozygous at a given locus has an equal chance of receiving either allele from
that locus. Where meiosis is fair, an allele can spread only if it benefits its
carriers (and therefore the genomes these individuals carry). Accordingly,
fair meiosis represents the common interest of the genome as a whole (Leigh
1971). What factors preserve the honesty of meiosis and render segregation
distortion so rare?

In species whose members inherit organelles only from the mother, some
cytoplasmic (organellar) mutants cause female-biased sex ratios (Hurst 1993a).
Where both egg and sperm contribute organelles to a zygote, conflicts between maternal organelles and their paternal counterparts may threaten the
zygote (Eberhard 1980). Such conflicts of interest between organelles
and their host cells—traces of a time when organelles were independent
organisms that had somehow entered host cells (Margulis 1993)—raise the
question, How is harmony between cells and their organelles normally maintained (Eberhard 1980)?

Cancers represent a conflict between an individual and certain of its cells.
Yet the human body is considered the very archetype of harmonious service
of the whole by its parts (1 Corinthians 12:20–25). How are conflicts between a multicellular organism and its constituent cells avoided or minimized (Buss 1987)?

These examples raise several general issues. What advantages did the
original entities derive by joining in groups? How did natural selection enforce the common interest of a group's members in their group's welfare?
What circumstances would lend this selection such power that the identity of
the individuals involved is almost lost in that of their group?

Community of Interest: Its Origin and Preservation

ADVANTAGES OF GROUP LIFE AND SYMBIOSIS

Joining others offers two kinds of advantage. The more familiar, which explains the origin of most animal societies, is safety in numbers: more eyes to
share the watch for predators, more teeth and claws to help defend resources
against competitors (cf. Kitchen and Packer, chap. 9). The other advantage is
the complementation of different functions or, if you will, a mutually beneficial division of labor, such as corals and their zooxanthellae gain from symbiosis, and plants and their mycorrhizae, pollinators, and seed-dispersers
gain from their partnerships (cf. Douglas 1994). The genes of a genome
share a common interest in each other's presence because each gene programs a process that benefits the carrier on whose reproductive success all
depend (Szathmáry, chap. 3). The community of interest between a cell and

its organelles, like that among the cells of a metazoan, is founded upon the advantageous way the functions of the different parts of the organism complement each other.

PROTECTING THE COMMON INTEREST

A mutualism, however, is simply a reciprocal exploitation (Herre, chap. 11). One or more partners to a mutualism might benefit, at least in the short term, by parasitizing the others, so that the relationship no longer serves the common good of the parties concerned. What factors can preserve the common good against such threats? I will outline several of the relevant processes in the context of the major transitions to which they presumably gave rise.

SELF-SUFFICIENT COMMON INTEREST

Mixed bird flocks—a characteristic feature of lowland tropical forest (Moynihan 1962)—remind us that common interest sometimes suffices of itself to maintain a mutualism (Maynard Smith 1991a; Leigh and Rowell 1995). In a mixed flock, a pair of each of several nuclear species, sometimes accompanied by their young, forage together in a jointly defended territory (Moynihan 1979; Gradwohl and Greenberg 1980). Such a group often progresses over a regular beat. As it does so, a bird on a smaller territory may join when the flock enters it. Some birds with larger territories may move from flock to flock (Gradwohl and Greenberg 1980; Munn 1985). The advantages of flocking are more eyes to watch for predators and the prospect of coordinated defense (mobbing) against smaller predators (Moynihan 1962, 1979; Willis 1972, pp. 135–147). Social relations within mixed flocks are sometimes complex (Moynihan 1962), and there appears to have been some coevolution among member species of some flocks to facilitate flocking behavior (Moynihan 1968). Nonetheless, mixed flocks are assemblages that are too loosely knit to function as units of selection. No mutual enforcement beyond the natural tendency of each bird to exclude conspecific strangers preserves this mutualism. In sum, a mixed-species bird flock simply expresses the mutual benefit of safety in numbers.

By permitting genes to be carried from one individual to another's offspring, sexual reproduction creates opportunities for a variety of conflicts: between mates (Lessells, chap. 5), between a mother and its fetus (Haig 1993b) or older young (Godfray, chap. 6), and among the genes of a genome (see below). Nonetheless, sexual reproduction usually reflects the unenforced common interest of two individuals in producing more varied offspring than either could unaided (Williams 1975).

Community of interest can maintain more elaborate mutualisms. In a monkey group in which coordinated defense against predators is essential, a monkey that causes the death of a group member by failing to play its part, endangers its own life (Leigh and Rowell 1995). To begin with, it has diminished the group on which it depends for its own safety. Moreover, insofar as predators return to groups that they have already raided with success, this monkey may have hastened its own death. Under such circumstances, the advantage of an individual and the good of its group coincide rather closely, at least if such deadbeat monkeys cannot escape the results of their misdoings by migrating to other groups. This proviso reminds us that social organization plays an essential role in aligning an individual's advantage with the good of its group.

SELECTION AMONG GROUPS AND THE COMMON INTEREST OF A GROUP'S MEMBERS

The circumstances that allow selection among groups to override selection within groups are quite restricted. Sometimes, however, groups are organized so that selection among groups enforces the common interest of their members. The effect of selection among individuals within a group on a metrical characteristic is the intensity of selection on that characteristic (the regression of log fitness on that characteristic's magnitude) times the variation available for selection, as measured by the group's heritable (additive genetic) variance in this characteristic (Price 1970). Likewise, the effectiveness of selection among groups is the intensity of selection among groups times the heritable variance among group means (Price 1972). Selection among groups is usually less effective than selection within groups because variance among groups is usually smaller than variance within groups. Exchange of migrants among groups and the joining of emigrants from two or more parent groups to found a new group both reduce variance among, relative to variance within, groups. Moreover, groups often live longer than their component individuals, making selection among groups slower. Selection among groups can contend with an opposing selection within groups on equal terms, that is, a given percentage increase in the average group's total number of offspring groups can counterbalance the same percentage increase in lifetime reproductive success of the average individual, only if the groups are organized in such a way that (1) less than one migrant is exchanged per two groups per group lifetime; (2) a new group nearly always has a single parent group where all its founders were born; and (3) there are more groups than individuals per group (Leigh 1983, 1991).

If we treat organelles of a particular kind within a cell as a group, they satisfy conditions that make selection among groups decisive (Leigh 1983). Organelles do not migrate from one cell to another. A mitotically produced cell obtains its organelles from its "parent." Finally, zygotes almost always inherit organelles of a given kind from one parent (Eberhard 1980). How might organelles have acquired the social characteristics rendering them susceptible to group selection? The tendency of mitochondria to defend their egg against conspecifics invading with sperm (Eberhard 1980) suggests that when protomitochondria originally invaded their ancestral hosts, presumably as parasites (Margulis 1993), they benefited more by caring for their current host than by increasing their transmission to other hosts at their current host's expense. This circumstance reflects the benefits of the complementation of functions between protomitochondria and their hosts (Blackstone 1995). Were it not for this fundamental community of interest between proto-organelles and their hosts, the circumstances permitting group selection on mitochondria would never have evolved. The territorial behavior of organelles, prompted by interests in common with their hosts, set the stage for a group selection in their host cells' interest. Once this stage was set, the effectiveness of this group selection was sealed by selection on their hosts for anisogamy, which facilitates uniparental transmission of organelles to zygotes (Cosmides and Tooby 1981; Hurst 1995, 1996a). The fate of organelles was further identified with that of their hosts by the transfer of certain organellar genes to the nuclear genome (Trench 1991).

Selection among groups also must have played a crucial role in the evolution of multicellular organisms. Metazoans, like vascular plants, all descend from sexually reproducing ancestors. Ancestral multicellular organisms presumably arose as aggregates developed mitotically from sexually produced zygotes. Therefore, variation within aggregates was lower than variation among aggregates—provided that cells could not migrate from one aggregate to another—so that selection among aggregates swamped selection within aggregates (Maynard Smith and Szathmáry 1995). Both sexual reproduction and the rarity of invasion of self by nonself were crucial to the evolution of multicellular organisms. Transforming these aggregates into genuine *individuals*, however, required the evolution of features suppressing conflicts between organisms and their cells, or limiting their consequences. Lowering somatic mutation rate and limiting the total number of cells per aggregate restrict genetic variation within aggregates (Michod 1997a). Buss (1987) has suggested that, in many metazoans, maternal control of early stages of development and sequestration of the germ line limit damage from rogue cell lines, and that in plants, which cannot sequester germ lines, rigid cell walls keep rogue cell lines from spreading. Selection among aggregates, however, had to be effective for such refinements to evolve.

A sort of selection among groups may also have played an essential role at the very origin of life (Boerlijst and Hogeweg 1991). "In the beginning," RNA sequences capable of both serving as catalysts and being replicated appear to have functioned as protogenes (Maynard Smith and Szathmáry 1995). Because proofreading enzymes were yet to appear, these sequences had to be short enough so that selection could keep replication errors from accumulating (Eigen 1992; Szathmáry, chap. 3). Thanks to the small size of each one, an array of different RNA quasi-species with complementary functions were needed to form a self-replicating network. This network was first imagined to be a *hypercycle*, in which A enabled B's production, B enabled C's, and so forth up to the final product, say F, which catalyzed A's production (Eigen et al. 1981). Szathmáry (chap. 3) now thinks such metabolic networks were unlikely to be hypercycles; he thinks of them as miniature ecosystems of mutually dependent catalysts. The compartmenting of hypercycles or metabolic networks into protocells allowed a selection among protocells to enforce the common interest of a network's constituent protogenes or quasi-species in the good of their network (Frank 1996; Szathmáry, chap. 3).

KIN SELECTION

It sometimes pays genes to program their carriers to sacrifice reproductive output in order to help a relative if this assistance enhances the reproduction of this related carrier. Such sacrifice is favored *if* the relative's gain in reproduction, times the correlation between the genotypes of the relative and the original carrier, exceeds the sacrifice in the carrier's reproduction (Hamilton 1964a; West Eberhard 1975; Bourke and Franks 1995). Students of kin selection love to cite Haldane's supposed willingness to lay down his life to save two (full) brothers, four half sisters, or eight first cousins (presumably the ages and future reproductive prospects of each of these relatives were comparable to Haldane's own). Indeed, selection among groups can be viewed as a form of kin selection, for selection among groups is efficacious only if the intraclass correlation among the genomes of fellow members of a group is high enough so that an organism benefits by helping other members of its group in contests or competition with outsiders (Crow and Aoki 1982).

Kin selection played a crucial role in the evolution of insect societies (West-Eberhard 1975, 1978; Keller and Reeve, chap. 8). Where there is safety in numbers, or benefit from reusing old nest cells or building new ones upon the old, wasps may benefit by nesting in groups. Group nesting among related individuals of formerly solitary species was the first step in the evolution of social wasps, and probably the first step in the evolution of all types of social hymenoptera (West-Eberhard 1978). Nesting in groups enhances competition among group members. Winners may eat losers' eggs or prevent losers from laying. Dangers of nesting alone may be such however, that if the loser is related to the winner, the loser may do best by staying and helping its winning relative reproduce (West-Eberhard 1975). The characteristic cycle of instincts in solitary wasps—egg development and nest building, followed by provisioning the egg and larva once the ovary is emptied—provides the foundation for an advantageous division of labor between winner and loser(s) (West-Eberhard 1987). If a subordinate's newly laid egg is eaten, or if a subordinate has resorbed an egg she was not allowed

to lay, the empty ovary may enhance that subordinate's instinct for provisioning eggs or larvae, which can be easily diverted to the young of the
dominant, neglected because the dominant mother is attending to aggressive
interactions and nest defense, attention that also protects the nest from potential predators. The more essential it is to nest in groups, the greater the
proportion of the reproductive output dominants can monopolize without
provoking losers to leave (Reeve and Nonacs 1992; Keller and Reeve 1994b,
chap. 8). The dominant must also be clearly stronger than subordinates, to
avoid the prospect of subordinates fighting to the death to take over the nest.
Other processes, however, must come into play before kin selection can favor the evolution of truly complex insect societies. As we shall see below,
both ecological constraints and social organization must be arranged so as to
suppress the possibility of appreciable reproduction by workers before a
clear-cut, complete division of labor can become consistent with the common interest of an insect society's members.

MUTUAL ENFORCEMENT

A group's members can enforce the common good by punishing members
who violate it (Trivers 1971). Axelrod (1984) modeled the feasibility of
mutual enforcement by the game of "iterated prisoner's dilemma." Consider
a group of two. At each play, a member can cooperate or defect. If both
cooperate, each earns three points; if both defect, each earns one point; if
one cooperates and the other defects, the defector earns five points, and
the cooperator, none. Here, we have a potential "tragedy of the commons"
(Hardin 1968). Whatever the opponent does, defecting earns more at any one
play; yet, if the participants play each other repeatedly, they share a common
interest in continual cooperation. Cooperation is best enforced by a strategy
related to tit for tat: Cooperate on the first play, and at play n, do as the
opponent did at play $n - 1$ (Axelrod and Hamilton 1981). In an error-prone
world, "win-stay, lose-shift" is more reliable. Here, the rule is, if the nth play
earned at least three points, repeat it the next time, otherwise shift (Nowak
and Sigmund 1993). Such strategies fail, however, if players can easily
change partners. To ensure cooperation, individuals must be penalized for
changing partners.

Hamlets, simultaneously hermaphroditic coral reef fish, avoid expending
half their reproductive effort on male functions by trading eggs for each
other to fertilize. Trading eggs avoids the need to produce excess sperm or to
fight for mates (Fischer 1981). Hamlets pair off at spawning time. Members
of a pair exchange sex roles in successive spawns (hence the egg-trading).
As the exchanges continue, each fish offers more eggs for its partner to
fertilize, as if it were becoming more confident of its partner's good faith. If,
however, a fish tries to play the cheaper male role for two successive

spawns, cooperation ends, and the cheater is penalized by the time required to find and inspire confidence in a new mate (Fischer 1988).

In Monera—bacteria and bluegreen algae—genes are arranged in a single circular chromosome, which expresses the common interest of its genes in each other's presence. This community of interest, however, is not unlimited. Genes for bacteriophage λ are sometimes part of the bacterial chromosome, where they are well-behaved, but sometimes they leave the chromosome and multiply explosively as independent virus particles, killing their host cell in order to infect others (Watson et al. 1987, pp. 517ff). A gene's ability to spread independently to other cells undermines its community of interest with the rest of its genome, just as the community of interest among a group's members would crumble if each could easily move to other groups.

A gene can create a conflict with its genome by becoming a "segregation-distorter," that is, biasing meiosis in its own favor, to spread itself through its population. The conflict becomes manifest when segregation distortion spreads an allele that harms its carriers (Lyttle 1977). How can the fairness of meiosis be enforced? Alleles on different chromosomes segregate independently of the distorter: None of these alleles can "ride the distorter's coattails." If the distorter inflicts a phenotypic defect on its carriers, all these alleles will suffer from it. At any of these unlinked loci, selection favors mutants suppressing the distorter, for they spare some of their descendants from the distorter's defect that they would otherwise have inherited (Prout et al. 1973). In this sense, a genome's genes have a common interest in fair meiosis (Leigh 1971, 1991). Selection for such suppressors is effective (Lyttle 1979). Suppression of successive distorters appears to have eliminated most of the possible means for biasing meiosis: How else are we to understand the notorious difficulty of selecting for changed sex ratios in species with chromosomal sex determination (Maynard Smith 1978; Williams 1979)? Nonetheless, other levels of selection must have been involved in the spread of honest meiosis. Surely those lineages whose species were less susceptible to segregation distortion lasted longer and radiated more successfully (cf. Nunney, chap. 12).

Truly complex insect societies can evolve only when subordinates cannot benefit by producing young of their own. This condition ensures that a subordinate's only hope of spreading its genes is to help its dominant relative reproduce (West-Eberhard 1975). Colonies of insects whose subordinates seldom or never try to reproduce on their own, such as honeybees, army ants and leafcutters, are marvels of self-organization, where each worker performs its appointed tasks automatically, without any trace of compulsion or even direction by the queen (E. O. Wilson 1980; Franks 1989; Seeley 1995). Honeybee queens create a situation in which workers make it unprofitable for each other to reproduce by mating with up to 20 males and mixing the sperm thoroughly. Thus, most of a worker's colleagues are half sisters. A worker is more closely related to its mother's eggs than to a half sister's, so it eats eggs laid by half sisters. Mutual policing by workers makes it pointless for them to *lay* eggs and thus creates a common interest among them in helping their queen (Ratnieks and Visscher 1989).

The common interest of chimpanzees, *Pan troglodytes*, in their group's welfare leads them to enforce the rudiments of morality (de Waal 1996). Chimpanzees can recognize each other and have the ability to imagine how they would act if in the circumstances of another individual, a self-awareness most clearly evidenced by the deceptions they practice on each other (de Waal 1996). These abilities enable chimps to do unto others as the others have done unto them, and to expect others to do to them as they have done to those others (de Waal 1996, p. 136). Chimps have a clear sense of gratitude for, and willingness to repay, those who have done them favors; a desire for retribution toward stingy troop members that do not share food; and a desire to exact revenge if wronged, a desire whose excesses the troop dominant is expected to restrain (de Waal 1996, p. 161). These attributes suggest that chimpanzees have the rudiments of a sense of justice.

The mutual enforcement of morality among chimpanzees is manifested in various ways. At the Arnhem Zoo, a whole troop of captive chimps undertook to chase and thrash two adolescent females who delayed the feeding of the troop for two hours by refusing to return when the troop was called back to its shelter one evening (no ape is fed until all return). They learned: The next evening they were the first to return (de Waal 1996, p. 89). Chimpanzees also expect their dominants to be fair, and to protect the weak. In this same troop, a young male who, with the help of an older colleague, became the troop's dominant, showed improper bias when he interceded in fights: He invariably sided with his colleague and with a few high-ranking female friends. Because this dominant was clearly not living up to expectation, coalitions of females prevented him from intervening in fights thereafter. His older colleague, who settled fights in a fair and restrained manner, was accepted as mediator instead. Thus, the chimpanzee group has a say in who functions as mediator and how he does it (de Waal 1996, p. 130). In another captive group, females restrained the dominant male from taking excessive revenge on a young male whom he had discovered mating with one of his favorite females (de Waal 1996, p. 91). More generally, chimps welcome and celebrate reconciliations between troop members who have been in a fight (de Waal 1996, p. 205). Indeed, even though there are squabbles for position, chimpanzees have a common interest in the hierarchical organization of their group, which is a framework for coordinating access to resources, keeping the peace, and organizing the group's response to predators or competing groups (de Waal 1996, p. 183).

Understanding Genetic Conflicts Clarifies the
Study of Evolution

GENETIC CONFLICTS, NATURAL SELECTION, AND THE
BIOLOGY TEACHER

One of the scandals of biology teaching is how the directing role of natural selection in adaptive evolution is usually argued. The student is told that evolution is driven by, or consists of, changes in allele frequencies in populations (some would find even this an unnecessarily contentious remark). Then the lecturer enumerates the possible causes of change in allele frequency: sampling error (genetic drift), mutation, migration, and natural selection (differential replication), and concludes that, because the only one of these four that can lead to adaptation is natural selection, natural selection is *the* cause of adaptive evolution. This logic leads to the extraordinary circumstance whereby "To buttress the theory of natural selection the same instances of 'adaptation' (and many more) are used, as in an earlier but not distant age testified to the wisdom of the Creator and revealed to simple piety the immediate finger of God" (Thompson 1942, p. 960). An exercise in "logic" has transformed a mechanistic hypothesis into a deus ex machina.

The analysis of genetic conflicts, however, enables one to look for footprints of the decisive role of natural selection in adaptive evolution. For selection among groups to achieve a major evolutionary transition, such as the evolution of eukaryotes or the transformation of multicellular aggregates into true metazoan individuals, certain conditions must be met (Leigh 1983, 1991). Migration among groups must be annulled. Either each group must be founded by migrants from a single parent group (as in the uniparental transmission of organelles of a given kind to a zygote, very probably the ancestral condition for all sexual eukaryotes; cf. Hurst 1995), or there must be some other means to ensure that among-group exceeds within-group genetic variance (such as the sexual production of zygotes, which then develop mitotically into organisms of many, genetically identical cells). The means by which conflicts between cells and their organelles (Eberhard 1980) or conflicts between individuals and their cells (Buss 1987; Maynard Smith and Szathmáry 1995) are suppressed or minimized provide unmistakable footprints of the decisive role of natural selection in these transitions. Each evolutionary transition has left traces of the genetic conflicts which that transition overcame and the means by which these conflicts were suppressed. For that reason, Maynard Smith and Szathmáry's study of the major transitions in evolution represents the first book in which evolutionary history testifies to evolutionary mechanism. Their book shows how to remove the argument for the directive role of natural selection in macroevolution from the domain of Cartesian analytic logic to that of empirical observation.

Hierarchies and Evolvability

The greatest stumbling block for laymen (and many biologists) trying to understand the theory of evolution by natural selection is seeing how natural selection of *random* mutations could lead to the complexity and precision of adaptation characteristic of living organisms. Quite distinguished minds find this idea an oxymoron (Polanyi 1958; Gilson 1971; Fabre 1989). What characteristics of organisms are responsible for "evolvability"? What features allow living things to evolve by natural selection of random mutations? One essential feature is modular organization, which allows mutation or selection to affect one feature of an organism without interfering with the others (Wagner and Altenberg 1996; Gerhart and Kirschner 1997).

THE VIRTUES OF MODULAR ORGANIZATION

Fisher (1958, pp. 41–44) argued that mutations of small effect contributed disproportionately to the raw material for adaptive evolution. He imagined a system whose actual state was specified by a point $\mathbf{X} = x_1, x_2, \ldots x_n$ in an n-dimensional space, whose optimal state was specified by a different point $\mathbf{O} = o_1, o_2, \ldots o_n$, and whose fitness was a monotonically decreasing function of the Euclidean distance $\|\mathbf{X} - \mathbf{O}\|$, the square root of the sum $(x_1 - o_1)^2 + (x_2 - o_2)^2 + \ldots + (x_n - o_n)^2$. Now imagine a mutation, a *random* change that shifts the system's state from $\mathbf{X}$ to $\mathbf{X} + \mathbf{r}$. The probability that $\mathbf{X} + \mathbf{r}$ is more fit than $\mathbf{X}$ is the proportion of the sphere of radius $\|\mathbf{r}\|$ about $\mathbf{X}$ enclosed within the sphere of radius $\|\mathbf{X} - \mathbf{O}\|$ about $\mathbf{O}$. This probability is the smaller the larger $\|\mathbf{r}\|$ relative to $\|\mathbf{X} - \mathbf{O}\|$. For a given ratio of these variates, the probability of improvement is smaller the larger n, that is, the more complex the characteristic affected by the mutation. The modular organization, both of genes themselves and the characteristics they affect, plays an essential role in making adaptive evolution possible, for modularity allows one feature to be changed without changing anything else (Lewontin 1978, p. 230; Wagner 1996; Wagner and Altenberg 1996). Indeed, episodic directional selection on one characteristic combined with stabilizing selection on the rest of the phenotype could favor transforming that characteristic into a relatively independent module (Wagner 1996). The principle here is analogous to Popper's (1991) argument that societies must be changed piecemeal, so that the effect of each change can be assessed with minimum ambiguity.

Moreover, the capacity for accommodating a variety of phenotypic insults can cause the various, independently programmed characteristics of an organism to adjust in extraordinarily appropriate ways to a major genetic change in one of their number. Maynard Smith (1958, pp. 279–280) tells the story of a goat, first studied by E. J. Slijper, whose front feet were rendered useless by a mutation. This goat walked bipedally, which led to an extraordinarily adaptive series of rearrangements in its skeleton and musculature. This capacity for accommodation among modules raises the question of what sorts of evolutionary jumps are possible when the environment is permissive enough.

HIERARCHIES, LEVELS OF SELECTION, AND EVOLVABILITY

Students of hierarchies like to tell the parable of the watchmakers (Leigh and Rowell 1995; Seeley 1995; Wagner 1995). One constructed his watches in a modular manner, forming subassemblies of ten basic parts apiece, assemblies of ten subassemblies apiece, and so forth until he had finished his watch. The other dispensed with subassemblies. Both watchmakers were subject to frequent interruptions. When interrupted, the one only had to start over on his current subassembly, whereas the other had to start from scratch. Naturally, only the modular watchmaker finished any watches. A more apposite story might be the organization of genetic algorithms, computer programs simulating natural selection of random mutation to solve complex optimization problems. The most workable of these programs successively evolve partial solutions serving as building blocks, which can then be combined to generate the final solution (Wagner 1995).

Evolution seems to have followed similar paths. Time and again, major evolutionary transitions occurred when larger wholes formed from smaller, "ready-made" parts that had already been tested by natural selection (Maynard Smith and Szathmáry 1995). Thus, organisms are modules of modules of modules . . . and so on toward Pascal's infinitely small. Even genes seem to be combinations of domains 100–300 nucleotides long (Eigen 1992, p. 22), which is about as big a gene as can be readily optimized by natural selection of chance mutations (Eigen 1992, p. 30). Eigen believes that genes acquired their essential features when they were much more mutable than now—*before* they cohered into chromosomes.

Modules that are themselves living creatures, or at least units of selection, have distinct advantages. When complementation of functions promoted mutualism between cells and their organelles, organelles played the role of self-designing macromutations for their host, macromutations with properties beyond the wildest dreams of a Goldschmidt (1940) or a Løvtrup (1976). An analogous capacity for accommodation applies to animals in societies. A previous section mentioned West-Eberhard's (1987) description of how the complex and beautiful division of labor is built on the varied reactions of solitary individuals to different environmental conditions. This same capacity for accommodation made it possible for Smythe (1991) to create in one generation a breed of social pacas from what is naturally a fiercely territorial species whose adults live in pairs, by suitable adjustments in the rearing of the newborn and very young. This capacity for self-design in the interests of their group, among parts that have not yet ceased to be units of selection (or evolution) in their own right, must have played a crucial role in many evolutionary transitions.

LEVELS OF SELECTION AND TRUTH, BEAUTY, AND GOODNESS

Some theologians, like Jaki (1983, pp. 60–63), and some biologists, like T. H. Huxley (1894) and George Williams (1989), place truth, beauty, and goodness utterly beyond the reach of natural selection. I agree readily enough with D'Arcy Thompson's (1942, p. 13) remark, "Consciousness is not explained to my comprehension by all the nerve-paths and neurones of the physiologist; nor do I ask of physics how goodness shines in one man's face and evil betrays itself in another." Nonetheless, our capacity to distinguish right from wrong, appreciate beauty, and know truly are related to how natural selection affects social beings. A previous section discussed de Waal's (1996) evidence for protomorality among chimpanzees, and its foundation on the troop's sense of justice, which, in good Aristotelean fashion, serves the troop's common interest.

Whence comes a sense of beauty is an odder issue. Sexual reproduction creates opportunity for members of one sex to compete for matings with the other. One consequence of such sexual selection is the evolution by males of characteristics that attract females: The peacock's tail, the blazing colors and gorgeous plumes of birds of paradise, and the like (Darwin 1859, p. 89). Darwin's assertion that male birds competed by appealing to the aesthetic sense of females of their species caused some offense. Nevertheless, human beings prize the colors, shapes, and sounds that many animals use to attract mates, and the colors, shapes, and scents by which flowers attract pollinators. Moreover, it is often essential for females to judge the beauty of males aright, for a male's beauty is often a good index of his health and suitability as a mate (Hamilton and Zuk 1982; Saino et al. 1997).

Organisms do not need complete knowledge, but what they do know, they must know truly enough to avoid predators and other dangers and to find food and mates (Lorenz 1978). Darwin's words of doubt about the efficacy of a mind descended from a monkey's, oft quoted by anti-Darwinians (cf. Jaki 1983, p. 57), miss the point. Even we may not know "things as they are," but we must know enough about some objects in our environment to predict their impact on us, or on each other, and to judge the impact of our actions on them. The same is true, to a lesser extent, of other animals. Even the animals that eat cryptic insects must generalize about the features that reveal their prey. The complicated measures by which cryptic insects disguise their head and legs are defenses against just this power to generalize (Robinson 1985). The capacity to predict and generalize must be equally important in social life. Even a chimpanzee's deceptions depend on its ability to predict how best to deceive its intended dupe, which appears to presuppose an ability to imagine how it would respond to its acts were it in the dupe's place (Jolly 1991).

Concluding Remarks

Adaptive evolution presupposes modular organization. Indeed, a precise understanding of the nature and history of the modular organization of living things is needed to assess their potential for adaptive evolution and may reveal that living things are organized to facilitate their evolution by natural selection of "random" mutations (Leigh 1987).

The most objective mark of evolutionary progress is the series of evolutionary transitions where parts combined to form larger, more effective wholes (Maynard Smith 1988). Each such transition involved potential conflict between different levels of selection. These conflicts, and the ways they are resolved, comprise one of the grand unifying themes of biology.

Parts join to form larger wholes only if there is a genuine community of interest among the parts, and if circumstances allow the enforcement of this common interest. In the major transitions of evolution, community of interest plays the same crucial role as in Aristotle's *Politics*.

The traces of the means by which conflicts between levels of selection are resolved in favor of the higher level represent unmistakable footprints of the decisive role of natural selection in macroevolution. These traces are instances where evolutionary history testifies to evolutionary mechanism.

Finally, evolutionary studies of social animals suggests that truth, beauty, and goodness are not totally beyond the reach of evolutionary biology.

3 The First Replicators

Eörs Szathmáry

The replicator concept of Dawkins (1976) has turned out to be extremely useful in analyzing evolutionary questions. Here I follow the definition of Hull (1980), who emphasized that replicators must pass on their structure largely intact. Although selection acts on them directly, interactors (such as organisms) do not qualify as replicators because their structures are not copied. I shall come back to this important conceptual issue at the end of this chapter because organisms usually qualify as reproducers.

My primary interest here lies in the origin of the earliest units of evolution. Entities qualify as units of evolution if they meet the following criteria (Maynard Smith 1987):

1. *Multiplication*. Entities should give rise to more entities of the same kind.

2. *Heredity*. Like begets like; A-type entities produce A-type entities; B-type entities produce B-type entities, etc.

3. *Variability*. Heredity is not exact; occasionally A type objects give rise to A' type objects (it may be that A' = B).

If objects of different types have a hereditary difference in their fecundity and/or survival, the population undergoes evolution by natural selection.

To explain the origin of life, we need to explain the origin of heredity in terms of chemistry. Heredity merely means that like begets like. This, in turn, requires variation: multiplication of an entity that can exist in only one form does not constitute heredity and could not form the basis of evolution by natural selection. I argue that mere heredity is not enough. Ongoing evolution requires "unlimited heredity," that is, the existence of replicators that can exist in an indefinitely large number of forms. Although, as I outline below, heredity with a small number of possible types can exist without copying, it seems very probable that unlimited heredity requires template copying of replicators with a modular structure.

The first experiment relevant to the origins of replicators (and life in general) was carried out more than a hundred years ago by Butlerov, a Russian chemist. He found that, if formaldehyde is kept in a reaction vessel for a few hours under moderately alkaline conditions, sugars readily form. Nowadays the "formose reaction" appears to be a formidable network of interconversions of sugars, among them ribose, which is a building block of RNA (e.g., Cairns-Smith and Walker 1974). Even more interesting is the fact that, given a sufficient amount of formaldehyde, the accumulation of sugars follows exponential kinetics, indicating that something is replicating in the solution. It is the sugars that replicate: The "hard core" of the reaction is the cyclic,

(a)

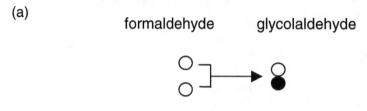

(b)

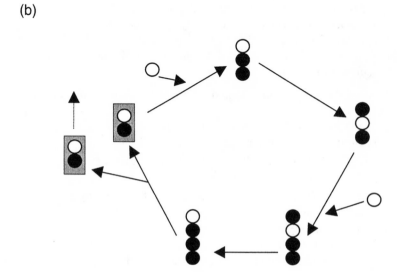

FIG. 3.1. The autocatalytic core, or seed, of the formose reaction. (a) The "spontaneous generation" of the core autocatalyst. (b) The central autocatalytic cycle. Each circle represents a chemical group including one carbon atom.

autocatalytic formation of glycolaldehyde at the expense of continuous formaldehyde consumption (fig. 3.1). (An autocatalyst catalyses its own formation.)

It takes some time until the first glycolaldehyde molecule is formed. The coupling of two formaldehyde molecules is the spontaneous generation, as it were, of the first replicator. This teaches us a general lesson: The first evolutionary units must have arisen, logically, by means other than evolution by natural selection.

There is practically no heredity in this system: We do not yet know of alternative, repetitively replicating variants of this network, but this may not be the final situation. It is noteworthy that the Calvin cycle, fixing carbon dioxide in plants, is also autocatalytic (Gánti 1979) at the level of the sugar phosphates: If one starts with three molecules of glyceraldehyde-3-phos-

phate, then after one turn of the cycle, one has four molecules. It is potentially confusing that this system is called autocatalytic, because every step in
it is catalyzed by a specific enzyme, not produced by the cycle itself. The
explanation is that there are two levels of catalysis in the system: one at the
level of small molecules (constituting the cycle as we draw it), and another
at the level of enzymes (also operating in cycles, of course).

For our topic we need examples of autocatalysts replicating without the
aid of enzymes, because the latter are evolved latecomers. Günter Wächtershäuser, chemist and patent lawyer, proposed a hypothetical system of this
sort, which may have been an archaic variant of the contemporary reductive
citric acid cycle. This cycle fixes carbon in a variety of bacteria (organic
acids are produced at the expense of the incorporation of carbon dioxide by
the cyclic transformation of the same organic acids; the amount of organic
acids thus increases). (The reductive citric acid cycle is almost the exact
reverse of the well-known Krebs [citric acid] cycle, producing carbon dioxide by breaking down activated acetic acid.) Note that Wächtershäuser produced an elaborate scenario of prebiotic metabolism on the surface of pyrite
(FeS_2, fool's gold; Wächtershäuser 1988, 1992). His primary motivation was
that many students in the field became disenchanted by the classical "primordial soup" scenario for the origin of life (see Maynard Smith and
Szathmáry 1995). He envisaged a protracted phase of evolution of what I
call replicators with limited heredity. Indeed, he suggested a series of alternative chemical cycles, mutants of his archaic reductive cycle. At one time,
evolution must have given rise to replicators that were copied digit by digit,
in the fashion of contemporary nucleic acid molecules.

The crucial difference in the mode of replication of the aforementioned
autocatalytic small molecules and polynucleotides can be visualized as follows. If one takes a snapshot of polynucleotide replication, one can easily
decide about the degree of completion. It is perfectly sensible to say, for
example, that replication is "half-way through"; this means that all nucleotides in the first half of the molecule pair with their complementary nucleotides. If one were to cut the polynucleotide in two halves, one could say
that one of the halves is completely copied, whereas the other one awaits
replication. Polynucleotides are copied digit by digit, or modularly; the nucleotides serve as chemical modules. This does not hold true for molecules
like the intermediates of the Calvin cycle. If one chooses an intermediate,
one can see all sorts of chemical transformations acting on it until after a
certain number of steps, the molecule suddenly falls into two identical
pieces. The molecule is continuously modified and processed until replication is complete. Wächtershäuser has called this mode of replication "processive." Replication in a sense is holistic: Half of the initial molecule cannot replicate at all.

A general problem arises here concerning heredity in autocatalytic networks of small molecules. One molecule of glycolaldehyde is indistinguish-

able from the other. If so, where is heredity? Maynard Smith and Szathmáry (1995) suggested a resolution. It makes sense to distinguish between replicators with limited heredity and those with potentially unlimited heredity (see also Szathmáry and Maynard Smith 1993b). In the former case, the number of types is smaller than, or roughly equal to, the number of objects (individuals) because objects exist only in a few alternative types. A didactic example is the set of possible hexanucleotides. In contrast, unlimited hereditary replicators have (many) more types than the number of objects (individuals) in any realistic system. What distinguishes even small oligonucleotides from other autocatalysts like glycolaldehyde is that, for the latter, the production of hereditary variation is much more difficult. One could argue that for replicators that are not modularly replicated, variants are allowed to arise only through macromutations (cf. Wächtershäuser 1988). Hence, as limited hereditary replicators, members of autocatalytic cycles lack the ability to undergo microevolution: Heredity is almost always exact. (Of course, there can be strong nonheritable fluctuations, but this is a different issue.)

It seems reasonable to assume that autocatalytic cycles only come in a relatively small number of types; their intermediates are limited hereditary replicators (Szathmáry and Maynard Smith 1993b; Maynard Smith and Szathmáry 1995). I must stress that, of course, these systems have replication at the molecular level. As Orgel (1992, p. 203) wrote, "All replicating systems are, by definition, autocatalytic and all autocatalytic systems result, in some sense, in replication." The reproduction of molecules leads to the growth of the population.

Note that this departure from the traditional "gene-chauvinistic" view of replicators (cf. Dawkins 1976) is absolutely essential if one wants to understand the origin of living systems, for the following reasons. First, genes are too complex to start with. Second, autocatalysis, and what has been called the Darwinian dynamic (Bernstein et al. 1983), transcend traditional genetics and even biology proper. This is why it is important to define units of evolution as generally as possible.

Recognizing the importance of replicators other than genes, Szathmáry (1995) offered the following classification of replicators:

1. Limited hereditary replicators
   a. Processive (holistic)
      Example: formose reaction, archaic reductive citric acid cycle
   b. Modular
      Example: oligonucleotide analogues
2. Unlimited hereditary replicators
   a. Processive
      Example: unknown, probably impossible
   b. Modular
      Example: genes of extant organisms

The information carried by processive and modular replicators can be termed *analogue* and *digital*, respectively (Wächtershäuser 1994), although the terms *holistic* and *digital* seem more appropriate (Maynard Smith and Szathmáry in prep.).

According to our current view, evolution proceeded from holistic, limited hereditary replicators through digital, limited hereditary replicators to digital replicators with unlimited heredity. In chemical terms, digital replicators are likely to have emerged as by-products of holistically replicating autocatalytic cycles, on which they initially act like parasites (Wächtershäuser 1992).

The first digital replicators must have been relatively small molecules because long templates do not easily dissociate from the copies made. Von Kiedrowski (1986) synthesized the first nonenzymatically replicating oligonucleotide analogue; there were more to follow. These results clearly show that digital replication is possible without enzymes, but do not yield, by themselves, a chemically realistic scenario. One of the main problems is that we still do not know what came before Gilbert's (1986) RNA world. RNA seems chemically too complex to originate by simple chemical evolution. People in the field seem convinced that some form of replicators, incorporating nucleic acid bases, must have preceded RNA, but they do not really know what the nature of the primordial backbone, instead of the contemporary ribose-phosphate-ribose-phosphate . . . , could have been. Schwartz (1997) gave a current account of the many relevant speculations and the few experimental approaches.

The dynamics of growth of such replicators turns out to be crucial. Indeed, from a chemical standpoint, natural selection is simply the dynamics of replicators. I review the reasons, empirical and theoretical, for thinking that the growth of the simplest digital replicators would have been parabolic, that is, slower than exponential. This is important because our traditional analyses of selection processes hinge on the assumption that growth would be exponential (Malthusian) without abiotic and biotic limitations. We shall see that parabolic growth does not, under selective conditions, lead to survival of the fittest.

At some time in early evolution, replicators with an exponential growth tendency must have appeared. Around that time, nucleic acids occupied their paramount status as carriers of genetic information. Again, the first nucleic acids could not have been very long. In the absence of specific replicases, copying would have been inaccurate, and large molecules would have accumulated errors. Thus, it is unlikely that the earliest nucleic acids had the length of even the smallest present-day chromosomes. Primordial genomes therefore must have consisted of several smaller pieces of nucleic acids. This raises the central problem of how cooperating groups of small replicators could have arisen, and how they could have protected themselves against invasion by molecular parasites. The answer lies in two phenomena that underlie all increases in complexity and cooperation in evolution, from the first populations of cooperating polynucleotides to the emergence of animal and human societies. These are synergism and genetic compartmentalization. That is, complementation of functions (division of labor) can result in strong synergistic fitness effects, and limited dispersal can result in different individuals bound to "sit in the same boat." I discuss the role of these two phenomena in the early evolution of life. Surface metabolism turns out to be important as a precursor of cellular organization in reducing dispersal and, hence, in favoring cooperation.

People have for a long time been puzzled by the proportion of traits that follow from some physical, chemical, or engineering constraint to those that are a result of historical contingency in biology. This question soon may become practical rather than purely theoretical. First, we may redesign life to some extent for biotechnological or mere scientific reasons. Second, we may find that life has arisen independently elsewhere in the solar system (e.g., on Europa, one of the Galileo moons of Jupiter). Would Europan organisms be running their heredity on digital replicators? Presumably yes. Would their hereditary material contain nucleotide bases or their analogues? Because such molecules are formed relatively easily under prebiotic conditions, the answer again is, presumably, yes. Would their genetic alphabet consist of four letters (analogous to our A, G, C, and U/T)? I argue that this is likely to be the case, if their hereditary material also had once been used for enzymatic purposes as well.

The last part of the chapter is concerned with the transition from replicators to reproducers. Existing organisms are not replicators; they do not reproduce by copying. Instead, they contain DNA that is copied, and that acts as a set of instructions for the development of the organism. Hence, reproduction requires both copying and development. Following Griesemer (1996), I outline the concept of a reproducer and show a particularly elegant model of it, conceived by the chemist Tibor Gánti (1971, 1975) more than two decades ago in order to understand what a minimum living system should look like. We shall understand the origin of life only when we are able to outline a convincing scenario for the origin of such a system—we are not quite there yet.

For obvious reasons, I lean heavily on the recent reviews of Maynard Smith and Szathmáry (1995; Szathmáry and Maynard Smith 1995, 1997).

Artificial Replicators and "Survival of Everybody"

ARTIFICIAL REPLICATORS

Von Kiedrowski (1986) and Zielinski and Orgel (1987) synthesized the first nonenzymatically replicating chemical species (in both cases, close chemical relatives to oligonucleotides). They found that, surprisingly, the growth dynamics of these replicators followed a slower than exponential time course.

An intuitive understanding of this phenomenon is possible if we understand why conventional replication of molecules is exponential. When RNA molecules are replicated in a test tube, a replicase enzyme efficiently separates the template and the copy so that both are ready to enter a subsequent round of replication. The same holds, in the absence of enzymes, for glycolaldehyde molecules in the formose network. In contrast, in the case of the von Kiedrowski-type replicators, template and copy do not fall apart so easily because they are bound together by hydrogen bonds. To the contrary, they associate spontaneously in solution at a certain rate. Now, it is important to realize that only the free templates can initiate a round of replication. Two templates (in this case identical to the template: copy complex) sticking together are replicationally inert. This amounts to self-inhibition: The net growth rate depends on the apparent equilibrium between association and dissociation of templates. In fact, the chemists found the growth rate to be proportional to the square root of the total concentration (all forms counted). In Malthusian growth, the rate is simply proportional to the total concentration.

PARABOLIC GROWTH: SURVIVAL OF EVERYBODY

The aforementioned self-inhibition leads to parabolic growth. Total concentration, were resources provided ad libitum, would reach infinity in infinite time, similar to exponential growth, but the growth curve follows a parabola. Contrary to the exponential case, parabolic growth entails "survival of everybody" in a competitive situation (Szathmáry and Gladkih 1989; Szathmáry 1991a; von Kiedrowski 1993) because e^{kt} (exponential) is a much steeper function than $(kt)^2$ (parabolic), where k is a rate constant, and t is time. As explained above, growth is slower than exponential because the replicators limit their own growth, but density reaches infinity in the limit because there are more and more replicators in the growing population. These two opposing effects lead to the parabolic nature of the growth curve.

Because self-limitation is based on molecular complementarity, AA and BB complexes (where A and B are two different replicators) are stronger than AB complexes. Hence, each species limits its own growth more strongly (by associating with itself). This condition for joint survival is also found in traditional Lotka-Volterra competitive systems. This is the ultimate cause for survival of everybody in parabolic systems (Szathmáry 1991a).

Since the pioneering work of von Kiedrowski, several replicators obeying the same type of growth dynamics have been constructed by Rebek (1994) and Sievers and von Kiedrowski (1995) among others. Growth of a recently synthesized self-replicating peptide (!) obeys similar kinetics (Lee et al. 1996). Von Kiedrowski (1993) worked out a detailed kinetic theory for parabolic growth of minimal replicators. It seems that the survival of everybody is a rather robust phenomenon among these replicators.

One of the important steps of prebiotic evolution thus must have been the emergence of replicators with exponential growth. An attempt to mimic such a process has been made in von Kiedrowski's lab (pers. comm.). In light of the general importance of surfaces in chemical as well as early biological evolution (see later), it is noteworthy that a certain surface plays a crucial role in this nonenzymatic, but exponential replication process: Sticking to the surface helps separate the strands.

Eigen's Paradox and the Importance of Population Structure in the Prebiotic Context

Serious considerations suggest that primordial nucleic acids (or their analogues) must have been rather short molecules because of excessive noise in their copying. If different replicators are thus needed to establish a primordial genome having the size of a few genes, some mechanism to ensure their dynamic coexistence must have operated. Various models show that if selfish mutants are taken into account, some form of population structure is mandatory for indefinite survival. As we shall see, in some cases, survival depends on genuine group selection.

THE ERROR THRESHOLD OF REPLICATION

Eigen (1971) called attention to the fact that the length of molecules (number of nucleotides) maintained in mutation-selection balance is limited by the copying fidelity. An intuitive appreciation of this point is readily obtained if one makes a simple calculation of the overall copying fidelity (Q) as a function of the copying fidelity per digit (in the concrete case, nucleotide, q) and the length (v) of the molecule to be copied. If we assume that mistakes made during replication are independent, then standard probability calculus gives $Q = q^v$. Let us take the numerical case of $q = 0.99$; that is, one mistake in a hundred is made per nucleotide per replication. Then for a molecule with $v = 100$ we obtain $Q = 0.37$; that is, only about one-third of the copies will bear no mutations. Whether the error-free copies can still be maintained depends on the strength of selection. Calculations and analysis of replication of viruses revealed that, realistically, for the value of q given above one cannot go beyond $v > 100$ (Eigen 1971). This phenomenon is Eigen's *error threshold of replication*. Whenever the mutation rate is lower than the critical one for the catastrophe to occur, a population of molecules will be maintained by mutation-selection balance. This is hardly surprising to a population geneticist. After all, the so-called quasi-species model of such a population (Eigen and Schuster 1979) is isomorphic to a population genetical model of a haploid, asexual population with many alleles coupled through mutations, subject to selection. In contrast, the deduction of the error threshold has been a genuine discovery. Manfred Eigen, who earned a Nobel Prize in chemistry, took a naive and fresh look at the mutational load, which enabled him to see something that the researchers within the field have missed.

Exponential replication, implying survival of the fittest, comes at a high price in early evolution. Early genomes must have consisted of independently replicating entities, but they would have competed with each other, and the one with the highest fitness would have won (Eigen 1971). Hence the "Catch-22" of molecular evolution: no enzymes without a large genome, and no genome without enzymes (Maynard Smith 1983b). We know, however, that evolution did not stop at the level of a few naked genes. Something that prevented this from happening must have occurred.

MOLECULAR HYPERCYCLES

Eigen (1971) thought to resolve this problem by proposing the hypercycle (fig. 3.2) as a model for molecular mutualists, coupled directly with mutual aid in replication. It is important to see that the hypercycle is a doubly autocatalytic system. First, each member serves as template in its own replication and is, therefore, autocatalytic. Second, each member receives help, through a replicase activity, from the one preceding it. Replication of each member thus depends on the product of its own concentration and that of the preceding one. This has been called second-order autocatalysis at the level of the system as a whole. Even if the replication rate constants of the members are different, dynamic coexistence is guaranteed because of the cyclic closure of replicational help (Eigen 1971).

HYPERCYCLIC ILLUSIONS

I must digress by discussing mistakes in the literature concerning which systems can be regarded hypercyclic. This is important because once a misidentification occurs, people think that hypercycle theory (Eigen and Schuster 1979) becomes readily applicable for the described systems. Conversely, such misinterpretations strengthen the perceived applicability of hypercycle theory to real cases. Considerable turmoil has already resulted from such mistakes.

Ricard and Noat (1986), for example, thought that any link between two chemical cycles results in a hypercycle. If a simple chemical cycle A (similar to the citric acid cycle) produces substance Z, which is consumed by cycle B, the two cycles are coupled, but are definitely nonhypercyclic. First, there is no autocatalysis whatsoever in the system; second, the coupling between the two cycles is not catalytic (Szathmáry 1988).

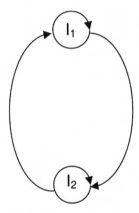

FIG. 3.2. The hypercycle, a model of molecular cooperation. Molecules I_i are auto- and heterocatalysts at the same time. Autocatalysis and heterocatalysis correspond to template and replicase activity, respectively. The number of members may be different from two.

The second case is much more confusing, because it is linked to an important experimental result, the publication of which bears the term "hypercycle" in its title, whereas the system made is not hypercyclic at all! Lee et al. (1997) presented fascinating experimental evidence of two self-replicating peptides that mutually aid the formation of each other. They made three claims.

1. Their system "constitutes a clear example of a minimal hypercyclic network."
2. "A large number of hypercycles are expected to be embedded within the complex networks of living systems."
3. Two previously presented experimental systems "may contain vestiges of a hypercyclic organization."

Unfortunately, none of these claims is correct.

Figure 3.3a shows a schematic diagram of the system synthesized by Lee et al. (raw materials are omitted throughout). Apparently, peptides I_1 and I_2 are both self-replicating, that is, autocatalytic. They are also heterocatalytic in that I_1 catalyzes the formation of I_2 and vice versa. This property suggested to Lee et al. that the peptides are molecular mutualists, and hence manifest two members of a minimum hypercycle. Like ecological mutualists, hypercycles are characterized by a heterocatalytic aid by one member of the system given to the *autocatalytic* replication, rather than mere formation, of the other, and vice versa. (Pollinators do not enhance the spontaneous generation of plants from inanimate matter, but they do help them in reproduction.) A truly hypercyclic variant of the system discussed is shown in figure 3.3b where first-order self-replication is combined with second-order hypercyclic coupling (the two processes run in parallel). When the first

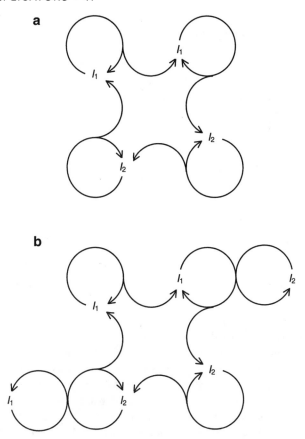

Fig. 3.3. A nonhypercyclic system (a) and a hypercyclic (b) system. (a) A schematic representation of the peptide network synthesized by Lee et al. (1997). *Open arrowheads:* stoichiometric chemical transformations.

type of process is omitted, one arrives at a truly minimum hypercycle, identical in schematic form to the one shown in figure 3.2. Note that filled arrowheads ($\longrightarrow$) in the symbolic version represent catalytic action, rather than stoichiometric transformation (open arrowheads, $\longrightarrow$). Consequently, the original system (fig. 3.3a) follows growth dynamics different from those of hypercycles. Although we are dealing with coupled replicators, the overall growth is still parabolic: The cycles are limited by product-inhibition because of the association between enzyme and product.

EVOLUTIONARY INSTABILITY OF NAKED HYPERCYCLES

It is true that the hypercyclic link ensures indefinite ecological survival of all member replicators. Problems arise, however, when mutations are taken into

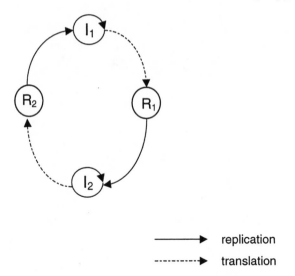

replication

-----▶ translation

FIG. 3.4. The hypercycle with translation. R_1 is a replicase protein enzyme coded for by gene I_1.

account. In order to consider them, it is worthwhile to look at a diagram in which auto- and heterocatalytic aids are functionally clearly separate, such as in a hypercycle with protein replicases (fig. 3.4). Mutants providing stronger heterocatalytic aid to the next member are not selected for. In contrast, increased autocatalysis is always selected for, irrespective of its concomitant effect on heterocatalytic efficiency. This is the well-known problem of parasites in the hypercycle (Maynard Smith 1979). As Eigen et al. (1981) observed, putting hypercycles into reproducing compartments helps, because "good" hypercycles (with efficient heterocatalysis) can be favored over "bad" ones. Two questions arise out of this. (1) Are there other means whereby parasites can be selected against? (2) Are there nonhypercyclic systems that function well in a compartment context? The answers to both of these questions are "yes." I discuss them below.

MOLECULES AND THE STRUCTURED DEME

Michod (1983) was the first to argue that a looser form of population structure could have been important in the selection against selfish genes in a prebiotic context. Szathmáry (1992a) showed that the same mechanism could ensure coexistence of competitive, useful templates as well. Below I follow the short account of these models as given in Szathmáry (1994). Let us imagine the following situation: Templates replicate at a surface, maybe on pyrite (Wächtershäuser 1992), where they grow and interact in semi-isolated groups. They are regularly washed away, become perfectly mixed,

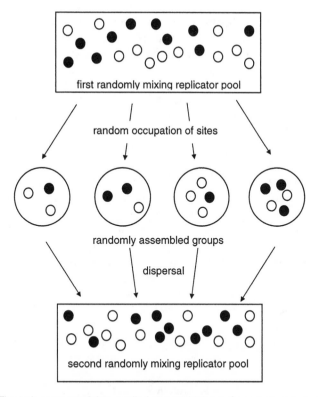

FIG. 3.5. The trait group model as applied to protobiology. *Open and closed circles:* two different replicators (an altruist and a parasite) or two complementing genes.

and are then readsorbed into the surface. Such a situation is similar to the one visualized in D. S. Wilson's (1980) trait-group (structured-deme) model (fig. 3.5).

The coupling among the templates is nonhypercyclic. Instead, we consider a so-called metabolic system (cf. fig. 45 in Eigen and Schuster 1978). The templates are assumed to contribute to metabolism via enzymatic aid, and metabolic products are in turn used up by the templates for replication at different rates. Although all templates contribute to metabolism ("the common good"), they are able to use it with different efficiency. Thus, in a spatially homogenous environment, competitive exclusion follows, despite the metabolic coupling (Eigen and Schuster 1978).

This is not so in the context of the structured-deme model. Local sites on which the relative proportions of templates are closer to optimal (hence metabolism is run more efficiently) yield more new templates, because more building blocks for reproduction are produced. Although differential replication rates of templates cause a shift in relative proportions, the shifted distribution will still not be too far from optimal. In contrast, local sites with a very unfavorable template composition will yield few new templates. In sum, sites closer to the optimal template composition contribute more to the template pool of the new homogeneous phase. The outcome is a "protected polymorphism" of all different templates, deviating strongly from the fatal homogeneous case (Szathmáry 1992a).

SELECTION DYNAMICS IN TWO DIMENSIONS: HYPERCYCLIC AND METABOLIC SYSTEMS

Interesting selection dynamics also occurs when molecules are bound to the surface without being washed away regularly. This problem is modeled by the use of "cellular automata." Without becoming too technical, it suffices to say that each square of a grid is assumed to be occupied by a single molecule (template) or to be empty. Templates can do two things: replicate (put an offspring into a neighboring empty cell, if available) or hop away into empty sites nearby. Replication may depend on the composition of the few neighboring cells. In the case of a hypercycle, for example, the template and a specimen of the preceding cycle member must be present in the same small area if replication of the former is to occur. This, of course, makes perfect chemical sense.

Boerlijst and Hogeweg (1991) simulated hypercycles on a surface exactly in this way. They found that rotating spirals on the surface do appear, provided the hypercycle consists of more than four members. This is linked to the fact that such a hypercycle without population structure shows sustained oscillation in time. Each wing of a rotating spiral looks a bit like the arm of a galaxy and is dominated by templates of the same membership in the hypercycle. Parasites are unable to kill the hypercycle in that system. This finding was attributed to the dynamics of spirals. Two questions emerge: Are spirals necessary? What happens if one models other systems in the same way (i.e., by cellular automata)?

Czárán and Szathmáry (1999) managed to show that, given such a spatial setting, nonhypercyclic systems are once again viable alternatives. The fundamental difference between their model and that of Boerlijst and Hogeweg (1991) is that the dynamical link among the replicators is realized through a common metabolism instead of direct, intransitive hypercyclic coupling. Using the cellular automaton model of the metabolic system, the aim was to show that (1) metabolic coupling can lead to coexistence of replicators in spite of an inherent competitive tendency; (2) parasites cannot easily kill the whole system; (3) complexity can increase by natural selection. The result— there is coexistence without any conspicuous pattern (i.e., something like spirals)—is robust and counterintuitive. It results from the inherent discreteness (i.e., the corpuscular nature of the replicator molecule populations) and spatial explicitness of the model, which grasp essential features of the living world in general, and macromolecular replicator systems in particular. An inferior (i.e., more slowly replicating) molecule type does not die out because there is an advantage of rarity in the system: a rare template is more likely than a common one to be complemented by a metabolically sufficient set of replicators in its neighborhood.

The general importance of surface dynamics seems more and more important for the origin of life in general. As Wächtershäuser (1992) pointed out, chemical evolution leading to more and more complicated networks, is likely to have taken place on the surface, especially on that of pyrite. Surface dynamics of replicators with indefinite heredity is a natural outgrowth of this "primordial pizza" dynamics (cf. Maynard Smith and Szathmáry 1995).

Protocells and Group Selection of Replicators

The phase of evolution just outlined refers to the precellular level. Later in evolution protocells must have appeared. Cellularization offers the most natural, and at the same time most efficient, resolution to Eigen's paradox. It also leads to the appearance of linkage, that is, the origin of chromosomes. The dynamics of genes encapsulated in a reproductive protocell are described by the *stochastic corrector* model (Szathmáry and Demeter 1987; Szathmáry 1989a,b), which rests on the following assumptions (fig. 3.6).

1. Templates contribute to the fitness of the protocell as a whole, and there is an optimal proportion of the genes. Concretely, we assume that genes encode enzymatic aid given to intracellular metabolism.

2. Templates compete with each other within the same protocell. As before, replication rates may differ from gene to gene.

3. Replication of templates is described by stochastic means. Since the number of genes in any compartment is small (up to a few hundred), their growth is affected by the plays of chance. Ecologists would express this as demographic stochasticity.

4. There is no individual regulation of template copy number per protocell.

5. Templates are assorted randomly into offspring cells upon protocell division.

I must emphasize that, in the stochastic corrector model, the templates are not coupled to one another through a reflexive (intransitive) cycle of replicational aid, because that would be a hypercycle. Instead, we assume that they contribute to the common good of the protocell by catalyzing steps of its metabolism. Within each compartment, the templates are free to compete, because they can reap the benefits of a common metabolism differently. (A

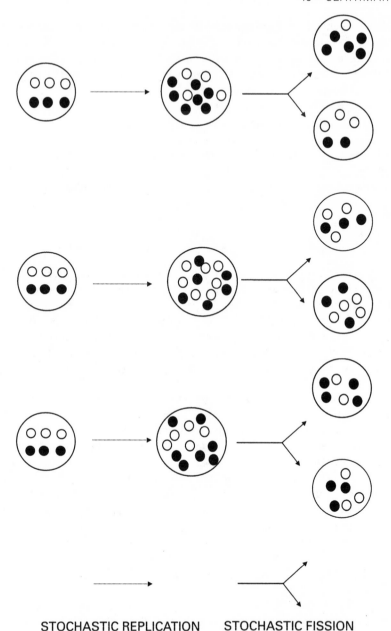

STOCHASTIC REPLICATION   STOCHASTIC FISSION

Fig. 3.6. The stochastic corrector model. *Open and filled circles:* two types of useful gene in the protocell, albeit with different within-cell replication rates. It is assumed that a 3:3 composition at the start of the protocell cycle is optimal for metabolism. Note that two such compartments (with bold cell envelopes) recur upon cell division, despite internal competition.

similar situation can arise among chromosomes and plasmids in contemporary bacteria.) Even though templates compete, the two sources of stochasticity generate between-cell variation in template copy number, on which natural selection (between protocells) can act. This is an efficient means of group selection of templates, because the protocells are the groups obeying the stringent criteria: (1) there are many more groups than templates; (2) each group has only one ancestor; (3) there is no migration between groups (cf. Leigh 1983). Grey et al. (1995) gave a fully rigorous reexamination of the stochastic corrector model. The two mentioned sources of stochasticity effectively lead to the correction of a malign within-protocell trend of harmful competition of the templates. It is interesting to see that genuine group selection is likely to have aided a major transition from naked genes to protocells. Group structure is provided by the physical boundaries of cells.

Within the same context, the origin and establishment of chromosomes (linked genes) in the population have also been analyzed (Maynard Smith and Szathmáry 1993). A chromosome consisting of two genes takes about twice as long to be replicated as do single genes. Chromosomes are strongly selected for at the cellular level even if they have this twofold within-cell disadvantage. Linkage reduces intracellular competition (genes are necessarily replicated simultaneously) as well as the risk of losing one gene by chance upon cell division (a gene is certain to find its complementing partner in the same offspring cell if it is linked to it). The molecular biology of the transition from genes to chromosomes has also been worked out (Szathmáry and Maynard Smith 1993a).

Simulations also show that sex is detrimental to protocells before the establishment of chromosomes because sex allows the horizontal spread of selfish genes (Szathmáry and Maynard Smith in prep.). Without sex, parasitic genes are passed on in a clonal manner and can be efficiently selected against. This is analogous to the problem of intracellular parasites and selfish organelles: There is selection for uniparental inheritance because it reduces intragenomic conflict (Eberhard 1980; Cosmides and Tooby 1981).

Hypercyclic versus Nonhypercyclic Compartments

Manfred Eigen aptly recognized the need for some kind of coupling among unlinked genes in early evolution. His concrete suggestion, the hypercycle, like other cooperative systems, is viable only with local interactions. One can justifiably ask the following questions. Do we need hypercycles or not? Did they play a role in prebiotic evolution?

The answer is that we do not know with certainty. For a long time, I have not favored hypercycles, partly because they have been oversold and sometimes misleadingly interpreted. More important, it seems that the stochastic corrector mechanism can solve the conundrum of dynamic coexistence of unlinked genes, the very problem that initiated the development of hypercy-cle theory. One must bear in mind that hypercycles are ecologically stable, and thus they could present a within-protocell copy-number regulation mechanism. Such a mechanism would be favored by natural selection in a population of compartments described by the stochastic corrector model.

The snag is that *any* such mechanism would be selected for. There are literally an infinite number of ways whereby copy-number control could be acheived through density-dependent growth. As I pointed out before (Szath-máry 1989a), the hypercycle is a particularly wasteful means of information integration (or copy-number control), for the following reasons.

1. It can produce shortcut mutations that make the cycle arbitrarily shorter. This type of mutation simply does not exist for nonhypercyclic systems.

2. Even hypercycles would have had to escape into compartments sooner or later (Eigen et al. 1981). For the compartment to thrive, the members of the hypercycle would have had to encode a function additional to the repli-case activity: This could have been enzymatic aid of metabolism (similar to the internal working of the stochastic corrector model). It is unlikely that in general the same gene could encode both functions. Thus, members of the hypercycle would have been twice as long as the genes of other systems. This would have entailed a much higher mutational load for hypercycles.

3. By the same token, the practical doubling in the number of genes would have led to an excessive metabolic burden also.

All in all, although hypercycles may have played an episodic, incipient role in early evolution, they are unlikely—in my opinion—to have had any decisive role, because alternative systems could have fulfilled the same role and at a reduced cost in terms of mutations and metabolites.

The Size of the Genetic Alphabet in a Metabolically Complex RNA World

An assumption of the stochastic corrector model is that the templates cata-lyze the reactions of intermediate metabolism, which in turn produces the monomers for template replication and the building blocks of the encapsulat-ing membrane. It could be true that this catalytic help is indirect: Like con-temporary nucleic acids, the templates could encode protein enzymes. This would, of course, require protein synthesis through translation. The problem is that translation is a highly evolved process: One does not easily see how a protocell could have had it to begin with. Happily, the idea of an "RNA world" (Gilbert 1986) brings us closer to bridging a gap between purely nonenzymatic systems and those based on protein enzymes. An extended version of this hypothesis—one that I favor—holds that the steps of intermediate metabolism were catalyzed by ribozymes (RNA molecules with enzymatic activity), an idea that goes back to suggestions by Woese (1967), Crick (1968), and Orgel (1968). If this hypothesis turns out to be true, it will bridge one of the most unpleasant gaps in evolutionary reasoning. Remarkably, it may also give an answer to the basic question, Why are there only four bases in the genetic alphabet?

THE RNA WORLD AND SELECTION OF RIBOZYMES

As I said, I favor the view that the RNA world was metabolically complex (Benner et al. 1987, 1989). Because there are almost no metabolic ribozymes known from contemporary organisms, what makes this suggestion nonetheless credible? Apart from arguments based on comparative biochemistry, it is the success of in vitro ribozyme genetics (for a review, see Szostak 1992) as was suggested for RNAs binding small ligands (Szathmáry 1984) and catalyzing reactions (Szathmáry 1989a, 1990a,b).

Very briefly, the experimental protocol yielding RNA molecules with specific functions consists of the following steps.

1. Choose a molecule (a so-called ligand) for which you want to generate RNAs that specifically bind to it. Simplifying somewhat, this molecule can be a substrate of a reaction that you want to be catalyzed by the RNAs generated.

2. Bind this molecule chemically to an indifferent material that will act as a vacuum cleaner collecting RNAs that bind the ligand.

3. Generate a pool of RNAs with different sequences.

4. Pour a solution of these RNAs over the material presenting the ligand.

5. Wash away RNAs that do not bind to the ligand.

6. Replicate those RNAs that bind the ligand.

7. If binding is strong enough, analyze the resulting RNA molecules; otherwise go to step 4.

By now, several different RNAs with enzymatic activity (ribozymes) have thus been selected (for a review, see Szathmáry and Maynard Smith 1997). It seems likely, therefore, that ribozymes were able to run a complex metabolism. Why then did protein enzymes replace most ribozymes? The consensus holds that the number of functional chemical groups provided by the 20 amino acids gives a definite advantage to proteins over ribozymes, the latter having only four building blocks (cf. Wong 1991). Obviously, one can make more versatile catalytic molecules with more building blocks.

There are two ways of increasing the catalytic potential of RNA-like molecules: (1) increasing the number of monomer types, and (2) post-synthetic modifications. I discuss these cases in turn.

NOVEL NUCLEIC ACID BUILDING BLOCKS

The beautiful results of Piccirilli et al. (1990) show that one can have as many as 12 different bases, forming six Watson-Crick-type base pairs, if one varies the hydrogen acceptor and donor groups on the monomers (bases). One of the theoretically possible novel base pairs was synthesized by Piccirilli et al. (1990). Polymerase enzymes do accept it in template-dependent reactions! Why is it then that we find only two base pairs in contemporary RNAs? The answer, as pointed out by Orgel (1990), is that (1) either Nature has never experimented with more than two base pairs, or (2) she decided that two were enough. Although we may never know whether (1) holds, rather forceful arguments have been put forward in favor of (2).

According to the suggestions of Fontana et al. (1991) two base pairs in natural nucleic acids is seen as a compromise between stability against mutations and thermodynamic stability. In general, the secondary structures of GACU sequences are richer and more variable than the corresponding structures built exclusively of GC or AU sequences. Calculated replication rate constants are maximal for much shorter chains for GC than for GACU. Because GC sequences form base pairs more readily than GACU sequences, the phenotype (in the context of the model the two-dimensional structure) of the former is less stable against random mutations. In contrast, three base pairs make it difficult for random sequences to fold into stable structures (for a review, see Schuster 1993).

A snag with the above explanation is that evolution of functional RNAs probably did not always proceed from random RNA sequences. An alternative, complementary, approach considers the fitness of ribo-organisms as a function of the size of the genetic alphabet, noting that catalytic efficiency of ribozymes increases with the number of letters, whereas copying fidelity of such molecules must decrease with it. (This decrease is intuitively obvious considering that it is easier to err when more, closely related molecules from a set are present.) The increase is slower than, and the decrease is faster than, exponential. Hence, there is an evolutionary optimum at a certain number of base pairs (Szathmáry 1991b, 1992b); some considerations even suggest that this optimum may indeed lie at 2. Thus, this trait may be a footprint of the decisive role of natural selection (cf. Leigh 1995) in molecular evolution.

Replicators and Reproducers: From Simple Autocatalysts to Chemotons

I presented the problems associated with the origin of life in the broader context of the major evolutionary transitions (Maynard Smith and Szathmáry 1995; Szathmáry and Maynard Smith 1995). How do higher-level evolutionary units emerge from the ones at lower levels? How did the storage and usage of hereditary information change? How did division of labor play an important role? As Griesemer (1996) aptly noticed, the frequent usage of the term "replicator" in the context of "comparative transitionology" (cf. Bonner 1995) is partly unwarranted. In many cases, the units that these authors were referring to were not replicators *sensu* Dawkins (1976) at all: Whole genomes, symbiotic organelles, cells within organisms, and sexual organisms within societies are certainly always vehicles, but rarely replicators. Their structure is usually not transmitted through copying. Reproduction of a whole mitochondrion is not replication.

In the present context, I prefer to use the term *reproducer* rather than *vehicle* (Dawkins 1976) or *interactor* (Hull 1980) because nothing in the latter concepts would suggest that organisms can be units of evolution. Concepts such as reproduction, heredity, and variability can be meaningfully applied to them. The term *reproducer* spells this out while emphasizing the difference to replication.

The origin of life itself (cf. Gánti 1997) is synonymous with the appearance of a certain type of chemical supersystem, the model of which is the chemoton (e.g., Gánti 1975, 1979, 1987). It consists of three autocatalytic subsystems: a metabolic network providing building blocks for the other two subsystems; a population of replicating templates; and an encapsulating membrane, essentially a protocell. The system as a whole is also autocatalytic and reproducing at the same time, but it is *not* a replicator. The only replicator playing a significant role from an orthodox Dawkinsian point of view is the template macromolecule. Although it is true that the intermediates of the metabolic cycle undergo processive replication at the molecular level, what is passed on to the offspring chemotons is a *population* of these replicators, that is by no means a replicator itself. By the same token, a similar reasoning applies to the genome of the chemoton: A bag of genes, undergoing random segregation into the offspring protocells does not undergo replication at the level of the bag. (Note, by the way, that when I conceived the stochastic corrector model, I had the chemoton in mind as a chemically detailed protocell model. Templates are imagined to catalyze the reaction steps of the metabolic cycle of A_i molecules.)

What would be the appropriate term to use in the context of transitionology? Griesemer suggests it is the reproducer. Reproducers often qualify as units of evolution, provided they have heredity and variation (for further discussion, see Szathmáry and Maynard Smith 1997). I think that the reproducer is a useful concept. It remains true that frequently a gene-centerd approach, like that of Williams (1966a) and Dawkins (1976), is extremely rewarding in the analysis of the spread of alleles in various contexts. It is also true, however, that (1) it is reproducers, rather than replicators, of a higher level that arose during the transitions; (2) when a higher-level reproducer appears, a novel type of development is worked out; and (3) rather old-fashioned replicators are packaged into novel reproducers.

One can summarize this chapter by considering the origins and functions of the subsystems of the chemoton. The origin of the membrane has not been dealt with here in terms of chemistry (see Maynard Smith and Szathmáry 1995), but we have seen that it provides the most stringent version of group selection by acting as a physical barrier to gene flow and forcing local interactions on the constituent genes. Local interactions were important before cellularization as well. Molecular cooperation of naked, unlinked replicators is inconceivable without such an effect.

The first replicators are likely to have emerged on mineral (probably pyrite) surfaces. The templates in the chemoton are assumed to gain ribozymic function aiding metabolism of the compartment. Obviously, evolution must have started before enzymes and templates. Simple autocatalysts, replicating in a holistic manner and having limited heredity (and thus limited evolutionary potential) are likely to have been the first replicators. Replicators carrying information in digital form appeared when replication became modular, rather than holistic. We know from experiments that such replicators can form a growing population in the absence of enzymes, but we do not know the evolutionary pathway to the appearance of the first RNA molecules. Nucleic acids are important because they are replicators with unlimited heredity. There can be so many types (sequences) that evolution may go on indefinitely.

Metabolism of autotrophic protocells hinges on the presence of an autocatalytic network of small molecules, modeled by the central cycle of the chemoton. At some point in evolution, nucleic acid replication became grafted onto such a metabolic network. Systems in which nucleic acids acted as ribozymes had an advantage: Metabolism and replication proceeded faster. The enzymatic function of RNAs in the has-been RNA world probably led to the fixation of the size of the genetic alphabet; accuracy decreases too fast if the number of base pair types is increased. The advantage from an increase in catalytic potential cannot compensate for this adverse effect.

Many experiments along the lines suggested by the current theories have not been carried out yet. This is an important task for the future. One would like to see a sensible scenario for the chemical origin of nucleic acids. Then one would like to see the spontaneous formation of protocells (chemotons) in the lab. Once we have achieved this, we shall have understood how life originated.

Acknowledgments

This work was supported in part by the Hungarian Scientific Research Fund (OTKA) and by a grant from the Ministry of Culture for research in higher education.

4 Individuality, Immortality, and Sex

Richard E. Michod

The emergence of new and higher levels of organization during evolution provides a compelling context for understanding the relations among certain fundamental properties of life, such as individuality, immortality, and sex. By *immortality*, I mean following Weismann (1890) the never-ending cycle of life; by *individuality*, I refer to the familiar levels in the hierarchy of life and their capacity to function as units of selection during evolution (genes, cells, organisms, societies, species); and by *sex*, I mean breakage and reunion of DNA molecules from different individuals.

The evolution of multicellular organisms is the premier example of the integration of lower evolutionary levels into a new, higher-level individual. Explaining the transition from single cells to multicellular organisms is a challenge for evolutionary theory. Sex and individuality are in constant tension as new units emerge, because sex mixes elements from different individuals and naturally threatens the integrity of evolutionary units. Yet, sex is fundamental to the continued well-being of evolutionary units and the immortality of life (Michod 1995). Although sex by creating mixis would seem to undermine individuality, history shows that sex is reinvented as each new level of individuality emerges in the evolutionary process.

Cooperation and Conflict

The benefits of cooperation provide the imperative for forming new, more inclusive evolutionary units. Increments in fitness are traded among levels of selection through the evolution of behaviors that are costly to individuals yet beneficial to groups. Cooperation is necessary for the emergence of new units of selection, precisely because it trades fitness from the lower level (the costs of cooperation) for increased fitness at the group level (its benefits). In this way, cooperation can create new levels of fitness and individuality (see table 4.1). This trade, if sustained through group selection, kin selection, and conflict mediation, results in an increase in the heritability of fitness and individuality at the new higher level. In this way, new higher levels of selection may emerge in the evolutionary process.

Although fueling the passage to higher levels, cooperation provides the opportunity for its own undoing through the frequency-dependent advantage of defection. Selfish interactions (defection) reap the benefits of cooperation while avoiding the costs and, for this reason, can be expected to spread

TABLE 4.1
Effect of Cooperation on Defection on Fitness at the Cell and Organismal Level

	Level of Selection	
Cell Behavior	Single Cell	Cell Group (organism)
Defection	+, replicate faster	−, less functional
Cooperation	−, replicate slowly	+, more functional

within the cooperating group. Selfish individuals typically stand more to gain than their selfishness costs any other individual, especially when rare. The "tragedy of the commons" (Hardin 1968) leads to conflict among lower-level units, which may sabotage the viability of cooperation and the creation of new higher levels of selection. By "conflict," I mean competition among lower-level units of selection leading to defection and a disruption of the functioning of the group.

Immortality, Mortality, and the Cycle of Life

From parent to offspring, from parent cell to daughter cell, from DNA strand to daughter DNA strand, the cycle of life continues. Life does not begin anew each generation, but is passed on through time like a family heirloom. Most evidence supports the view that life began once around 4 billion years ago and has been passed down through the eons. Each of us can, in principle, trace our ancestry back in time to this ancient founder. This immortal ancestry or cell lineage Weismann referred to as the germ line. (Used in this way, the term *germ line* refers to the cell lineage that can be traced backward in time from any living thing, whether or not it is a multicellular organism and possesses a germ line, in the sense of a sequestered and differentiated cell type specialized in forming gametes, and a somatic line with terminal differentiation.) In 1890, Weismann first defined the immortal in biological terms (Weismann 1890, p. 318). Weismann contrasted immortality with eternity; immortality is a state of activity and change in which the cycle of life continues indefinitely through time.

And what is it, then, which is immortal? Clearly not the substance, but only a definite form of activity . . . the cycle of material which constitutes life returns even to the same point and can always begin anew, so long as the necessary external conditions are forthcoming . . . the cycle of life, i.e., of division, growth by assimilation and repeated division, should [n]ever end; and this characteristic it is which I have termed immortality. It is the only true immortality to be found in Nature—a pure biological conception, and one to be carefully distinguished from the eternity of dead, that is to say, unorganized, matter.

The continued well-being of life resides in the information encoded in genes, and life's immortal potential requires that genes be passed on in good repair. Sex functions to repair damaged genes and otherwise cope with genetic errors such as mutations (Michod 1995). In so doing, sex is an integral part of the well-being and immortality of life, both in unicellular and multicellular organisms.

In multicellular organisms, immortality requires totipotency. Sex, totipotency, and immortality became special characteristics of certain differentiated cells termed germ cells, in contrast to the somatic cells, which replicate by mitosis and are terminally differentiated into the various cell types that make up the tissues and organs of multicellular organisms. Why did immortality and reproduction become the function of one group of cells, the germ cells, with the other somatic cells becoming terminal? Again it was Weismann who first spoke convincingly on this subject.

According to Weismann, the germ-soma differentiation was invented in multicellular organisms because of the advantage to the fitness of the organism of specialization and division of labor among its cells. Somatic cells specialize in making bodies adapted to the contingencies of existence, and germ cells specialize in making good gametes. Furthermore, once somatic cells began specializing in making bodies, they naturally would lose their immortality and capacity to divide forever. Why? Because, unnecessary but costly structures or activities should be lost in evolution. If the germ cells specialize in immortality, there is no longer any need—from the point of view of the whole organism—for the soma to maintain this capacity. Sounds reasonable, at least when you think in terms of the needs of the whole organism, and this was how Weismann approached the matter.

The problem with Weismann's approach is that multicellular organisms do not always exist as evolutionary units and consequently organismal needs are not always recognizable to selection. Before and during the transition from solitary cells to multicellular organisms, cells cannot be counted on to behave in the interests of the organism. After all, cells have been evolutionary individuals in their own right for billions of years before the first multicellular organism emerged. Even now, with our individuality well protected by such marvelous adaptations as a germ line, immune system, and programmed cell death, humans are threatened by the evolutionary potential of extant microbes (witness the recent antibiotic-resistant forms of bacteria and other microbes).

Why would cells relinquish their evolutionary rights in the interests of the organism? Although there can be no question that division of labor among cells is important to the functioning of an organism, evolution must first settle the question of individuality. Upon which evolutionary unit will selection focus: the cell, the multicellular organism, or some mixture of the two? We have clearly gotten ahead of ourselves by depending on the needs of the organism to explain the ancient differentiation of the germ and the soma.

According to our studies on the evolution of individuality reviewed below, the evolution of segregated and differentiated germ cells may be an example of conflict mediation. Conflict mediators may arise during the transition between single cells and multicellular organisms, and resolve, in favor of the organism, the multilevel selection process that must have been responsible for the origin of cell groups in the first place. Conflict mediation probably played a role in the origin of the ancient differentiation between the immortal (germ) and the mortal (soma) cells. By preserving the fitness gains at the level of the cell group, or organism (these fitness gains resulting from cooperation among cells), the germ line served to increase the heritability of fitness at the level of the multicellular organism and allowed it to emerge as an evolutionary individual (Michod 1996, 1997a,b, 1999; Michod and Roze 1997). To understand the basis for this claim, let me consider the question of organisms more systematically. Where did multicellular organisms come from? To get our bearings on this question, let us go back to the very beginning and sketch out a plausible scenario of the first 3 or 4 billion years of life on earth.

The First Individuals

IN THE BEGINNING

As far as we know, life first originated as simple replicating molecules, probably similar to extant single stranded RNAs. These ancient replicators could both encode information as a sequence of nucleotides and fold up upon themselves to act like proteins. Replication was a little sloppy at first because the copying of information from parent strand to daughter strand relied heavily on the free energies of base-pair formation between complementary nucleotides. Proteins that aid in this process and make it more faithful in extant life forms had not been invented yet. For thermodynamic reasons, in DNA, the nucleotide A pairs with T and G with C. This complementarity provides the basis for replication of an RNA or DNA strand and also provides the basis for reproduction at higher levels (cell, organism, etc.).

COOPERATIVE GENE NETWORKS

Genes began cooperating because two genes can do more than a single gene alone. Perhaps one sequence could serve as a kind of catalytic surface that facilitated the replication of another sequence. These cooperative interactions led to cooperative networks of genes, termed *hypercycles* by Eigen and his colleagues (Eigen and Schuster 1979), in which each gene contributed to the replication of other genes and also shared in the beneficial effects of the products produced by its neighbors. In time, proteins would be produced as a way of mediating these cooperative interactions. Although beneficial to others, proteins are produced at some cost, if only for the time and energy put into their production. Costly acts that are beneficial to the group provide a (short-term) advantage to cheating: There is a "temptation" (read immediate selective advantage) to use the benefits produced by others and not contribute to the group. Before considering the consequences of cheating for group living, I would like to consider another aspect of these early gene networks: their lack of individuality and how mixing of genes from neighboring networks could serve the function of recovery from genetic damage.

SEX AND PROMISCUITY AMONG THE NAKED GENES

These early networks of genes were a poorly defined lot, with few barriers to the flow of genes between gene groups. The lack of individuality may have had at least one advantage (Michod 1995). It would be very easy for genes to become damaged, exposed as they were without a cell membrane to the ultraviolet radiation and damaging reactions that must have been frequent on the primitive earth. Breaks, loss of nucleotides, and loss of methyl groups are just a few of the kinds of damage that occur to DNA in modern organisms and probably threatened the existence of these early gene networks. Damaged genes would just fail to replicate. As long as there was at least one undamaged copy of the same gene in the vicinity, this good copy could replicate and replace its damaged brethren. The result would be a kind of mixing of genes, or sex, with undamaged genes replicating to take the place of damaged neighbors. Sex (mixing for purpose of recovery from genetic error) came easily to these early replicators. There was much promiscuity and little individuality at this early stage. However "mixed up" we imagine the early gene networks to be, without any cell membrane to trap the errors, a kind of repair (or recovery from genetic error; Bernstein et al. 1984) could occur spontaneously with little "effort" or design on the part of the gene network.

TRAGEDY OF THE COMMONS

Now let us return to the costs of cooperation and the immediate selective advantage of defection. Hardin identified the tragedy of the commons as the fundamental problem of group living (Hardin 1968). The tragedy of the commons occurs when an individual stands to gain more by behaving selfishly than his selfish behavior costs each member of the group. For this reason, in cooperative groups there is always a temptation for individuals to defect. What keeps defection from arising and taking over the group, ruining the network of cooperation on which its very advantage depends? Human societies have laws and police forces to reduce this temptation. What happens during evolution? How can behavior beneficial to the group ever evolve? In short, the group must become an individual, but how does this happen? Conflict mediation underlies the transition to new levels of individuality.

CONFLICT MEDIATION AND INDIVIDUALITY

In time, the cooperating networks of genes would become encased in a cell. The cell was a wonderful invention, protecting the genes from the damaging effects of the environment and allowing resources and proteins to be kept close at hand, instead of diffusing away to be used by others. By encapsulating gene networks into a cell-like structure, one's proteins would be available only to nearest neighbors, those sharing the same cell. If one of these neighbors turned selfish—say, it used its neighbor's proteins but did not take time to make its own—all the genes in the same cell would be threatened, including the selfish gene. By putting everybody in the same boat, so to speak, everybody's self interest becomes more closely aligned with the interest of the group.

The cell is an example of a device that reduces conflict because it more closely aligns the interests of the genes with the interests of the group. For all the genes to replicate, the cell must replicate. Therefore, each gene has an interest in promoting the replication and well-being of the cell and in policing any selfish renegade genes. To understand the evolution of individuality and new levels of organization, we need to identify and explain evolutionarily those mechanisms and structures that serve to align the interests of the lower-level units with the interests of the group. More inclusive individuals must regulate the selfish tendencies of their components—genes in the case of gene networks; component cells in the case of multicellular organisms. How is individuality created at a new level (cell or organism) so that the lower-level units may be regulated, especially when no controlling organizer sits outside of the system? Conflict mediation is necessary; otherwise, new adaptations at the new level cannot evolve.

Sex and Individuality

Individuality has costs. Once the cell was invented, genetic errors were trapped on the inside. Sex had come easy to the free-living molecular replicators because genetic redundancy was always available in the form of gene copies in neighboring groups. With the gene group now encapsulated in a cell, new forms of sex had to be invented between cells as a means to obtain backup copies of genes for genetic repair.

Sex between cells involved mating (fusion) of two cells followed by mutual repair (recombination) and then splitting into daughter cells again. Diploidy (just stay fused and carry a set of genes in tow) was another possible strategy for coping with genetic error, but it has costs in terms of the extra resources and time needed to replicate additional genes. The costs of diploidy can make sex the preferred strategy under certain conditions (Michod 1998).

Sex and individuality are in constant tension, because sex involves fusion and mixis of genetic elements and thus naturally threatens the integrity of evolutionary units. Yet, sex is fundamental to the continued well-being of evolutionary units too. Sex and its antithesis in the evolution of reproductive systems, parthenogenesis, provide different options for the reproduction of evolutionary units. Although sex seems to undermine individuality, sex has been rediscovered as each new level of individuality emerges in the evolutionary process. Sex holds the promise of a better future and a more whole and undamaged individual.

Theories discussing the benefit of sex are discussed in three collections of papers on the topic (Stearns 1987; Michod and Levin 1988; American Genetics Society Symposium for the Evolution of Sex 1993). According to the repair hypothesis, genetic redundancy and repair occur during the sexual cycle and are the key to greater wholeness and well-being for the individual (Michod 1995). Cloning, on the other hand, offers ease and efficiency of reproduction at the expense of future generations and the well-being of the individual.

But with the successful cloning of a sheep recently announced in Scotland, has biological science found a more direct means to perpetuate what makes us the individuals we are (Michod 1997c)? The possibility, however faint, that a person might create offspring without the benefit of a partner has brought that question and others about sexual reproduction into unusual prominence. After all, sex extracts high costs in energy, time, and resources. Would it not be more efficient to make copies of ourselves asexually, as some think? Does generating one new person by combining the genes of two aging parents make any more sense than a one-for-one exchange? Would begetting a clone bring about a closer approximation of immortality than procreating in the usual fashion?

For those who have fantasized—and the fantasy seems all too common—that cloning could lead to the endless renewal of individual lives, the biological evidence suggests otherwise. In fact, it turns out that sex leads to a kind of immortality by repairing the genes of the egg and sperm cells so essential for the continuation of life (Michod 1997c). Far from being rejuvenating, cloning, on the contrary, could threaten the continuing evolutionary well-being of genes, cells, organisms, and even the very nature of species (see Michod 1991; 1995, chap. 9).

Major Transitions in Levels of Selection

The major transitions in evolutionary units are from individual genes to networks of genes, from gene networks to bacterialike cells, from bacterialike cells to eukaryotic cells with organelles, from cells to multicellular organisms, and from solitary organisms to societies (Buss 1987; Maynard Smith 1988, 1990, 1991a; Maynard Smith and Szathmáry 1995). These transitions in the units of selection share two common themes: (1) the emergence of cooperation among the lower level units in the functioning of the new higher level unit, and (2) regulation of conflict among the lower-level units.

Eigen and Schuster proposed the hypercycle as a way to keep individual genes from competing with one another so that cooperating gene networks could emerge (Eigen and Schuster 1979; Eigen 1992). Localizing genes in the cell keeps selfish parasitic genes from destroying the cooperative nature of the genome (Michod 1983; Eigen 1992; Maynard Smith and Szathmáry 1995). Chromosomes reduce the conflict among individual genes (Maynard Smith and Szathmáry 1993, 1995). Meiosis serves to police the selfish tendencies of genes and usually insures that each of the alleles at every diploid locus has an equal chance of ending up in a gamete. As a result of the fairness of meiosis, genes can increase their representation in the next generation only by cooperating with other genes to help make a better organism. Uniparental inheritance of cytoplasm may serve as a means of reducing conflict among organelles either through the expression of nuclear genes (Hoekstra 1990; Hurst 1990; Hastings 1992), or organelle genes (Godelle and Reboud 1995), or both. Finally, concerning the final transition—from organisms to societies of cooperating organisms—the theories of kin selection, reciprocation, and group selection provide three related mechanisms for the regulation of conflict among organisms: genetic relatedness, repeated encounters, and group structure. These are just a few of the ways in which the selfish tendencies of lower-level units are regulated during the emergence of a new higher-level unit.

As initially conceived, the field of sociobiology focused on the transition from solitary organisms to groups of organisms, or societies, and the emergence of cooperative functions at the social level, the level of the colony, say, in the case of eusocial behavior in insects (Wilson 1975). However, the set of tools and concepts used in studying conflict and cooperation during the transition from organisms to societies has proved useful for studying the other major transitions.

What happened during the transition between solitary cells and multicellular organisms? Organisms can be thought of as groups of cooperating cells.

Selection among cells could destroy this harmony and threaten the individual integrity of the organism. For the organism to emerge as an individual, or unit of selection, ways must be found of regulating the selfish tendencies of cells while at the same time promoting their cooperative interactions. The purpose of our work is to use multilevel selection theory to study the transition from cells to cell groups to multicellular organisms. More generally our goal is to develop a theoretical framework to study the emergence of individuality and new levels of fitness.

Evolution of the Organism

INDIVIDUALITY RECONSIDERED

Natural selection requires heritable variations in fitness. Levels in the biological hierarchy—genes, chromosomes, cells, organisms, kin groups, groups, societies—possess these properties to varying degrees, according to which they may function as units of selection in the evolutionary process (Lewontin 1970). Beginning with Wilson (1975) and the transition from solitary animals to societies, then Buss (1987) with the transition from unicellular to multicellular organisms, and more recently Maynard Smith and Szathmáry (Szathmáry and Maynard Smith 1995; Maynard Smith and Szathmáry 1995), attention has focused on understanding transitions between different levels of selection.

To understand the origin of individuality, therefore, we must understand how the properties of heritability and fitness variation emerge at a new and higher level from the organization of lower-level units, these lower-level units being units of selection in their own right initially. As already mentioned, unicellular organisms enjoyed a long evolutionary history before they merged to form multicellular organisms. In so doing, single cells relinquished their evolutionary heritage in favor of the organism. Why and how did this occur?

A SCENARIO

To help fix ideas, let us consider a scenario for the initial stages of the transition from unicellular to multicellular life. We may assume that reproduction and motility are two basic characteristics of the early single-celled ancestors to multicellular life, and these single cells were likely able to differentiate into reproductive and motile states (Margulis 1981, 1993; Buss 1987). Cell development was probably constrained by a single microtubule organizing center per cell, and, consequently, there would have been a tradeoff between reproduction and motility, with reproductive cells being unable to develop flagella for motility, and motile cells being unable to develop

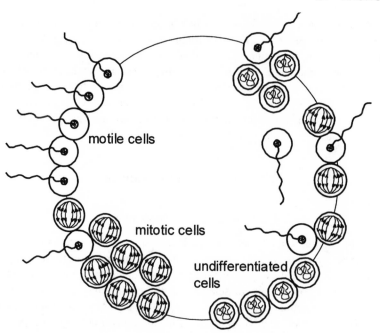

motile cells

mitotic cells

undifferentiated cells

FIG. 4.1. Scenario for the first organisms (groups of cells). Adapted from Michod and Roze (1999).

mitotic spindles for cell division. Single cells would switch between these two states according to environmental conditions. Finally, the many advantages of large size might favor single cells coming together to form cell groups. At this point, our investigations begin. Figure 4.1 shows three kinds of cells: motile cells with a flagella; nonmotile mitotically dividing cells; and cells that have yet to differentiate into either motile or mitotic states. Because of the constraint of a single microtubule organizing center per cell, cells cannot be motile and divide at the same time. Motile cells are an example of cooperating cells, and mitotically reproducing cells are an example of defecting cells.

If and when single cells began forming groups, the capacity to respond to the appropriate environmental inducer and differentiate into a motile state would be costly to the cell but beneficial for the group (assuming it was advantageous for groups to be able to move). Because having motile cells is beneficial for the group, but motile cells cannot themselves divide, or divide at a lower rate within the group, the capacity for a cell to become motile is a costly form of cooperation, or altruism. Loss of this capacity is then a form of defection, as staying reproductive all the time would be advantageous at the cell level (favored by within-group selection), but disadvantageous at the group level (disfavored by between cell-group selection). We are led, according to this scenario (and many others), to consider the fate of cooperation and defection in a multilevel selection setting during the initial phases of the transition from unicellular life to multicellular organisms.

A Model for the Origin of Multicellular Organisms

During the past several years, we have developed mathematical models of the evolutionary transition between single cells and multicellular organisms using the methods of population genetics and multilevel selection theory for the purpose of evaluating the levels of variation created within cell groups and studying the effect of this variation on the levels of cooperation and individuality attained.

Mathematical models show what is possible, based on assumptions about how the world works. By themselves, they cannot prove a hypothesis is true. They can, however, rule out poorly formulated or illogical hypotheses as well as suggest new hypotheses and fruitful lines of inquiry. By guiding experiment and observation, models are an integral part of scientific discovery. I primarily use simple population genetics models because they have great predictive and heuristic value in the understanding of complex evolutionary dynamics (Provine 1971, 1977, 1986; Ruse 1973; Wimsatt 1980; Michod 1981, 1986).

To understand the origin of organisms, it is helpful to think about them as groups of cooperating cells related by common descent (often from a single cell, the zygote). Selection among cells—below the level of the organism—could destroy the harmony within the organism and threaten its individual integrity. Competition among cells might favor cancerlike defecting cells that pursue their own interests at the expense of the organism. For the organism to emerge as an individual, or evolutionary unit, ways must have been found of regulating the selfish tendencies of cells, while at the same time promoting their cooperative interactions for the benefit of the organism. In addition, ways must have been found to ensure the heritability of the properties of these ensembles of cells so that the organism could continue to evolve as an evolutionary unit.

Consider a multicellular organism without a well-developed germ line, like a tree, coral, or hydra. An overview of the model life cycle is given in figure 4.2. The subscript j indexes types of zygotes. After zygote formation, organisms grow by replication of cells during development. This proliferation of cells during development is indicated by the solid vertical arrows in figure 4.2. During this proliferation, deleterious mutation may lead to the loss of cell function. Cell function is represented by a single cooperative strategy, and mutation leads to loss of cell function and thus is assumed to produce defecting (selfish) cells from cooperative cells. Mutant cells use less time and resources to cooperate and, as a result, may survive better or repli-

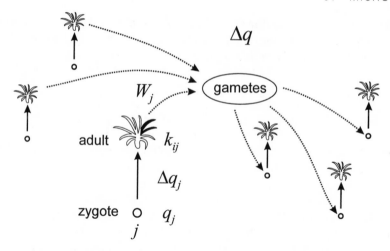

FIG. 4.2. Model life cycle of organisms. Adapted from Michod (1997b).

cate faster than cooperating cells (as occurs in the development of human cancers). Deleterious mutation will also produce uniformly deleterious cells that are impaired at both the cell level and the level of the group. Both kinds of mutations (selfish and uniformly deleterious) select for the modifiers of within-organism change discussed in subsequent sections (see Michod and Roze 1999). We assume no back mutation from defection to cooperation, because it is much easier to lose a complex function like cooperation and synergism between cells than it is to gain it. Because of mutation and different rates of replication of different cell types, gene frequencies change within the organism during its development. This is within-cell group or within-organism change, represented by the solid vertical arrow and Δq_j in figure 4.2 (mutant cells in the adult are represented by the solid black color).

After development, the adult organism contributes gametes to start the next generation, and the genetic makeup of these gametes will reflect the new gene frequencies in the adult form (after within-organism change). In the case of sexual reproduction, the gametes produced by the cell group fuse randomly to form a diploid zygote. In the case of asexual reproduction, the gametes become zygotes and develop directly into the adults of the next generation. Gene and genotype frequencies also change in the total population of organisms because of differences in fitness (gamete output) among the adult organisms. Adult fitness is a function both of adult size (number of cells in the adult stage) and functionality, represented here by the level of cooperation in the adult stage. Differences in adult fitness lead to between-cell group, or between-organism, change, and is represented by the dashed arrows in figure 4.2. The two components of frequency change—within organisms and between organisms—give rise to the total change in gene frequency in the population, Δq.

Cooperation trades fitness between levels in a selection hierarchy (table 4.1). In the case of cells within organisms, cooperation benefits the organism while detracting from the fitness of cells. Because it takes time and energy to help the group, cooperating cells may replicate more slowly or survive worse than mutant defecting cells. I assume that adult size is indeterminate and that organism fitness is a function of both adult size as well as the level of cooperativity among the organism's cells.

In modern multicellular organisms, there is a dual inheritance system: genetic and epigenetic (Maynard Smith 1990; Maynard Smith and Szathmáry 1995). During development, differentiated cell types are generated by turning on and off different genes in different cells. This epigenetic state is passed on during cell reproduction so that, say, liver cells (once differentiated) give rise to liver cells. Deleterious mutation during development may then involve, not just mistakes at the loci-determining cell type (as I have considered), but also errors in the epigenetic state, that is, turning on or off the wrong genes. The situation we have considered to date only involves genetic inheritance. Because we assume that only a single locus determines cell type, our model cannot allow for different genes to be turned on and off in different cells. A more complicated multilocus model determining cell type with epigenetic inheritance is under development. Presumably, allowing for epigenetic effects will increase the levels of within-organism variation, but this variation may or may not be heritable, depending on the epigenetic inheritance systems in place in the single cells (Jablonka and Lamb 1995).

KIN SELECTION REDUCES CONFLICT AMONG CELLS

In the models described here, all cells are clonally derived from a single cell zygote and related genetically (for consideration of propagule reproduction, see Michod and Roze 1999). This provides a kind of worst case for the study of the evolutionary effects of within-organism variation and conflict, because the zygote should restrict the opportunity for within-organism variation. If the organism began as a mixture of cells of different ancestries, as is probably the case for a migrating slug in the cellular slime mold *Dictyostelium discoideum*, the levels of within-organism change and conflict would likely be greater. By often reproducing through a single cell stage—the zygote— organisms insure close genetic relatedness among their component cells. Maynard Smith and Szathmáry argued that close kinship among cells should be sufficient to regulate the selfish tendencies of cells in an organism (Maynard Smith and Szathmáry 1995). In the limit, if cells are absolutely genetically identical, then their interests are one and the same. Is the kin structure created by the zygote stage sufficient to regulate the selfish tendencies of cells?

Levels of cooperation can be low in organisms even when reproduction occurs through a single-cell zygote stage (Michod 1997a,b). Depending on the level of mutation, intensity of selection, and time for development, mutation and selection can create sufficient within-organism change and, hence, genetic variation within the cell group. This suggests that there is a problem in coping with within-organism variation and selection that the zygote stage (and the resulting kinship among cells) does not adequately deal with. We have used two-locus modifier theory to show that this variation and conflict can select for "conflict modifiers," genes that restrict the opportunity for within-organism change, for example, modifiers that create a germ line or police the selfish tendencies of cells (Michod 1996; Michod and Roze 1997).

CONFLICT MODIFIERS

For the organism to emerge as a new unit of selection, within-organism change and interaction must be controlled so that heritability of fitness may increase at the organismal level. We model this by considering a second modifier locus that modifies the parameters of within-organism change at the first cooperate/defect locus possibly at some cost to the organism. In our model, a modifier allele is introduced at an equilibrium that is polymorphic for cooperation and defection (cooperation can never reach fixation because of recurrent mutation leading to defection). At this equilibrium, there exist two kinds of groups, those stemmimg from cooperating zygotes and those stemming from defecting zygotes. By definition, cooperative genotypes bias selection toward the group or organismal level (because cooperation takes fitness away from the cells and gives it to the group), while defecting genotypes do the opposite and bias selection toward the cell level. To maintain cooperation at the equilibrium before the modifier is introduced, cooperative groups must be more fit (produce more gametes) than defecting groups, because the fitness benefits of cooperation at the group level must compensate for mutation at the cellular level toward defection. The modifier allele increases by virtue of being associated with the more fit cooperating genotype. As a result of increase of the conflict modifier, both the level of cooperation within the organism and the heritability of fitness at the organism level increase (see figs. 4.3–4.5 below).

How might evolution modify the parameters of within-organism change so as to increase the fitness of the organism? According to Buss (1987), the individual integrity of complex animal organisms is made possible by the germ line, the sequestered cell lineage set aside early in development for the production of gametes. By sequestering a group of cells early in development, the opportunity for variation and selection is limited. As a consequence, evolution depends on the fitness of organisms, and the covariance of adult fitness with zygote genotype, and not the fitnesses of the cells that comprise the organism. The heritability of organismal traits encoded in the zygote is thereby protected. The trait of interest here concerns the level of specialization and differentiation among cells within organisms, which is represented here by the level of cooperativity among the cells.

The essential feature of a germ line is that gamete-producing cells are sequestered from somatic cells early in development. Consequently, gametes have a different developmental history from cells in the adult form (the soma) in the sense that they are derived from a cell lineage that has divided for a fewer number of cell divisions with, perhaps, a lower mutation rate per cell replication. The main parameters affected by germ line modifiers are the developmental time and mutation rate per cell division in the germ line relative to the soma.

Such germ line modifiers that lower the developmental time or mutation rate may be selected in our studies. Maynard Smith and Szathmáry (1995) suggested that germ line cells may enjoy a lower mutation rate but do not offer a reason why. Bell interpreted the evolution of germ cells in the Volvacale as an outcome of specialization in metabolism and gamete production to maintain high intrinsic rates of increase while algae colonies got larger in size (Bell 1985; see also Maynard Smith and Szathmáry 1995, pp. 211–213). I think there may be a connection between these two views.

As metabolic rates increase, so do levels of DNA damage. Metabolism produces oxidative products that damage DNA and lead to mutation. It is well known that the highly reactive oxidative by-products of metabolism (e.g., the superoxide radical O_2^-, and the hydroxyl radical $\cdot OH$ produced from hydrogen peroxide H_2O_2) damage DNA by chemically modifying the nucleotide bases or by inserting physical cross-links between the two strands of a double helix, or by breaking both strands of the DNA duplex altogether. The deleterious effects of DNA damage make it advantageous to protect a group of cells from the effects of metabolism, thereby lowering the mutation rate within the protected cell lineage.

This protected cell lineage, the germ line, may then specialize in passing on the organism's genes to the next generation in a relatively error-free state. Other features of life can be understood as adaptations to protect DNA from the deleterious effects of metabolism and genetic error (Michod 1995): Keeping DNA in the nucleus protects the DNA from the energy-intensive interactions in the cytoplasm; nurse cells provision the egg so as to protect the DNA in the egg; and sex serves to repair genetic damage effectively while masking the deleterious effects of mutation. The germ line may serve a similar function of avoiding damage and mutation—by sequestering the next generation's genes in a specialized cell lineage, these genes are protected from the damaging effects of metabolism in the soma.

As just mentioned, according to Bell (1985), the differentiation between the germ and the soma in the Volvocales is correlated with increasing colony size, with true germ soma differentiation occurring only when colonies reach about 10^3 cells as in the *Volvox* section *Merillosphaera*. Although Bell inter-

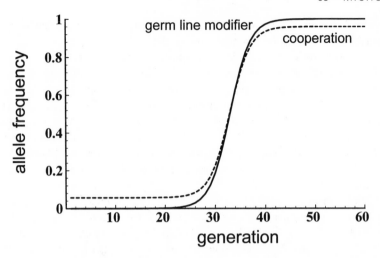

FIG. 4.3. Allele frequencies of cooperation and germ line modifier during an evolutionary transition. Adapted from Michod and Roze (1997) in which the underlying model is explained. See also Michod (1996).

preted the dependence of the evolution of the germ line on colony size as an outcome of reproductive specialization driven by resource and energy considerations (as did Weismann over a century ago), this relation is also explained by the need for regulation of within-colony change (see Michod and Roze 1999).

Another way to reduce conflict among cells is for the organism to actively police and regulate the benefits of defection (Boyd and Richerson 1992). How might organisms police the selfish tendencies of cells? The immune system and programmed cell death are two possible examples. For an introduction to the large and rapidly developing area of programmed cell death, or apoptosis, see Carson and Ribeiro (1993), Ameisen (1996), and Anderson (1997). This hypothesis is studied more in Michod and Roze (1999) and Michod (1999).

EFFECT OF TRANSITION ON THE LEVEL OF COOPERATION

Mediation of conflict among lower-level units is an essential feature of transitions to new higher levels of organization. I now consider the consequences of the evolution of conflict modifiers for the level of cooperation among cells and the heritability of fitness at the cell group, or organismal level. For reasons of space, I only consider the evolution of the germ line modifiers and our results for asexual reproduction, but we have obtained qualitatively similar results for sexual reproduction and the other forms of conflict mediation such as self-policing modifiers and modifiers that regulate group size.

In figure 4.3, the evolution of a germ line modifier dramatically increases the level of cooperation in the organism. The level of cooperation always increases during the evolution of conflict modifiers. To understand the forces that lead to the evolution of the modifier, we have used covariance methods.

INCREASE OF FITNESS AT ORGANISMAL LEVEL

An especially useful and illuminating method for representing selection in hierarchically structured populations is Price's covariance approach (Price 1970, 1972, 1995). Price's approach posits a hierarchical structure in which there are two selection levels—in our case, (1) between cells within organisms, viewed as a group of cells, and (2) between organisms within populations. Both levels of selection can be described by the single equation 4.1, Price's equation for organisms:

$$\Delta q = \frac{\text{Cov}[W_i, q_i]}{\overline{W}} + \text{E}[\Delta q_i]. \tag{4.1}$$

Variables q and q_i are the frequencies of a gene of interest in the total population and within zygotes; $\text{Cov}[x,y]$ and E $[x]$ indicate the weighted covariance and expected value functions, respectively. The first term of the Price equation 4.1, $\text{Cov}[W_i, q_i]$, is the covariance between fitness and genotype and reflects the heritable aspects of fitness. The second term of equation 4.1, $\text{E}[\Delta q_i]$, is the average of the within-organism change resulting from mutation and selection among cells.

In figure 4.4, the two components of the Price covariance equation 4.1 are plotted during the increase in frequency of the germ line allele given in figure 4.3. These components partition the total change in gene frequency into heritable fitness effects at the organism level (solid line) and within-organism change (dashed line). In the model studied here, within-organism change is always negative, because defecting cells replicate faster than cooperating cells, and there is no back mutation from defection to cooperation. At equilibrium, before and after the transition, the two components of the Price equation must equal one another, or the population could not be in equilibrium (fig. 4.4). During the transition, however, we see that the covariance of fitness with genotype at the emerging organismal level (fig. 4.4, solid curve) is greater than the average change at the cell level (fig. 4.4, dashed curve). This greater heritable covariance in fitness at the higher level forces the modifier into the population.

In figure 4.4, we see that modifiers of within-organism change evolve by making the covariance between fitness at the organismal level and zygote

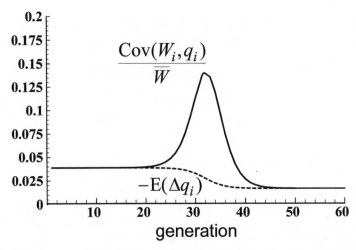

FIG. 4.4. Study of evolutionary transition by Price equation. Same model and parameter values as figure 4.3. The x-axis is the same as in figure 4.3. Adapted from Michod and Roze (1997). See text for explanation.

genotype more important than the average within-organism change. This implies that modifiers increase the heritability of fitness at the level of the new organism.

HERITABILITY OF FITNESS AND THE EVOLUTION OF INDIVIDUALITY

Before the evolution of cooperation among cells, the population is assumed to be genetically homogeneous (all cells defect, and there is no modifier allele). In such a population, the heritability of fitness equals unity. When the cooperation allele appears in the population, evolution (directed primarily by kin and group selection) may increase its frequency, leading to greater levels of cooperation. However, within-organism change increases along with co-operation. Deleterious mutation produces nonfunctional defecting cells that have an advantage over cooperating cells at the cellular level. As a consequence of within-organism change, the heritability of fitness must decrease as soon as cooperation evolves.

The basic problem for the evolution of a new unit of selection—in this case, the multicellular organism—is that the organism cannot evolve new adaptations, such as the traits enhancing cooperation, if these adaptations are costly to cells, without increasing the opportunity for conflict within and thereby decreasing the heritability of fitness. Deleterious mutation is always a threat to new adaptations because it produces cells that go their own way. By regulating within-organism change, there is less penalty for cells to help the organism. Without a means of regulating within-organism change, the "organism" is merely a group of cooperating cells related by common de-

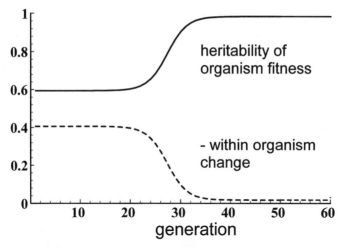

FIG. 4.5. Heritability of organism fitness and within organism change during evolutionary transition. Same parameter values as figure 4.3. The x-axis is the same as in figure 4.3. Adapted from Michod and Roze (1997).

scent. Such groups are not individuals, because they have no functions that exist at the new organism or group level.

Conflict modification is the first, uniquely organismal level function. Does heritability of fitness, the defining characteristic of an evolutionary individual, increase during the evolutionary transition mediated by conflict modifiers? Heritability of fitness may be defined as the regression of the fitness of offspring on fitness of the parents. It can be shown that the evolutionary transition mediated by the evolution of conflict modifiers always leads to an increase in the heritability of fitness. In addition, the eigenvalues of the different equilibria involve ratios between products of fitnesses and heritabilities (Michod 1999; Michod and Roze 1999). This illustrates clearly that what determines whether a new characteristic can increase in frequency in the population is the heritability of fitness of the new evolutionary individuals with this feature.

Before the evolution of modifiers restricting within-organism change, the "organism" is just a group of cooperating cells related by common descent from the zygote. Because of high kinship, heritability is initially significant at the group (organism) level ($h_w^2 \approx 0.6$ for the particular model studied in fig. 4.5), but this value is still low for asexual haploidy. (Heritability at the organismal level should equal unity in the case of asexual organisms when there is no environmental variance.) Low heritability of fitness at the new level resulting from within-organism change poses a threat to the continued evolution of the organism. In the case considered here, developmental time, and hence organism size, could not increase without the evolution of conflict modifiers. Indeed, the continued existence of cell groups at all is highly unlikely before the evolution of the conflict modifier, because the cooperation allele is at such a low frequency (fig. 4.3), and stochastic events would probably lead to its extinction. As the modifier begins increasing, the level of within-organism change drops (fig. 4.5, dashed curve), and the level of cooperation among cells increases dramatically (fig. 4.3, dashed curve), as does the heritability of organism fitness (fig. 4.5, solid curve).

Transitions in Individuality

EFFECT OF SEX AND DIPLOIDY ON EVOLUTIONARY TRANSITIONS

Sex and diploidy have profound effects on the evolutionary transitions in our models (see Michod 1999). With diploidy, there are two dominance relations to consider at the cellular and organismal levels. Diploidy may facilitate the initial increase of cooperation through the masking of the advantage of defection in heterozygous zygotes. Diploids may also reach larger organism sizes than haploids, although at much reduced levels of cellular cooperation. If adult size is held constant, however, these advantages of diploidy no longer pertain. The buffering effect of increasing the heritability of fitness at the organismal level, whereby the level of harmony and cooperation within the organism is maintained in the face of increasing within-organism change, still pertains under diploidy, although it is affected by dominance.

There are quantitative effects of recombination in breaking up the association between the modifier and the cooperate/defect locus. In the case of germ line modifiers, sex requires larger decreases in developmental time in the germ line (for the effect of recombination on evolution of the germ line modifier, see Michod 1996, fig. 2). With sex, it also takes longer for the transition to occur (results not shown here for reasons of space). The modifier increases by virtue of being more often associated with cooperative alleles in gametes, and recombination breaks apart this association. Although sex can retard the transitions modeled here, because of the effects of recombination in breaking up the genetic associations needed for the modifier to increase, I do not see these quantitative differences as presenting any real barriers to the evolution of conflict modification and evolutionary transitions in sexual progenitors. More important, I think, is the way in which sex organizes variability and heritability of the traits and capacities that affect the fitness of the new emerging unit.

Sex helps diploids maintain a higher heritability of fitness under more challenging conditions, especially when there is great opportunity for within-organism variation and selection. With sex, as the mutation rate, and concomitantly the amount of within organism-change, increases more and more of the variance in fitness is heritable, regardless of dominance. Sex allows the integration of the genotypic covariances in a way not possible in asexual populations.

The genome-wide mutation rates are vastly different in modern multi-cellular organisms (≈ 0.5 on haploid genome basis) and in modern microbes (≈ 0.003). Once multicellularity evolved, the continued evolution of multi-cellular organisms required new gene functions with a corresponding increase in genome size. With increasing genome size came the problem of increased rates of deleterious mutation. It is often noticed that diploidy helps multicellular organisms tolerate this increase in mutation rate by the masking of recessive or nearly recessive deleterious mutations (see, e.g., Michod 1995, fig. 1 and associated discussion). Once the diploid species reaches its own mutation selection balance equilibrium, however, the mutation load in diploid species actually increases beyond what it was under haploidy (Haldane 1937; Hopf et al. 1988). There must be another factor that allows complex multicellular diploids to tolerate a high mutation rate.

We have found that, as the mutation rate increases in sexual diploid organisms, the regression of fitness on zygote gene frequency actually increases (Michod 1999). In other words, as the mutation rate increases, and along with it the amount of within-organism change, more of the variance in fitness in sexual diploids is heritable, that is, explained by the alleles carried in the zygote. How can this be? The greater mutation rate must result in greater levels of within-organism change. At equilibrium, this within-organism change must be balanced by a larger covariance of fitness with zygote frequency. This is what Price equation (4.1) says. In haploid and asexual diploid populations, this is accomplished by a greater variance in zygote gene frequency, whereas in sexual populations, this can be accomplished by a greater regression of organism fitness on zygote frequency. Sex allows a greater precision of mapping heritable propensities of the zygote onto adult fitness under more challenging conditions.

These conclusions are based on equilibrium statistics before and after the evolutionary transition, and it is unclear whether these conclusions can be extended into the nonequilibrium realm of the transition and if the results will hold up under more realistic genetic models. If so, the greater precision in the mapping of cooperative propensity onto organism fitness should allow sexuals to make the transition from cells to multicellular organisms more easily under more challenging circumstances. This result is consistent with the view that the protist ancestor of multicellular life was probably sexual (Maynard Smith and Szathmáry 1995).

COMPONENTS OF EVOLUTIONARY TRANSITIONS

Our results suggest that, even in the presence of high kinship among cells, within-organism change can be significant enough to lead to the evolution of a means to regulate it. Examples of such conflict mediators may be the segregation of a germ line during the development and the evolution of a means of policing cells, such as the immune system or programmed cell death. The germ line functions to reduce the opportunity for conflict among cells and promote their mutual cooperation both by limiting the opportunity for cell replication (Buss 1987) and by lowering the mutation rate (Maynard Smith and Szathmáry 1995). Mutual policing (Boyd and Richerson 1992) is also expected to evolve as a means of maintaining the integrity of the organisms once they reach a critical size. Any factors that directly reduce the within-organism mutation rate are also favored.

Once within-organism change is controlled, high heritability of fitness at the new organismal level is assured. Individuality at the organismal level depends on the emergence of functions allowing for the regulation of conflict among cells. Once this regulation is acquired, the organism can continue to evolve new adaptations at the new level, without increasing the conflict among cells, as happened when cooperation initially evolved.

Development evolves so as to restrict the opportunity for conflict among cells. The evolution of modifiers of within-organism change lead to increased levels of cooperation within the organism and increased heritability of fitness at the organismal level. The evolution of these conflict mediators are the first new functions at the organismal level. An organism is more than a group of cells related by common descent; to exist, organisms require adaptations that regulate conflict within. Otherwise, continued improvement of the organism is frustrated by the creation of within-organism variation and conflict. The evolution of modifiers of within-organism change are a necessary prerequisite to the emergence of individuality and the continued well-being of the organism.

In summary, what happens during an evolutionary transition to a new higher-level unit of individuality, in this case, the multicellular organism? While taking fitness away from lower-level units, cooperation increases the fitness of the new higher-level unit (cell to organism). In this way, cooperation may create new higher-levels of selection. However, the evolution of cooperation sets the stage for conflict, represented here by the increase of defecting mutants within the emerging organism. The evolution of modifiers restricting within-organism change are the first higher-level functions at the organismal level. Before the evolution of a means to reduce conflict among cells, the evolution of new adaptations (such as the underlying traits leading to increased cooperation among cells) is frustrated by defecting mutants. Individuality requires more than just cooperation among a group of genetically related cells, it also depends on the emergence of higher-level functions that restrict the opportunity for conflict within and ensure the continued cooperation of the lower-level units. Conflict leads—through the evolution of adaptations that reduce it—to greater individuality and harmony for the organism.

5 Sexual Conflict in Animals

Catherine M. Lessells

Whatever the reasons for the evolution of sexual reproduction, it leads to a situation in which there are two individuals who may have no genetic interest in each other's future, the parents, but who nevertheless have a joint genetic interest in another individual or group of individuals, their offspring. If investment to improve the prospects of the current offspring reduces the parent's own prospects (i.e., there is a "cost of reproduction"; Williams 1966b; Lessells 1991), then there will be a conflict between the two parents (Trivers 1972). This conflict arises because each parent's fitness is generally maximized if it invests less and the other parent invests more than would maximize the other parent's fitness. Although this conflict has been referred to as "the battle of the sexes" (Dawkins 1989), selection for each parent to exploit the efforts of the other parent relies only on sexual reproduction and not the existence of separate sexes.

The Evolution of Sexes

Conflict between parents may even be responsible for the evolution of sexes. Theoretical models show that conflict between initially sexually undifferentiated parents over gamete size can lead to the evolution of gamete dimorphism, and subsequently to disassortative mating between large and small gametes (Parker et al. 1972; Parker 1978). A parent with a fixed quantity of resource for the production of gametes can only produce more gametes at the expense of gamete size. When the fitness of zygotes depends strongly on zygote size, there may be disruptive selection for gamete size: Parents producing large gametes have the advantage of zygotes with high fitness, and parents producing small gametes have the advantage of numerous zygotes. Once gametes are dimorphic, there is conflict between the two sizes of gamete over fusion partners. Small gametes are selected to fuse disassortatively with large gametes, but large gametes are selected to fuse assortatively among themselves. Small gametes appear to have won this evolutionary conflict. Possible reasons include the stronger selection on small gametes to mate disassortatively; the larger number of mutations occurring each generation in the more numerous small gametes (Parker et al. 1972); the lack of suitable fusion partners for mutant ova that could fuse only with other ova (all wild-type ova having been rapidly removed from the gamete pool by the vastly more numerous sperm; Parker 1978, 1982); and the cost for larger gametes of maintaining the motility necessary for fusion with other large gametes exceeding the consequent benefits (Parker et al. 1972; Parker 1978).

Alternative theories for the evolution of sexes envisage intragenomic conflict rather than parental conflict as the selection pressure for two disassortative mating types (Cosmides and Tooby 1981; for more detail, see Pomiankowski, chap. 7). For instance, the initial evolution of large gametes may have been due to selection on cytoplasmic genes for increasing gamete size (which increases the probability of those genes being in the majority and hence winning out in competition in the resulting zygote; Cosmides and Tooby 1981). Alternatively, small gametes may have been an easily evolved mechanism for the exclusion of cytoplasmic genes from gametes (Hurst 1990, 1992b; Hastings 1992; Law and Hutson 1992). However, none of these theories rule out parental conflict as an additional selection pressure for the development and maintenance of anisogamy.

Sexually Antagonistic Genes

Once sexes exist, a gene that is expressed in both sexes may be beneficial to one sex, but deleterious to the other (Rice 1984). Such "sexually antagonistic genes" may exist without conflicts of interest between males and females. For example, tarsus length in collared flycatchers (*Ficedula albicollis*) is perfectly genetically correlated in males and females, implying that the same genes determine tarsus length in both sexes. Selection acts in opposite directions in the two sexes, however, with small males and large females tending to survive better from shortly before fledging until capture as breeding adults in the following year (Merilä et al. 1997).

Sexually antagonistic mutations are able to invade populations, depending on the relative selection pressures in the two sexes and the proportion of generations in which they are selected in each sex (Rice 1984). Autosomal genes have equal probabilities of finding themselves and being expressed in males or females (but not necessarily in alternate generations). Hence, sexually antagonistic autosomal mutants will invade when the advantage to one sex outweighs the disadvantage to the other. A mutant allele that is linked to a sex-determining locus will be expressed unequally in the two sexes, however. In a species with XY sex determination and an inactive Y chromosome, recessive alleles on the X chromosome will be expressed more frequently in the heterogametic (XY) sex (because all recessive alleles will be expressed in the hemizygous XY individuals who carry only a single copy of the gene, whereas recessive alleles will not be expressed in heterozygous XX individuals). Similarly, dominant alleles will be expressed more frequently in the homogametic sex. As a result, recessive alleles that favor the heterogametic sex and dominant alleles that favor the homogametic sex may spread even when their advantage in that sex is smaller than their disadvantage in the other sex (Rice 1984). When there are functional genes on the Y chromosome, a mutant allele on that chromosome will be transmitted (except when there is recombination) only to offspring of the same sex as the parent. Thus, the gene will be found (and selected) predominantly in one sex, and the net selection on antagonistic alleles may allow them to spread even when they are lethal to the other sex (Rice 1987).

The theoretical prediction that sexually antagonistic alleles can accumulate if they are linked to sex-determining alleles has been used to demonstrate the widespread existence of sexually antagonistic alleles in *Drosophila melanogaster* (Rice 1992). Genetic markers were artificially selected in the experimental line so that they were transmitted as if they were novel female-determining genes. As a result, alleles that were linked to the genetic markers will have been passed from mother to daughter for most or all of the experiment. In a control line, the genetic markers were artificially selected so that they were found in males and females in alternate generations. After twenty-nine generations, the effect of the alleles linked to the genetic markers on male fitness was tested. Males carrying the genes that had previously been confined to females showed a drastic reduction in fitness compared with those carrying genes that had alternated between males and females. The rapid accumulation of sexually antagonistic alleles in a small randomly chosen part of the genome (the part linked to the artificially selected genetic markers) suggests that such alleles are present at low frequency at many loci widely dispersed through the genome.

The existence of sexually antagonistic genes implies that each sex holds back the adaptation of the other (Rice 1984; Slatkin 1984; Lande 1987), with selection in one sex undoing changes in gene frequency made in response to selection in the other sex. This has been shown for tarsus length in collared flycatchers (Merilä et al. 1997, 1998; see above), and it also seems to be true for beak color in zebra finches (*Taeniopygia guttata*; Price 1996). In the laboratory, male and female beak color show a high genetic correlation (Price and Burley 1993; Price 1996), as would be expected if the same genes pleiotropically control beak color in the two sexes. There is also sexually antagonistic selection in the laboratory (Price and Burley 1994): Males with the reddest bills had the highest fitness, because they started nesting more quickly at the start of the experiment and had a higher rate of reproduction (but did not survive better). In contrast, females with the orangest (least red) bills had the highest fitness, because they had a higher rate of reproduction and survived longer (but did not start nesting more quickly). Thus, the alleles that pleiotropically affect beak color in both males and females have antagonistic effects on male and female fitness. This explains why beak color does not evolve, despite the directional selection on beak colour in each of the sexes separately.

In the long-term, the existence of sexually antagonistic alleles will select for modifier genes for sex-limited expression of the alleles (Fisher 1958). The existence of the modifier genes will in turn reduce selection against the sexually antagonistic allele, so that both genes will eventually evolve to fixation (Rice 1984), although the process may be slow (Lande 1980, 1987). At this point, the originally sexually antagonistic allele is expressed only in one sex and transmitted silently through members of the other sex. When there is sexual conflict, the behavior of the two sexes will be determined by different loci that can evolve independently. It thus becomes possible to model such situations in terms of male and female strategies without the need for explicit population genetic models: Sexually antagonistic alleles have been replaced by sexually antagonistic behavior.

Sexually Antagonistic Behavior

THE EXTENT OF CONFLICT

Trivers (1972) emphasized that each parent is selected to exploit the parental investment of its partner: An individual's fitness will be maximized if its partner behaves in a way that maximizes the first individual's fitness rather than the partner's own fitness. The degree of conflict therefore depends on the extent of common interest between the pair in current and future breeding attempts. The most intense conflict will occur when there is no shared interest in the current offspring, as occurs when one of the pair is not the biological parent of the brood. Human children are more at risk of being murdered by stepparents than by biological parents (Daly and Wilson 1988), and infanticide occurs in a number of primate (Hausfater 1984; Hiraiwa Hasegawa and Hasegawa 1994) and carnivore species (Packer and Pusey 1984) when males from outside a social group supplant the incumbent breeding males. Infanticide also occurs in an eresid spider *Stegodyphus lineatus*, in which males attempt to destroy the egg sacs of females with whom they have not previously mated (Schneider and Lubin 1996, 1997). It is clear that there is a conflict in reproductive interests because females provide suicidal care to the young, which kill and consume her a few weeks after hatching. Males that do not destroy egg sacs therefore have no expectation of producing offspring with a female who already has eggs. Even without infanticide, males of other species may reduce their investment in broods where they have reduced paternity, as occurs in reed buntings (*Emberiza shoeniclus*; Dixon et al. 1994) and dunnocks (*Prunella modularis*; Burke et al. 1989), although such a reduction is not always expected (Westneat and Sherman 1993; Lessells 1994; Kempenaers and Sheldon 1996) or found (for a list of studies, see Møller and Birkhead 1993). The possibilities of infanticide and reduced paternal care imply that females should conceal the paternity of their offspring (Trivers 1972). Female chimpanzees (*Pan troglodytes verus*) mate furtively outside their social groups: Although a genetic analysis revealed that more than half of the young were fathered by males from outside their group, females were never observed approaching males from neighboring groups except during confrontations (Gagneux et al. 1997).

The extent of conflict also depends on the extent to which the future reproductive success of the members of a pair is interdependent. If there were lifelong obligate monogamy, then the current and future reproductive interests of the partners would coincide entirely, and there would be no conflict. However, it is not sufficient that monogamy is usually lifelong or long-lasting: If an individual can replace a lost mate without cost to its own future reproductive success, then there will be as much conflict as if different individuals formed pairs for each breeding attempt. Factors that can impose a cost on the loss of a mate, and hence reduce conflict, include strong territoriality in which both members of a pair are required to defend the territory and widowed individuals are evicted, or an increase in reproductive success with increasing breeding experience with the same partner, as occurs in Bewick's swans (*Cygnus columbianus bewickii*; Scott 1988) and Cassin's auklets (*Ptychoramphus aleuticus*; Emslie et al. 1992). At the other end of the scale, when a pair never breeds together more than once, selection for complete exploitation will be limited if the costs of parental care by the partner are paid during the current breeding attempt ("a current cost"; Lazarus and Inglis 1986) and hence threaten an individual's own reproductive success.

THE SUBJECT OF CONFLICT

PARENTAL INVESTMENT

Parental investment by either parent can take diverse forms but usually involves provisioning or protection of the young (Trivers 1972; Clutton-Brock 1991). By definition (Trivers 1972; but see also Clutton-Brock 1991), parental investment reduces future reproductive success, whether with a social partner or as part of a mixed reproductive strategy involving extrapair copulation or brood parasitism (e.g., Petrie and Møller 1991). An individual may therefore maximize its lifetime fitness by restricting the amount or duration of investment in the current family. Because this restriction will often decrease the lifetime fitness of the partner, there will frequently be sexual conflict over investment. An obvious example is a male insect that will fertilize most or all of the eggs that a female lays in the first clutch after she mates with him, but few or none of the eggs in succeeding clutches (Parker 1970). The male will have highest fitness when the female lays the largest possible clutch, even though this may damage her survival prospects and hence her lifetime fitness. The reverse situation is found when males partition their sperm reserves between copulations in a way that maximizes their reproductive success but may lead to less than 100% fertilization success. For example, male bluehead wrasse (*Thalassoma bifasciatum*) with the highest daily mating rate release the fewest sperm per mating and fertilize a smaller proportion of the females' eggs (Warner et al. 1995). Sexual cannibalism, in which females consume their mates during mating, is another example of conflict over investment by males (Elgar 1992; Johns and Maxwell 1997). Although the behavior of males of some species makes them apparent accomplices in their own demise (e.g., the "copulatory somersault" of the Australian redback spider [*Latrodectus hasselti*] into the jaws of the female; Forster 1992), other males approach the female with extreme caution, suggesting a conflict with the female over their fate (Lawrence 1992; Johns and Maxwell 1997).

In conflict over the duration of care, one parent may be able to put the other into a "cruel bind" by being the first to desert (Trivers 1972). The deserted parent's decision to stay or leave should be based only on the future effects on its own fitness (Dawkins and Carlisle 1976; Boucher 1977), even when the other parent's fitness would be damaged more by the desertion of the second parent. Conflict over the amount of parental investment leads to the range of variation that occurs between species in patterns of parental care (Trivers 1972; Clutton-Brock 1991). In hermaphrodites, conflict over the amount of investment is expressed as conflict over the amount each individual reproduces in the male or female role (Charnov 1979).

In some species, the conflict over which parent will invest in the young is played out on a behavioral rather than an evolutionary time scale. In penduline tits (*Remiz pendulinus*), only one parent (usually the female) incubates the eggs and rears the young, while the other parent pursues further breeding opportunities (Persson and Öhrström 1989). Which parent cares for the brood seems to depend on which is able to desert first at the end of laying. About a third of clutches laid are deserted by both parents (Persson and Öhrström 1989; Valera et al. 1997), reinforcing the idea that there is indeed conflict over which parent assumes parental care for the brood.

Conflict over the amount of investment by each parent is usually waged between genes expressed in the parents, but can also be mediated via paternally and maternally derived alleles expressed in the offspring (Haig and Westoby 1989; Haig 1992b, 1997). For example, in embryonic mice, a gene promoting embryonic growth, and hence increasing maternal investment, is expressed when paternally but not maternally derived (Haig and Graham 1991). Such conditional expression of genes, known as genomic imprinting, is allowed by molecular labeling that reveals the parental origin of an allele. Selection for differential expression of maternally and paternally derived alleles occurs whenever alleles expressed in the offspring can influence the outcome of conflicts between the parents (Haig and Westoby 1989; Haig

1997), although such selection can also be regarded as the outcome of intra-genomic conflict (Pomiankowski, chap. 7) or parent-offspring conflict (God-fray, chap. 6).

LIFE-HISTORY VARIABLES

Once males and females have different patterns of investment, there is abundant opportunity for further conflict over almost every aspect of their joint breeding attempt, including such basic considerations as family size and the timing of breeding. Optimal family size may differ between the parents, even when egg production itself is costless (Houston and Davies 1985; Winkler 1987). Similarly, some birds appear to breed later than opti-mal in terms of the availability of food for raising the chicks, apparently because of limited food availability for the laying female (Perrins 1970, 1996). This implies that earlier breeding might decrease the fitness of fe-males but increase the fitness of their partners who do not pay the costs of egg laying. There may even be sexual conflict over the degree of hatching asynchrony (the spread of hatching dates within an avian brood, which can be controlled by the timing of onset of incubation relative to egg laying; Lack 1947; Magrath 1990). Female blue tits (*Parus caeruleus*) have higher overwinter survival when the degree of synchrony of their brood is experi-mentally increased, but the reverse is true for the males (Slagsvold et al. 1994). Conversely, it has been suggested that females may benefit from asyn-chronous broods because this allows them to extract more parental care from their mates (Slagsvold and Lifjeld 1989).

In polygamous species, conflict can occur over the number of mates that one sex acquires. In polygynous species, the arrival of a secondary female may increase the male's fitness but decrease the primary female's fitness (Verner 1964; Verner and Willson 1966; Orians 1969), creating conflict between the male and primary female over the acceptance of additional fe-males (Slagsvold and Lifjeld 1994). This conflict may be manifest in aggres-sion by primary females toward secondary females (review in Slagsvold and Lifjeld 1994; Kempenaers 1995). Experimental manipulation of the distance between primary nests and potential secondary nest sites of European star-lings (*Sturnus vulgaris*) suggests that such aggression can prevent secondary females from settling (Sandell and Smith 1996). Aggression between fe-males even extends to infanticide by secondary females of the eggs of pri-mary females in house sparrows (*Passer domesticus*; Veiga 1990) and great reed warblers (*Acrocephalus arundinaceus*; Hansson et al. 1997). Conflict can also occur in the reverse direction (Houston and Davies 1985; Harada and Iwasa 1996; Houston et al. 1997). Female dunnocks on territories with two males benefit if they mate with the beta male in addition to the alpha male because the beta male then provisions the young instead of making ovicidal attempts, but the alpha male then suffers reduced fitness (Houston and Davies 1985; Davies 1992).

Conflict between the sexes also often occurs over relative investment in sons and daughters. When the relative costs of raising a son or a daughter differ between the parents (e.g., in polygynous species where males invest little more than sperm, but females rear the sexually dimorphic young), the evolutionarily stable sex ratio will also differ between the parents (Fisher 1958; Trivers 1974). Similarly, helping at the nest acts as partial "repayment" of the costs of producing the helping sex of offspring and hence alters the equilibrium sex ratio (Emlen et al. 1986), but only when the helper is the offspring of the breeding individual that it helps (Lessells and Avery 1987). As a result, the equilibrium sex ratio will differ between the parents when helpers more frequently help one sex of parent. In a number of cases, the relative amounts of care provided by the two parents varies with offspring sex or brood sex ratio (Stamps et al. 1987; Gowaty and Droge 1991; Clotfelter 1996; Nishiumi et al. 1996), although the selective pressures for this are not entirely clear (Gowaty and Droge 1991; Lessells 1998).

Lastly, males and females may differ in their willingness to accept helpers. "Secondary" helpers in pied kingfishers (*Ceryle rudis*) appear to benefit through the chance of acquiring the breeding female as a future mate. Breeding pairs only accept secondary helpers under conditions of food shortage, and then males, who risk losing their mates in the future, are more reluctant than their mates to accept the aspiring helper (Reyer 1980, 1984, 1986).

MATE CHOICE

None of the above forms of conflict require variation between the individuals of one sex. When such variation occurs, however, members of one sex may vary in their value as mates to members of the other sex. When an individual's reproductive success is not limited by the availability of mates, it may be selected to choose among potential partners. Such choice gives rise to sexual conflict over mating and leads to the evolution of such traits as the abdominal spines of female water striders (*Gerris incognitus*; Arnqvist and Rowe 1995) and the abdominal "gin trap" of male sagebrush crickets (*Cyphoderris strepitans*; Sakaluk et al. 1995) to avoid or impose copulations.

Mate choice is a source of sexual selection on the chosen sex (Darwin 1871; Andersson 1994). As first noted by Bateman (1948), because males (usually) invest less than females in each breeding attempt, males can increase their reproductive success by mating more, whereas females cannot. As a result, females limit male reproduction and are usually the choosy sex. Their choices impose sexual selection on male traits (Trivers 1972). For species in which neither sex is strongly limited by the availability of mates (e.g., monogamous species), both sexes may practice mate choice and hence be subject to sexual selection. Experiments manipulating the crest length of models made from mounted skins suggest that this is the case in the crested auklet (*Aethia cristatella*; Jones and Hunter 1993).

Mate choice and the sexual selection that it engenders can be categorized according to the kinds of traits that are selected. Females may choose on the basis of direct benefits that will increase their reproductive success (Kirkpatrick 1987). Such direct benefits might include access to increased resources or help from the male in raising the young (Andersson 1994). Alternatively, preferences for good genes can evolve through indirect selection (Pomiankowski 1988): If there is genetic variation in both the preference in females and the preferred trait in males, the tendency for choosy females to mate with males with the preferred trait will bring genes for the preference and trait together in the offspring, so that the preference and trait become genetically correlated. Sexual selection on the male trait then also acts indirectly on female preference and causes female choosiness to evolve. The good genes that are chosen by females may have an effect on either the viability or the mating success of their offspring. In the latter case, selection for the preference and the trait may become self-reinforcing and lead to runaway sexual selection in which the male trait becomes extremely exaggerated (Fisher 1958; Pomiankowski et al. 1991).

Although mate choice may often be accomplished through a direct decision by the female to mate based on the traits displayed by a male, more subtle mechanisms may also be at work. Mate choice may be achieved indirectly by provoking competition between potential mates or sperm and then accepting the victor for copulation or fertilization (Harvey and Bennett 1985; Birkhead et al. 1993; Keller and Reeve 1995; Wiley and Poston 1996). Alternatively, females may mate more or less indiscriminately with males but then exercise direct but cryptic choice between the ejaculates of different males through anatomical or physiological adaptations of the reproductive tract (Eberhard 1996). For example, there is evidence that females exert a choice between the sperm of different ejaculates in certain ascidians (Bishop 1996; Bishop et al. 1996; Wirtz 1997). Lastly, divorce may be a means of choosing, or rather rejecting, previous mates. In contrast to the earlier view that divorce was a means of breaking up incompatible pairs from which both ex-partners benefit (Coulson 1966), recent studies suggest that divorce is a form of mate choice in which one individual is the unwilling victim (Ens et al. 1993; Orell et al. 1994; Otter and Ratcliffe 1996).

Evolutionary Outcomes of Sexual Conflict

PLAYING THE GAME . . .

In the conflicts described above, the benefit to an individual of behaving in a particular way usually depends on the behavior of other individuals. In such situations, game theory is the appropriate approach (Maynard Smith 1982),

TABLE 5.1

Payoffs to the Male/Female in Maynard Smith's (1977) Game Theory Model of Parental Investment.

Male	Female	
	Guard	Desert
Guard	eS_2/eS_2	ES_1/ES_1
Desert	$eS_1(1 + p)/eS_1$	$ES_0(1 + p)/ES_0$

NOTE: The female lays e eggs if she guards and E if she deserts, which have a survival of S_0, S_1, or S_2, depending on whether 0, 1, or 2 parents guard. Males who desert have a probability p of remating. There are four possible ESSs. (1) *Both desert*. This requires that (a) $ES_0 > eS_1$, the female's payoff is higher if she lays more eggs and leaves them unguarded than if she lays fewer eggs and cares for them alone, or she will guard; and (b) $S_0(1 + p) > S_1$, the male's payoff is higher if he leaves the eggs unguarded and seeks a second mate than if he guards the eggs alone, or he will guard. (2) *Stickleback*. The female deserts, and the male guards. This requires that (a) $ES_1 > eS_2$, the female's payoff is higher if she lays more eggs and leaves them in the care of the male than if she lays fewer eggs and helps him care, or she will guard; and (b) $S_1 > S_0(1 + p)$, the male's payoff is higher if he guards the eggs alone than if he leaves them unguarded and seeks a second mate, or he will desert. (3) *Duck*. The female guards, and the male deserts. This requires that (a) $eS_1 > ES_0$, the female's payoff is higher if she lays fewer eggs and guards them alone than if she lays more eggs and leaves them unguarded, or she will desert; and (b) $S_1(1 + p) > S_2$, the male's payoff is higher if he leaves the eggs to be guarded by the female alone and seeks another mate than if he helps her care, or he will guard. (4) *Both guard*. This requires that (a) $eS_2 > ES_1$, the female's payoff is higher if she lays fewer eggs and helps the male to guard than if she lays more eggs and leaves them to be guarded by the male alone, or she will desert; and (b) $S_2 > S_1(1 + p)$, the male's payoff is higher if he helps the female guard than if he leaves her to guard the eggs alone and seeks another mate, or he will desert. (After Clutton-Brock and Godfray 1991).

and the evolutionary outcomes predicted are evolutionarily stable strategies (ESS). A set of behaviors by interacting individuals is an ESS when each individual is maximizing its fitness *given the behavior of the other individuals*. When this is true, no individual is selected to change its behavior. In game theory models that represent conflict between males and females the fitness benefit (payoff) of a particular strategy may depend on the behavior of the (potential) mate, the behavior of individuals in the population as a whole, or both.

PARENTAL CARE

Maynard Smith (1977) was the first to propose a model analyzing conflict over investment in offspring (table 5.1). The model allows each parent a choice of two strategies: invest ("guard") or not ("desert") in the current brood. The survival of the current brood depends on the number, but not the sex, of parents caring for it. The cost of guarding by a male is expressed as a reduction in the probability of his obtaining a mate for a second breeding attempt, and the cost of guarding by a female as a reduction in her fecundity in the current breeding attempt. Under these assumptions there are four possible ESSs: (1) neither parent guards; (2) the male guards alone (referred to by Maynard Smith as "stickleback"); (3) the female guards alone ("duck"); (4) or both parents guard. Biparental care will be evolutionarily stable when two parents can raise at least twice as many offspring as one, or the chance that a deserting individual can remate is small. If one parent is almost as effective as two in caring for the young, one parent will care for the offspring. This will tend to be the male when care by females reduces their fecundity, and the female when the chances for a male to remate are good. In addition there are sets of conditions in which both stickleback and duck are alternative ESSs, and which of them evolves depends on the evolutionary starting point.

This model, while providing many of the elements of later models, has three obvious shortcomings. First, it takes the amount of investment by guarding individuals as a fixed value. Second, members of the pair do not react to whether their mate has deserted. Third, the probability of remating is assumed to depend only on whether the individual male has guarded or not, whereas this probability must in reality depend on the number of males competing for second mates and hence on the guarding behavior of all other males in the population. Two groups of models have addressed these shortcomings to different extents.

The first group of models (Maynard Smith 1977; Grafen and Sibly 1978; Lazarus 1990) asks the question of how long a parent should invest in a breeding attempt, but assumes that the amount of care provided when a parent does care is fixed. Thus, the fitness of the offspring depends only on the amount of time for which each parent guards. Maynard Smith's (1977) and Grafen and Sibly's (1978) models take into account the effect of the desertion strategies of individual males and females on the expected time for individuals of each sex to find a new mate. The predictions recall those of the guard/desert model: Desertion is more likely when one parent is nearly as effective as two in providing parental care, and is more likely by the parent who is less effective at providing parental care and whose remating chances are greater (the minority sex in the population as a whole) (Grafen and Sibly 1978). Lazarus' (1990) model is the only one in which the strategy of each individual depends on whether the other parent has already deserted and hence is the only one to consider Trivers' (1972) cruel bind. It shows that this can lead to preemptive desertion, in which each parent is selected to desert progressively earlier in an attempt to avoid being deserted itself and left in a cruel bind (Lazarus 1990).

The second group of models asks how much a parent should invest in a given interval of time (Chase 1980; Houston and Davies 1985). In these models, the payoff of a given level of investment depends only on the behavior of an individual and its mate, and not on the behavior of males or females in the population as a whole. The models assume that the fitness

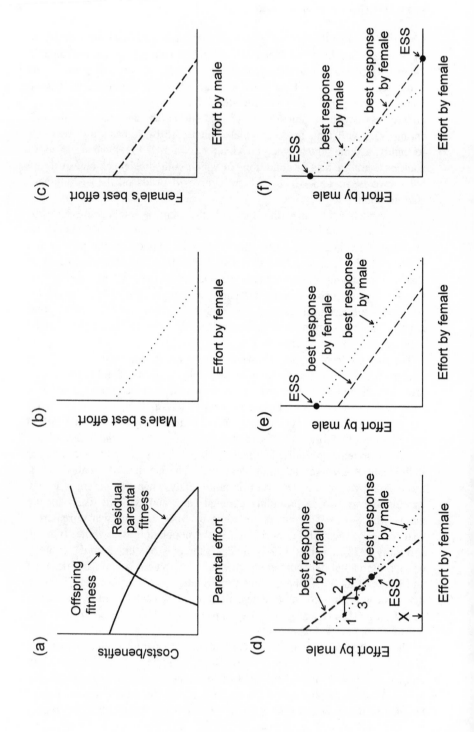

(a)

Costs/benefits

Offspring fitness

Residual parental fitness

Parental effort

(b)

Male's best effort

Effort by female

(c)

Female's best effort

Effort by male

(d)

Effort by male

best response by female

best response by male

2

4

3

1

ESS

X

Effort by female

(e)

Effort by male

ESS

best response by female

best response by male

Effort by female

(f)

Effort by male

ESS

best response by male

best response by female

ESS

Effort by female

benefit through the current family depends on the total care given to them by the two parents (fig. 5.1a), and that the fitness cost of care to each parent depends only on the care given by itself and not by its mate (fig. 5.1a) (Houston and Davies 1985). Under these circumstances, the optimal amount of care by each parent depends on the amount of care being given by the other parent (fig. 5.1b–c). The model predicts the same range of outcomes as the previous models (fig. 5.1d–f; Houston and Davies 1985; Motro 1994; Ratnieks 1996). In order for biparental care to be the ESS, any change in care by one parent must be "compensated" to some extent by the other parent (fig. 5.1d). If the two parents are equivalent, changes must be under-compensatory, a decrease (or increase) in care by one parent being met by a smaller increase (or decrease) by the other parent.

There is experimental evidence for compensation in nestling provisioning by European starlings (Wright and Cuthill 1989) and orange-tufted sunbirds (*Nectarinia osea*; Markman et al. 1995, 1996). Parental care by one parent was experimentally decreased either by loading it with small weights or temporarily removing it. In each case, the unmanipulated parent increased its feeding rate, but without completely compensating for the decrease in care (or absence) of the other parent. This undercompensation was confirmed by the lower weights of chicks in experimental nests. These experiments also show that adjustments to the mate's parental effort take place on a behavioral rather than evolutionary time scale. Such bargaining may well take place indirectly through the begging behavior or condition of the chicks rather than by direct monitoring of the mate's behavior (Chase 1980), as

FIG. 5.1. A game theory model of the amount of parental investment by each parent when both parents care (Houston and Davies 1985). The evolutionarily stable amounts of care depend on the relationship between the residual fitness of each parent and the amount of care that it gives (here these two relationships are assumed identical, but they need not be) and the fitness of the offspring and the total amount of care that they receive from the two parents (a). From these two relationships, it is possible to predict for the male the level of care that will maximize his fitness for any given level of care by the female [(b), the male's "reaction curve"]. Similarly, it is possible to plot the female's best level of care for any given level of care by the male (c). The ESS combination of care by the two parents can be discovered by plotting the reaction curves of the two parents on the same axes (d–f). When the curves cross as shown in (d), an intermediate level of care by both parents is the evolutionarily stable strategy (ESS): if the female invested X, the male's best level of care would be represented by the point 1; if he invested at that level, the female's best response would be represented by the point 2; the male's best response is then point 3, and the female's 4, and so on. At the ESS, each parent is making its best response to the other's level of care, and neither can benefit by changing. If the male's reaction curve lies entirely above that of the female (e), only the male cares. The reverse would be true if the female's reaction curve lay entirely above that of the male (not shown). When the curves cross as shown in (f), the intersection is an equilibrium, but it is not stable. Changes in the level of care by one parent will be overcompensated by the other parent and lead to one of the two alternative ESSs, in which either only the male cares or only the female. (After Clutton-Brock and Godfray 1991; with permission of Blackwell Scientific Publications).

has been shown by experiments on pied flycatchers in which the relative provisioning effort of the members of pairs was manipulated by playing back chick begging calls only during visits to the nest by one of the parents (Ottosson et al. 1997).

The model above assumes that the fitness of the offspring depends on the total amount of care given by the two parents, but not on the proportion of the care given by each. Motro (1994) has considered the outcome of the fitness of the offspring being higher ("super-additivity") or lower ("sub-additivity") when both parents care than when one parent provides the same total amount of care. In general, biparental care is more likely to be the ESS as the degree of super-additivity increases. This recalls the results of the mate-desertion models showing that biparental care is favored when two parents are more than twice as effective as one in providing care.

Potentially, the above two groups of models can be combined to model the duration and amount of care simultaneously. Unless care in successive periods has independent additive effects on fitness (Godfray 1995b), dynamic programming models (Mangel and Clark 1988) will be needed to determine the evolutionary stable pattern of care throughout the period of care. A comprehensive model should also include the possibility of strategies that are conditional on whether the other parent has already deserted.

MATE CHOICE

If all mates were of equal quality in terms of the expected fitness consequences from a breeding attempt paired to them, there would be no advantage to mate choice, and individuals of both sexes should accept the first potential mate that they encounter. When potential mates vary, however, an individual will increase its fitness by rejecting low-value mates when the gain from mating with a mate of higher value more than offsets the extra costs of continuing to search for such a mate.

The simplest situation to model is when mates of low value are rarely encountered, because then the choices made in these encounters have no appreciable effect on the composition of the pool of individuals searching for mates. This might be the case when populations are divided into ecotypes or incipient species that rarely meet and hybrid disadvantage reduces the fitness of offspring from hybrid matings (Parker 1979; Parker and Partridge 1998). The model assumes that each breeding attempt incurs a time cost (e.g., in parental care or replenishing gametes) that is fixed for each sex. Search costs are also paid in units of time and depend on the number of individuals of the opposite sex that are searching. This in turn depends on the sex ratio of the population as a whole and the time invested by males and females in a breeding attempt. Mates of high value should always be accepted, but mates of low value should be accepted only when they offer a better rate of gain of fitness than continuing to search for a higher-value

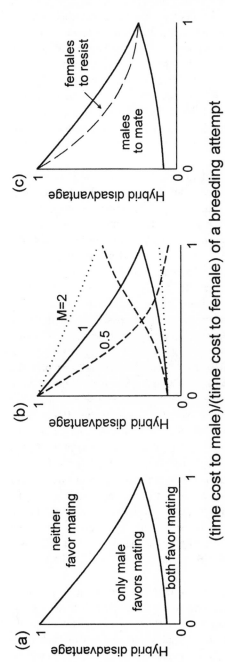

(time cost to male)/(time cost to female) of a breeding attempt

FIG. 5.2. Mate choice in rare encounters between individuals of different ecotypes (Parker and Partridge 1998; with permission of the Royal Society). An individual is selected to accept a potential mate of the other ecotype when the rate of fitness gain from so doing is higher than that of rejecting that mate and instead searching for and mating with an individual of its own ecotype. Mating is more likely to be favored when hybrid disadvantage is low (a), the time cost of mating for the individual is low compared with that for the other sex (a), and the sex ratio in the population as a whole (M, number of males per female) is skewed toward individuals of the same sex (b). In general, there is a range of values in which only one of the potential pair is selected to mate, and where there is thus conflict over mating. When the population sex ratio is unity, it is the sex that has the lower time cost for a breeding attempt (shown as males in the figure) which are selected to mate and the other sex to resist. Over much of the range of values where there is conflict, the sex that favors mating is more strongly selected to mate than the reluctant sex is to resist (c); the broken line separates the area of conflict (from (a)) into areas where the male is more strongly selected to mate, and the female is more strongly selected to resist. This is because the sex that favors mating stands to gain a breeding opportunity by winning the conflict, whereas the reluctant sex stands only to improve the quality of its mate.

mate. Mates of the other ecotype therefore will be accepted only when hybrid disadvantage is low or search times are high relative to the time costs of a breeding attempt. If the time cost of a breeding attempt differs between the sexes or if the sex ratio is skewed, there are conditions when there is conflict between the male and female over whether a mating should occur (fig. 5.2a,b): When hybrid disadvantage is high and the two sexes commit equal time, it will pay neither sex to mate. When hybrid disadvantage is low and the two sexes commit equal time, it will pay both sexes to mate. But if hybrid disadvantage is intermediate and the two sexes commit unequal time, there will be conflict. If the sex ratio in the population as a whole is unity, the sex that commits more time to each breeding attempt is selected to refuse the mating.

The above model delineates the "battleground" (Godfray and Parker 1991) where there is conflict between the sexes, but does not predict the outcome. If neither sex can impose a mating that is not to the selective advantage of the other sex, then no matings will take place in the conflict zone. Otherwise, the outcome will depend on the intensity of selection on the two sexes (the difference in the rates of fitness gain from mating and not mating) and the costs of achieving the selected outcome. In general, males have more to gain than females have to lose from mating (fig. 5.2c), which implies that they should more often win the conflict. However, the costs of imposing a mating for a male will frequently be higher than the costs of resisting a mating for a female, so the outcome is not as clear as it first appears (Parker 1979, 1983; Parker and Partridge 1998).

Conflicts over mating may be settled by a "war of attrition" in which the sex that is prepared to persist the longer "wins" (Maynard Smith 1982), either males giving up and leaving the female in order to search for other mates, or females giving up and mating with the male (Parker 1979, 1983; Clutton-Brock and Parker 1995a; Parker and Partridge 1998). Alternatively, conflicts may be settled using weapons to impose or resist matings (Parker 1979, 1983; Parker and Partridge 1998). Provided that there are differences between individuals, either in the ratio of benefits from winning to costs of persistence (and the opponent cannot perfectly judge this ratio), or in the level of armament that can be achieved for a given cost, there will be an evolutionarily stable distribution of persistence times or levels of armament that results in the sex that generally has the higher benefit-to-cost ratio usually, but not always, winning (Clutton-Brock and Parker 1995a; Parker and Partridge 1998).

More complicated models of mate choice consider the situation in which both sexes vary in value as mates and encounter each other at random (Parker 1983, analyzed further by Johnstone et al. 1996). The ESS must specify the threshold value of acceptable mate for males and females of each mate value. This is not straightforward because each threshold value depends on

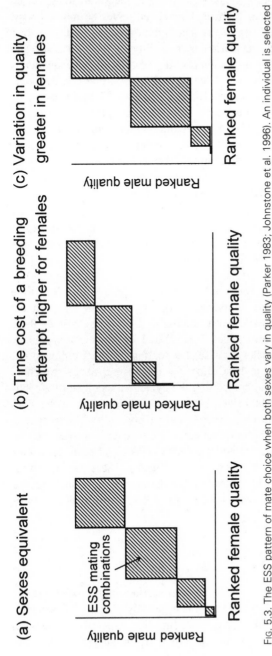

(a) Sexes equivalent

(b) Time cost of a breeding attempt higher for females

(c) Variation in quality greater in females

Ranked male quality

Ranked female quality

ESS mating combinations

FIG. 5.3. The ESS pattern of mate choice when both sexes vary in quality (Parker 1983; Johnstone et al. 1996). An individual is selected to accept a mate of a given quality when its fitness gain from so doing is higher than that of rejecting that mate and searching for a higher-quality mate. The ESS takes into account the range of quality of members of the other sex that are prepared to accept the individual as a mate, and assumes that mating does not take place when there is conflict. When the time cost of a breeding attempt and the variance in quality is the same for the two sexes, and the sex ratio is unity, the ESS gives symmetrical assortative mating (a). When the sexes are not equivalent, the poorest-quality individuals of the sex with the lower time cost for a breeding attempt (b) or with the higher variance in quality (c), or the majority sex in the population as a whole (not shown) are rejected as mates by all individuals of the other sex. The rather strange pattern of assortative mating generated can be explained as follows. All males will be prepared to mate with the highest-quality female. She, however, will only be prepared to mate with males above some acceptability threshold of quality. Because males above that threshold are acceptable to the highest-quality female, they are acceptable to all females. Hence, all males above the acceptability threshold for males have the same distribution of mates who are prepared to accept them (i.e., all females) and should all have the same acceptability threshold for females. The reciprocal argument starting with the highest-quality male results in the conclusion that males and females above their respective acceptability thresholds will mate exclusively with each other, but that mating will be random within this group of individuals. These individuals are therefore removed from the pool of potential mates for the remaining individuals, and the whole argument can be repeated for those remaining individuals.

the benefit of searching for a higher-value mate. This in turn depends on the range of potential mates that are prepared to mate with that kind of individual. In general, higher-value individuals will be acceptable to higher-value individuals of the other sex, and hence should themselves be more choosy, leading to assortative mating (fig. 5.3a). The tightness of assortative mating should increase with increasing variation in quality of the two sexes and increasing time costs of breeding attempts. If variation in mate quality is higher, or the time cost of a breeding attempt is lower, in one sex, the poorest-quality individuals of that sex may be rejected by all individuals of the other sex (fig. 5.3b–c). When breeding is not continuous, the pool of potential mates will change, and the risk of not finding a mate will increase as the season progresses. In this case, models also predict assortative mating, but with the level of choosiness varying through the season (Johnstone 1997).

GAME THEORY MODELS WITH CYCLICAL DYNAMICS

Although many game theory models predict static ESSs, others predict cyclical ESSs with never-ending self-regenerating changes in the strategies of the players (Maynard Smith 1982). For example, Dawkins (1989) described a model of conflict over investment in which the female can play "coy" (demanding a lengthy courtship period from a male before mating) or "fast," and the male can play "faithful" (courting the female for a long time and helping raise the young) or "philanderer" (giving up on the female if she demands a lengthy courtship and never helping to raise young). When there are many coy females, faithful males do well and increase in numbers. A preponderance of faithful males then selects for fast females (who do not pay courtship costs), which in turn select for philanderer males, which in turn select for coy females, so completing the cycle. An example of a cyclical ESS that might in part be driven by female choice occurs in the side-blotched lizard (*Uta stansburiana*; Sinervo and Lively 1996). The frequency of three male morphs differing in throat color cycled over a six-year period. Males with dark blue throats are nonaggressive and defend small territories and are replaced by aggressive orange-throated males. These are in turn replaced by "sneaker" males with yellow striped throats who resemble females and do not defend territories, who are in turn replaced by the mildly territorial blue-throated males.

. . . BUT BENDING THE RULES

Game theory models like those in the previous section often predict a stable outcome to conflict in which neither sex can gain by altering its behavior. ESSs, however, depend on the strategy set (the list of possible behaviors) and associated payoffs, that constitute the "rules of the game." Conflict between the sexes will promote the fitness of individuals who bend the rules in their own favor.

MANIPULATION

Game theory models usually assume that each participant controls only its own behavior, and that an evolutionary equilibrium results when neither participant can gain by changing its behavior. However, an individual that is able to control the behavior of its opponent can directly impose its evolutionary optimum.

One of the best understood examples is the manipulation of female *Drosophila melanogaster* by the males during mating. Because of a cost to egg production (Partridge et al. 1987) and because a female remates throughout her adult life, male reproductive success is maximized by a higher rate of egg laying and a longer interval to remating by the female than would maximize the female's lifetime reproductive success. The males are able to manipulate the females' reproductive physiology with peptides in the seminal fluid produced by the accessory glands (Chapman et al. 1995). These peptides raise the egg-laying rate (Chen et al. 1988), reduce the receptivity of the female to further matings (Chen et al. 1988), and destroy or disable sperm from previous matings (Harshman and Prout 1994; Chapman et al. 1995; Clark et al. 1995). Although it is known that these effects are produced by peptides in the seminal fluid, the mechanism by which this is brought about is not clear, at least for egg-laying rate. One possibility is that the peptides act directly on the physiological mechanism controlling laying rate. Alternatively, any chemical that reduced the female's survival prospects would favor a reallocation of resources from her own maintenance (and hence future reproduction) to a burst of "terminal reproductive investment" (Clutton-Brock 1984). Such manipulation of egg-laying rates via allocation decisions has been suggested as an explanation for the injuries caused by hypodermic insemination in hermaphrodites (Michiels, 1998). The potential for manipulation when there is mating conflict is particularly high because manipulatory chemicals may be passed to the female along with sperm.

Manipulation also appears to occur in mate choice when the males of some species have evolved courtship signals that exploit preexisting sensory biases in the females' perceptual systems (sensory exploitation; Ryan et al. 1990; Ryan and Keddy-Hector 1992). For example, males of the túngara frog (*Physalaemus pustulosus*) produce a mating call consisting of a whine that is sometimes followed by a chuck that increases the attractiveness of the call to females. The preference for calls with chucks is also shown by females of *P. coloradorum*, a closely related species whose males do not produce the chuck. The pattern of occurrence of chuck production by males and preference for chucks by females within the species group suggests that the preference evolved before the mating signal (Ryan and Rand 1993). The reasons for the preexisting sensory bias in this case are unknown. Habituation could be a ubiquitous explanation for a preference for increasing complexity (Searcy 1992), and theoretical models suggest that recognition mechanisms may frequently result in hidden preferences for stimuli outside the normal range (Arak and Enquist 1993; but see also Dawkins and Guilford 1995). Alternatively, the sensory bias may have been selected in a different context, as seems to be the case for the water mite (*Neumania papillator*; Proctor 1991). Copepod prey are detected by the vibrations that they produce when swimming. Males perform courtship trembling with their legs, which generates vibrations of a similar frequency, and to which the females orient as if to prey. The pattern of occurrence of this feeding technique and courtship trembling in closely related species suggests that, as in the frogs, the sensory bias evolved before the courtship signal (Proctor 1992; although the evidence is not unequivocal).

COERCION THROUGH PUNISHMENT

Even when an individual is unable to enforce its evolutionary optima on its mate or potential mate, it may still be capable of discouraging the other from repeating the action by physical punishment (Clutton-Brock and Parker 1995a,b). By doing so, it alters the pay-offs that the other individual gains from different actions, and hence its evolutionarily stable behavior. Because inflicting punishment is assumed to carry a cost, punishment will only be evolutionarily stable when the same individuals interact repeatedly, and the punished individual changes its behavior only (or primarily) toward the individual that has punished it. If individuals interact only once, the punisher has no chance to gain from the punishment, and if punished individuals change their behavior toward all other individuals, punishers have lower fitness than nonpunishers because only they pay the cost of the punishment, but all individuals gain from its effects (Clutton-Brock and Parker 1995a).

One situation in which punishment may be used to coerce another individual is in mate choice. For example, males of many primate and ungulate species attack females that refuse to mate with them (Clutton-Brock and Parker 1995a). Models in which the same male and female interact more than once, and in which females may learn to submit to the mating advances of males who have punished them after previous refusals, show that punishment by males and submission by females may be an ESS (Clutton-Brock and Parker 1995a,b). For females, the cost of being punished must be greater than the benefit of refusing to mate (perhaps through the acquisition of a better-quality mate), or it will never pay for them to submit to mating attempts that are not in their own interest. For males, the total cost of inflicting punishment before the female learns must be less than the total benefit through submission of the females to mating attempts after she learns, or it will not pay to punish females who do not submit (Clutton-Brock and Parker 1995a).

In general, the conflict must be played out on a behavioral rather than evolutionary time scale in order for coercion through punishment to be an ESS. Conflict over the amount of investment in a brood may be played out repeatedly between the same pair of individuals during the course of one breeding attempt, giving ample opportunity for punishment of the partner for insufficient investment to be rewarded by an increase in subsequent investment. This kind of punishment is meted out by breeding individuals to potential helpers within the social group in several species with cooperative breeding (Clutton-Brock and Parker 1995b). For example, dominant male superb fairy wrens (*Malurus cyaneus*) harass helpers that have been temporarily removed at a time when they would normally be helping with chick provisioning, but not if they are removed outside the breeding season (Mulder and Langmore 1993). It is a puzzle why such intimidation over parental investment apparently does not occur in any species between the members of a pair.

DECEPTION

When the conflict between the sexes is played out on a behavioral rather than evolutionary time scale, the most profitable behavior for a male or female will often depend on factors such as the value of an individual as a mate. Individuals, however, will often have more accurate information about themselves than do other individuals, opening up the possibility of deceit as a way of achieving an outcome in their own evolutionary interest. Individuals may deceive another by withholding information or by providing false information.

The polyterritorial system of pied flycatchers (*Ficedula hypoleuca*) is a possible example of deception by withholding information. Males sometimes become polygynous by attracting a second mate (secondary female) at about the time or shortly after their first mate starts laying. Males clearly increase their fitness by acquiring a second mate, but secondary females would gain higher fitness if they were the sole mate of a male (Alatalo and Lundberg 1990). One peculiar aspect of polygyny in this species is that males move to a new territory that may be several territories distant before attempting to attract a second mate. One hypothesis (among others) for such polyterritoriality is that it is a means for males to hide the fact that they are already mated (Haartman 1969; Alatalo et al. 1981). This idea is supported by the higher remating success of males who move farther (Slagsvold et al. 1992; Rätti and Alatalo 1993; Slagsvold and Dale 1994).

Females may also deceive males by withholding information. Female penduline tits often bury their eggs during the laying period in the soft material that forms the base of their hanging woven nests. This seems to be a means for females to hide from males the fact that they are laying for long enough to complete the clutch and desert, leaving the male "holding the baby" (Valera et al. 1997). Males do not usually desert until after the female leaves the clutch uncovered for the first time, and when eggs were experimentally uncovered males deserted on the same day. Conversely, females were able to desert first only if they had buried the eggs for at least part of the laying period, and if the male was experimentally removed early in the laying period, the female usually uncovered the eggs within a day (Valera et al. 1997).

In comparison with the above examples of deceit by withholding information, examples of deceit by providing false information are hard to find. The evolutionary reasons for this are obvious: Individuals will be selected to ignore unreliable signals. Nevertheless, many species have mating signals that reliably advertise an individual's quality (for a review, see Johnstone 1995b). Two mechanisms can account for the honesty of such signals. First, the signal may simply be uncheatable. For example, many species use carotenoids to produce the coloration in signals, but cannot synthesize them themselves so their foraging success sets an unbreakable limit to the intensity of the signal that they can produce (e.g., Hill and Montgomerie 1994). Second, if the signal is costly to produce and the cost is higher for poor-quality individuals, then better-quality individuals are selected to produce more intense or elaborate displays ("the handicap principle"; Zahavi 1975; Grafen 1990a). Although cheating is possible, it is selected against because, for instance, a poor-quality male that produced a deceptively elaborate display would incur an increase in costs that more than offset his increased mating success. Theoretical models show that signaling does not have to be completely honest to be evolutionarily stable (Johnstone and Grafen 1993). If the cost of display is lower than normal for some individual males, they will be able to produce a more exaggerated display than other males of the same quality. Females will then be misled as to their quality, but provided that such males are rare and the cost of being misled is small, the signal may still be a reliable enough indicator to females of the fitness value of a male for them to continue to use the signal in mate choice. Occasional deception is therefore possible.

RUNNING WITH THE RED QUEEN

Conflict implies that an evolutionary step forward by one sex is a slide backward for the other, leading to a potentially never-ending coevolutionary arms race between the sexes ("Red Queen" coevolution; Van Valen 1973; Dawkins and Krebs 1979). The bending of rules by manipulation, coercion, or deceit is expected to be met with countermeasures. Manipulation that leads to a loss of fitness in the exploited individual selects for a reduction in sensitivity to the manipulatory stimulus. However, when sensitivity to the stimulus is useful in other contexts, such as the detection of predators or food (as in the case of the water mite, *Neumania papillator*; Proctor 1991), the balance of selection pressures may favor continued sensitivity and manipulation. In some cases, a mating signal that has evolved through sensory exploitation may be selectively maintained either because it reduces the costs of mate search for females (Dawkins and Guilford 1996), or leads to the selection of high-quality mates. For example, a sensory bias for lower-frequency chucks in túngara frogs results in females mating with larger males that fertilize more of their eggs (Ryan 1983). Harassment of females by males also leads to counteradaptations. Females of some species, including those of the African swallowtail butterfly (*Papilio dardanus*; Cook et al. 1994) and the damselfly (*Ischnura ramburi*; Robertson 1985), have reduced the excessive attentions of males by evolving malelike coloration. Deception leads to selection on the deceived party to collect information in the way that is most likely to reveal the true state of affairs. Female pied flycatchers make many short visits to a prospective mate's territory in the few hours before choosing him as a partner (Dale and Slagsvold 1994). Because already-mated males are absent from their secondary territory more than unmated males from their primary territory (Stenmark et al. 1988), but remain on their territory once they have detected the presence of a female (Searcy et al. 1991), repeated visits maximize the probability of a female detecting that a male is already mated (Dale and Slagsvold 1994; Getty 1996). Deception is more likely to be successful when time is limited for the duped individual to acquire reliable information. In pied flycatchers, females must choose a mate quickly because of a steep seasonal decline in reproductive success (Alatalo et al. 1981) and competition from other females for mates (Dale et al. 1992). In penduline tits, females have only to conceal the clutch until the completion of laying for the tactic to be successful.

Conflict not only creates an evolutionary arms race in its original context but may aggravate antagonistic coevolution by engendering new conflicts of interest. For example, the seminal fluid products that benefit male *Drosophila melanogaster* by increasing female fecundity and reducing competition from other males also reduce female survival, presumably as an unselected side effect (Chapman et al. 1995). They thus create a cost of mating per se in females (Fowler and Partridge 1989) and provoke a new conflict between males and females over the mating rate. Similar costs of mating, as distinct from sperm transfer, have been shown in the nematode *Caenorhabditis elegans* (Gems and Riddle 1996). Adaptations for sperm competition in males—such as persistent guarding, prolonged copulation, and multiple mating—often generate costs to females—including reduced foraging success or increased predation risk—and thus provoke mating conflict (Magurran and Seghers 1994; Stockley 1997). However, the extent to which these behaviors are adaptations to intramale competition or to conflict with females over mate choice is not clear.

Support for the idea that sexual conflict promotes rapid coevolution comes from both observations and experiments. Seminal fluid proteins in *Drosophila* evolve rapidly (Aguadé et al. 1992; Tsaur and Wu 1997), although intramale competition may also be an important driving force. Rapid coevolution in response to sexual conflict has been demonstrated experimentally in *D. melanogaster* (Rice 1996). Males in an experimental line were given the evolutionary upper hand by being allowed to adapt to females from a control line, but the control females were denied the opportunity to adapt to the experimental males. Each new generation of the experimental line was produced by pairing males from the experimental line with females from the control line, while the control line was perpetuated by pairing control males and females. As a result, sex-limited genes expressed in experimental males could adapt to the control female phenotype, whereas sex-limited genes expressed in control females could not adapt to the experimental male phenotype. Secondly, genes entering the experimental line from the control females were removed in the next generation by artificial selection on genetic markers. As a result, sexually antagonistic alleles in the experimental line had been selected only in males, whereas sexually antagonistic alleles in the control line had been selected in approximately equal numbers of generations in males and females. After fewer than 40 generations, the experimental males had a clear reproductive advantage over control males when mated to control females. They produced 24% more offspring, had a higher rate of mating with already mated females, and were more successful in preventing the females with whom they had mated from producing offspring by later mates. The sexually conflicting nature of these adaptations is emphasized by the lower survival of control females when paired with experimental rather than control males.

The evolutionary change that conflict engenders is not confined within species boundaries but can also act as an "engine of speciation" (Rice 1996). Sexual conflict promotes speciation by contributing in various ways to both pre- and postzygotic isolation (see Parker and Partridge 1998).

SEXUAL CONFLICT AND COOPERATION

Because sexual conflict is so obviously an outcome of sexual reproduction, it is easy to overlook more cooperative aspects. Biparental care is often associated with some level of role specialization in which the care provided by each parent is complementary. This state of affairs concurs with models of evolutionarily stable parental investment predicting that role specialization (represented in the models by two parents being much better than one or by "super-additivity") will select for biparental care. Unfortunately, however, contemporary selection pressures may not tell us much about the selection pressures during the evolution of a particular pattern of parental care (Clutton-Brock 1991). In the case of role specialization, the extent to which role specialization drove the evolution of biparental care or was simply allowed to evolve by the occurrence of biparental care is not clear.

Members of a pair may also cooperate to defend their common interests when these are threatened by a third party. One possible example is parent-offspring conflict (Godfray, chap. 6), in which, because of asymmetries in relatedness, offspring are selected to acquire more than their fair share of parental care, while the interests of the parents coincide in favoring a more equitable division of resources. Even though some models imply that the proportion of care received by a particular offspring from each parent is selectively neutral (although the total care given or received by an individual is not) (Lessells 1998), it is not unusual for each member of a pair to direct their attention toward different young. In some bird species, females preferentially feed the smaller chicks (e.g., Stamps et al. 1985; Gottlander 1987), whereas in others, the parents divide the brood after fledging, and care only for their own part of the brood (e.g., Harper 1985; McLaughlin and Montgomerie 1985; Price and Gibbs 1987; Byle 1990). In great tits (*Parus major*), each parent feeds consistently from a particular position on the nest rim, (Kölliker et al. 1998). One explanation for these observations is that by allocating their attention to different parts of the brood, the parents may actually be cooperating in preventing individual chicks from monopolizing care and hence attaining closer to an equitable distribution of care overall.

Acknowledgments

I wish to thank Nick Colegrave, Charles Godfray, Mark Ridley, Ben Sheldon, Henrik Smith and Rebecca Timms for their helpful comments on this chapter.

Parent-Offspring Conflict

H. Charles J. Godfray

An important consequence of sexual reproduction is that parents and their young are not genetically identical and hence natural selection has the potential to act on genes expressed in parents and young in subtly different ways. Since the pioneering work of Robert Trivers in the 1970s, these differences have become known as parent-offspring conflict. In many ways, parent-offspring conflict is just a special case of a much broader class of conflict between relatives, although interactions between parents and young have sufficient distinct features to justify their separate consideration. Parent-offspring conflict is also an interesting example of selection acting at different levels: The main players are individuals, yet a consideration of purely individual-level selection is apt to be misleading, and a gene-level view much more informative.

I first describe the classical theory of parent-offspring conflict as developed by Trivers (1974). Despite many attempts to topple it, Trivers's idea is still fundamental to the modern subject. I then discuss the first generation of studies spurred by Trivers, taking a loosely historical perspective and attempting to explain why parent-offspring conflict was slower to be accepted than many other areas of the new science of sociobiology, and why it initially generated rather few experimental studies. In the third section, I contrast models of the battleground between parents and young with more recent models that attempt to predict how the battle is resolved. Here I concentrate on what might be called the classic problem first defined by Trivers: how are resources allocated to current offspring as opposed to being used to enhance parental survival and hence future reproductive success? The problem is most easily phrased with birds in mind, though it also applies to mammals and to other species with parental care. In the following section, I discuss related issues such as brood reduction and the intermeshing of parent-offspring and sibling conflict, issues that again largely apply to birds and mammals. Finally, I mention briefly other problems to which the theory of parent-offspring conflict has been applied, drawing parallels between this subject and the other areas of genetic conflict discussed in this book.

Trivers (1974)

Ten years before Trivers's paper, Hamilton (1964a,b) initiated a revolution in our understanding of social evolution. Hamilton realized that a gene could increase its representation in the future gene pool in two ways: first by enhancing the survival and reproduction of the individual in which it found itself, and second by enhancing the survival and reproduction of relatives who may carry copies of the gene through inheritance from a common ancestor. However, whether it is worth switching resources from producing your own descendant young to the nondescendant young of relatives, depends on how distantly related they are. This distance can be measured by the coefficient of relatedness, r (technically the probability that a gene present in one individual is also present in the other, identically by descent). Hamilton deduced that an act that "cost" an individual an amount of fitness c (measured, e.g., in terms of a reduction in offspring), would be favored by natural selection if the costs were less than the "benefit" to relatives, b, multiplied by the coefficient of relatedness: In symbols: if $c < rb$. Hamilton's ideas went a long way toward explaining the bounded altruism found within animal social groups (Kitchen and Packer, chap. 9) and were especially successful in explaining patterns in haplodiploid social insects (Keller and Reeve, chap. 8).

Hamilton did briefly discuss the application of his ideas to parents and offspring, but it was Trivers, ten years later, who fully explored the ramifications of the new thinking to parent-offspring interactions. We can follow Trivers's argument by considering an animal in which one parent looks after the single offspring that is born each breeding season. There will be a trade-off for the parent between continuing to invest resources in the current offspring, and in retaining those resources to increase survival until the next breeding season or to invest in future offspring. Assume that the population is stationary, neither growing nor decreasing in size. In these circumstances, natural selection should act on the parent's resource-allocation strategy to maximize the total number of offspring produced over its lifetime. Resources will only be allocated to the present offspring, if the benefits (b) in terms of its fitness are greater than the costs (c) to future offspring measured in terms of their fitness: $b > c$. To lapse into anthropomorphism, the female is making an economic decision between investment in the present and in the future, where she weights equally the value of offspring produced now and in subsequent breeding seasons. The weighting is equal because the gene influencing resource allocation in the parent will be present in any of her offspring with the same probability.

Now let us suppose that the offspring can in some way influence resource allocation. Exactly how this may happen is discussed below. From the offspring's point of view there is also a trade-off between causing resources to be diverted to itself and in allowing the parent to retain those resources for future young. The benefits (b) to the offspring of obtaining those resources are clear; the costs (c) are measured in terms of lost fitness to siblings, which, although not descendant relatives, are significant because they carry copies of the focal offspring's genes. Following Hamilton, we devalue the costs to relatives by the coefficient of relatedness between them, r. The gene influencing resource allocation is certainly present in the focal offspring, but the probability that it is present in any particular future offspring is r. Thus, the offspring will favor diversion of resources toward itself when $b > rc$, a different criterion from that of the parent $b > c$. There will be a genetic conflict of interest, which I call a battleground, in the region $rc < b < c$.

Trivers pointed out that parent-offspring conflict might arise at any time during the period of parental care. As long as the amount of resources received on any particular occasion influences both the fitness of the current offspring and that of future offspring, then the battleground exists. Trivers, however, thought that the greatest potential for parent-offspring conflict was at about the time of independence. It is at this time, when young can virtually fend for themselves, that the benefits of further parent resources are most likely to hover in the battleground of $1 < b/c < r$. Indeed, observation of many mammals show that there does seem to be an increase in aggressive interaction between parents and young at about this time, and Trivers referred to this as weaning conflict. Note, however, that weaning and the time of offspring independence need to coincide. Bateson (1994) has pointed out that various physiological constraints may lead to relatively little conflict over the timing of the switch from milk to solid foods in mammals: true weaning. Trivers's arguments, however, still apply to the transition from parental care to full independence. In a very elegant experimental study, Evans et al. (1995) determined optimal incubation temperatures for parent and offspring herring gulls and found them very similar. This study reinforces Bateson's point that physiological constraints may markedly reduce the scope of parent-offspring conflict. An unanswered question is whether such conflict reduction is itself an evolved adaptation.

Thus, Trivers's thesis was that the application of gene-centered and kin-selection thinking to parent-offspring interactions leads to the identification of a potential genetic battleground, of circumstances in which natural selection acts in different ways on genes expressed in the parent and in the offspring. But there is a big difference between showing that a battleground exists—potential genetic conflict—and showing that the resolution of the conflict influences the behavior of animals as observed in the wild—actual phenotypic conflict (for a discussion of these issues in a social insect context, see also Ratnieks and Reeve 1992; Keller 1997; Keller and Reeve, chap. 8). I return to this point in detail below but mention here that, although Trivers did not construct a formal theory of the resolution of parent-offspring conflicts, he suggested verbally how young might influence parental behavior and how the conflict might be resolved. Trivers used the metaphor of psychological manipulation to argue that young would misrepresent their resource requirements by, for example, pretending to be hungrier or younger than they actually were. Parents would evolve means of detecting such misrepresentation, but there would be an arms race, with selection acting on both sides to shift resource-allocation decisions toward their own optimum.

I finish this section with a series of notes, some essentially technical, on the assumptions involved in Trivers's argument. First, the concentration on a single parent rearing a single offspring was made for convenience. The arguments extend to both larger families and to situations in which both parents care for the young. However, the picture is complicated because now "straight" parent-offspring conflict intermeshes with conflict between parents (see Lessells, chap. 5) and between siblings (see below).

The assumption that the population was stationary was also made for convenience. In a constant population, fitness normally equates with the total number of offspring produced (or total number of grandchildren, when it is important to account for different quality offspring). In a growing population with overlapping generations, young produced earlier in life make a greater contribution to the future gene pool because they can start reproducing themselves earlier. In consequence, we have to measure costs and benefits in a more subtle way, which leads to considerably more complicated mathematical expressions, but with Trivers's qualitative arguments remaining unchanged (Godfray and Parker 1991; Lundberg and Smith 1994).

The Reaction

Trivers's theory of parent-offspring conflict attracted immediate criticism from Alexander (1974) who made two main points. The first was that even if a battleground existed, it was irrelevant as the parent was in a position to impose its optimum by *force majeure*. Alexander thus rejected Trivers's idea that the offspring could manipulate the parent into giving it more resources. This argument did not contradict the theory of parent-offspring conflict but rendered it irrelevant for people trying to understand the behavior of animals in the wild. One can also view this argument as a simple and straightforward theory of the resolution of parent-offspring conflict—the parent always wins—and I return to this point in the next section when discussing resolution mechanisms.

Alexander's second point attacked the logical basis of parent-offspring conflict itself. Parent-offspring conflict could not occur because any advantages that accrued to the young in conflict with their parents were nullified when the young grew up and became parents themselves—for the simple reason that their offspring would inherit the very same genes for conflict.

This is a population genetic argument and five years later was shown to be incorrect, at least as a general criticism, by the explicit genetic models of Parker and Macnair (1978). Because the argument has great intuitive appeal, it is often resurrected today and has, I believe, been in large part responsible for the comparatively slow acceptance of ideas about parent-offspring conflict compared with other areas of kin-selection thinking. The problem with Alexander's argument is that it adopts an individual-centered perspective when parent-offspring conflict is a situation where a gene-centered view is essential, a point originally made (unsurprisingly) by Dawkins (1976). Consider a rare allele for a "conflictor" gene that allows the offspring somehow to get more food (when $rc < b < c$), in a population largely consisting of "nonconflictors" (that obtain food only when $b > c$). The reason why this gene—expressed in the offspring—spreads, is that in each generation it redirects resources from individuals not carrying that allele to those that do. In each generation there will be an overrepresentation of young carrying the conflictor allele among the progeny of parents carrying both alleles.

As mentioned above, this argument was put on a firm quantitative footing in a series of papers by Parker and Macnair (1978, 1979; Macnair and Parker 1978, 1979; Parker 1985). They focused on the problem of a bird chick or chicks begging in the nest and made the assumption that the parent would respond to increased solicitation by providing more resources for the offspring, at the expense of future siblings. Offspring paid a price for increased begging: A reduction in fitness that might be due to increased energetic expenditure, or an increased risk of attracting predators. The advantage of obtaining more resources was higher fitness, although the benefits were obtained with diminishing returns. Parker and Macnair analyzed this problem using explicit population genetic models, which was important in answering the genetic criticisms of Alexander. However, nearly all their results can be obtained by using equivalent kin-selection arguments (Clutton-Brock and Godfray 1991; Godfray and Parker 1992), which simplifies matters as one can use the coefficient of relatedness to describe different breeding systems. I describe the results of their analysis in these terms. First, Parker and Macnair showed that the optimum resource allocation for the parent occurred at the point where the marginal benefits to the current young equaled the marginal costs to the future young. Marginal benefits and costs reflect the fitness changes that occur when a very small amount of resources is transferred from future to current young. They represent the slope of the function relating benefits to resources evaluated at the optimum. Call these benefits and costs B and C, the capital letter signifying that they are marginal benefits and costs to be measured in the vicinity of the optimum. Parker and Macnair thus showed that the parental optimum occurred when $B = C$. They then considered what would happen if the offspring controlled resources. The offspring optimum was again set by a balance of marginal costs and benefits, but, as in Trivers's original argument, future offspring were relatively devalued. The offspring optimum occurred when $B = rC$, and hence the battleground exists in the region $1 < B/C < r$ (fig. 6.1).

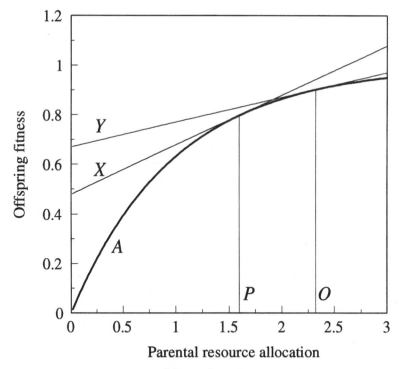

FIG. 6.1. A geometrical interpretation of optimal resource allocations for parent and young. Curve A represents the relationship between offspring fitness and parental resource allocation measured in arbitrary units. The marginal benefits of more resources (B in the text) is simply the slope of this curve. I assume here that the costs to the parent of providing more resources (in terms of lost future offspring) rise linearly, as described by a line with the same slope as X. The marginal cost (C in text) is simply the slope of this line. The parental optimum (P) is given by the point at which the marginal benefits and costs are equal, that is, when a line with slope C is tangent to the curve A. The young devalues future offspring and so its inclusive fitness costs rise less steeply, as described by a line with the same slope as Y (= rC, I assume here full sibs, r = 1/2). The offspring ESS (O) is given by the point at which a line with slope rC is tangent to the curve A.

Thus, Parker and Macnair (1978; Macnair and Parker 1978) essentially confirmed Trivers's arguments about the existence of a battleground of parent-offspring conflict in the case of a parent feeding a single young. Their measures of cost and benefit are slightly different from those of Trivers, but this is because they are predicting ESS levels of resource allocation, rather than only whether a gene resulting in a change of allocation will be favored. In subsequent papers, they extended the analysis to the more complicated case of larger families (Macnair and Parker 1979), to biparental care (Parker 1985), and, in a model discussed below, they moved from defining the battleground to predicting the resolution of the conflict (Parker and Macnair 1979).Their work, along with a number of other roughly contemporaneous genetic studies (Stamps et al. 1978; Metcalf et al. 1979), firmly established the potential for parent-offspring conflict.

From Battleground to Resolution

Models and theorizing about parent-offspring conflict fall into two camps: those that seek to define and identify circumstances when selection acting on genes expressed in parents acts differently from that expressed in young, and those that seek to predict how the potential conflict is resolved. This is the difference between battleground and resolution that we have touched on already.

It is important to realize that battleground models do not themselves make predictions about what we should expect to see in nature. A number of authors have been critical of the logic that identifies apparent behavioral conflict as observed between parents and young with the genetic conflict demonstrated in battleground models (Mock and Forbes 1992; Bateson 1994). To some degree, this identification rests on the shared word "conflict." What is needed by experimental and field workers are models that predict how the genetic conflict demonstrated in battleground models is resolved. The presence of the battleground is a logical necessity given simple assumptions of Mendelian genetics and Darwinian natural selection. Predicting the resolution of the conflict is much more complex and typically requires more assumptions to be made and more biology to be incorporated.

No Parental Care

Predicting the resolution of parent-offspring conflict is probably most simple when parents do not remain with their young for a period of extended care. Consider a hypothetical case of a gregarious insect, a moth or butterfly, that lays its eggs in clumps on the host plant. Alternatively, we might think of a gregarious parasitoid wasp that oviposits a batch of eggs into its host. In both cases, we assume that the larvae develop on a fixed amount of resources (in the herbivore case, we imagine the eggs laid on an isolated specimen of the food plant). The more resources an individual larva obtains, the greater its size as an adult, and hence—if reproductive success is correlated with size—the greater its fitness. Such natural history assumptions apply to many insect species. Consider now two aspects of the insect's life history: its clutch size and the speed with which it consumes resources. The optimum clutch for the parent will be influenced by different factors, such as the availability of other oviposition sites, and the relationship between the number of eggs placed in the clutch and the fitness of the individual offspring. There is a large literature exploring these issues (e.g., Godfray et al. 1991), and using standard techniques, one can calculate the optimum clutch size from the viewpoint of both the parent and the offspring. Unsurprisingly, the offspring would "prefer" smaller clutch sizes than the parent.

Now consider the speed with which the offspring consume the limited resources. Physiological studies have shown that a trade-off often exists between speed of consumption and assimilation efficiency. To maximize the total fitness returns from the clutch, the larvae should feed relatively slowly so that the limited resources are used in the most efficient manner. This is what would maximize parental fitness. But selection acting on genes expressed in the larvae is likely to favor faster feeding if this allows certain individuals to preempt resources and thus ultimately become larger than their competitors. True, the severity of competition expected between siblings is considerably less than would be expected in no-holds-barred competition between unrelated individuals, yet simple models easily show that the evolutionarily stable level of feeding results in a reduction in total brood fitness below that achievable in the absence of sibling conflict.

We have identified two battlegrounds: clutch size and consumption speed. How are they resolved? Surely the parent will always win the clutch-size conflict and the offspring the consumption-speed conflict. The young can hardly influence parental behavior before they are born; whereas, by the time the offspring hatch and begin to compete for resources, the parent is long gone or even dead. The resolutions here appear to be both trivial and one-sided: Either the parent or the young is the absolute winner. If such arguments carried over to species with parental care, then Alexander's first criticism would be true: Parent-offspring conflict, even if it existed, would be irrelevant to explaining the behavior of animals in the wild as one side (in Alexander's view, the parent) always wins.

We thus have a hypothetical but reasonable example of parent-offspring conflict where the resolution is both one-sided and simple. Although this is true if we consider either conflict in isolation, there is a twist if we analyze their joint evaluation (Godfray and Parker 1992; for a related model, see Forbes 1993). If offspring consumption rate responds to local levels of sibling competition, then a parent may be selected to lower its clutch size to reduce sibling competition and hence to increase the efficiency with which the brood uses the limiting resources.

EARLY RESOLUTION IDEAS

Return now to the much more difficult problem of the resolution of parent-offspring conflict in birds and mammals with extended periods of parental care. Trivers's ideas about resolution have already been mentioned. He envisaged parents monitoring the needs of their offspring while the offspring systematically tried to deceive the parent by misrepresenting their resource requirements. Trivers also pointed out the self-destructive tantrums shown by many young mammals. Zahavi (1977a) suggested that these may involve young "blackmailing" their parents. The offspring engage in activities that reduce their own fitness (and hence the fitness of their parents), and the parents provide more resources to stop the young from further harming themselves.

Psychological manipulation and parental blackmail are highly charged metaphors that have raised the hackles of some biologists. Ideally, we need to formalize these ideas and analyze them within a quantitative framework. This is far from straightforward and has not yet really been achieved within an abstract model, let alone one quantified with a particular animal system in mind. Nevertheless, the last 10 years have seen the development of a number of approaches for studying resolution models, which I describe before returning to the challenges that lie ahead. My approach is largely nonmathematical, and interested readers should refer to the original papers for full details of the very different models and modeling approaches or to my attempt (Godfray 1995a) to interpret the different ideas within a single modeling framework.

BLACKMAIL AND BEGGING

The first formal resolution model was in the fourth of the series of papers by Parker and Macnair (1979) analyzing the population genetics of parent-offspring conflict. Recall from the earlier discussion of their work (reinterpreted in a kin-selection framework) that they established that the parental optimum occurred when $B/C = 1$ and the offspring optimum occurred when $B/C = r$ (B and C are marginal costs and benefits with changing resource allocation). The problem is to predict where in the region $1 < B/C < r$ the ESS resolution will occur.

Parker and Macnair began by assuming that parents respond to increased offspring solicitation through providing more resources, whereas offspring respond to increased resource supply by decreasing the level at which they beg. In the model, these two responses by the parent and the young are in a sense hard-wired and are not themselves allowed to evolve. The particular functions used by Parker and Macnair to model the parental response assumed that a fixed change in solicitation led to a relative rather than an absolute, change in resources, and that the fixed change would produce less response when begging levels were already high. Similar assumptions were incorporated into the offspring response. With these assumptions in place, the simultaneous parent and offspring ESS was found to occur at $B/C = (1/2)(1 + r)$. In other words, exactly midway between the parental and offspring optima.

The resolution occurs precisely halfway between parent and offspring optima because of the symmetrical way in which the parent and offspring responses were modeled, but for a broad class of possible functions, the optimal resolution will be somewhere near the halfway point. This model can be thought of as partially embodying the ideas in Zahavi's verbal theory of parent-offspring blackmail. The model assumes no variation in underlying offspring need, but it also assumes that parents gain by providing extra resources for their young in order to encourage them to desist from begging or other activities that lower their fitness. The main problem with the model is that a critical part, the responses of parents and young, must be assumed. If these responses themselves were allowed to evolve, would one still get an intermediate resolution ESS? This question has not yet been addressed and would probably require a change in modeling framework to incorporate the dynamic interplay between parents and young.

Yamamura and Higashi (1992) have developed a general approach to conflict resolution among kin that can be applied, at least in principle, to parent-offspring interactions (Godfray 1995a). Both parents and young are assumed to be able to influence resource share, but they pay a price for trying to manipulate the outcome. Deviation from the ESS by either parent or offspring is assumed to lead to a conflict that entails costs for both parties. A critical parameter is k, the relative costs that the parent and young experience when attempting to determine resource level. When $k < 1$, the parent finds it relatively cheap to manipulate offspring behavior, and the reverse is true when $k > 1$. The ESS resource allocation is always intermediate between the optima for the parent and young, though nearer (but not equal) to the parental optima when $k < 1$ and nearer (but not equal) to the offspring optima when $k > 1$. This is a general model of the resolution of conflict between relatives that has not yet been applied specifically to parent-offspring conflict. The assumptions it makes about the nature of the behavioral interaction between parent and young need to be assessed within the framework of a more mechanistic model of parent-offspring conflict.

A rather different approach to conflict resolution has been taken by Eshel and Feldman (1991). Recall that the parental optimum is given by the condition $B/C = 1$. The ratio of the marginal costs and benefits of changing resource allocation should be unity. What determines the marginal benefits to the current offspring is the extent to which its fitness increases as it receives more resources. Conceivably, the offspring could change this relationship. Take an extreme example. Consider a young bird whose fitness is 1 (in arbitrary units) if it gets one insect an hour, 1.5 if it gets two insects an hour, 1.75 of a unit if it gets three insects an hour, and so on. The parental optimum might then be to provide resources of two insects. What would happen if the young changed this relationship so that its fitness was zero if it received only one or two insects an hour? The parent might then find it optimal to supply three insects an hour, and offspring fitness would be increased. More generally, the offspring changes the function relating fitness to resources received such that at the new parental optimum, $B'/C' = 1$, offspring fitness is higher than at the old parental optimum, $B/C = 1$.

The extent to which Eshel and Feldman's mechanism operates in nature will be determined by how readily the young can manipulate the fitness-resource function. This will be influenced by a variety of physiological and ecological constraints. The playing field on which parent-offspring conflict is contested may have already been influenced by cumulative changes in the fitness-resource function over the evolutionary history of the species. Eshel and Feldman (1991) concentrated on how the young might manipulate the rules of the game, but the parent might be able to adopt the same strategy, perhaps by changing the fitness consequences to future offspring of selfishness by current young. These ideas need further investigation.

SIGNALING RESOLUTION

The feature shared by all the resolution models discussed in the last section is that they assume no variation in the young's requirement for resources (or at least variation in the young's needs are not an essential part of the model). Yet before Hamilton, most biologists would have interpreted the noisy begging and bleating of young birds and mammals as an attempt to communicate need to their parents. After Hamilton and Trivers, such a cozy interpretation was harder to sustain. How could communication of need be evolutionarily stable? What was to stop an individual offspring from cheating and misrepresenting its needs for its own benefit? Because of these problems, Trivers and many of the people who first worked on parent-offspring conflict emphasized resolution mechanisms that involved blackmail, psychological manipulation and other skulduggery.

However, there are evolutionarily stable ways by which offspring can communicate variation in need to their parents, and which provide a different type of resolution from those discussed in the last section. The origin of these ideas lies in the work of Zahavi (1977a), who argued that most if not all biological signals in situations of potential conflict are costly, and that these costs are essential for the evolutionary stability of the signaling system. For many years, Zahavi's ideas lay fallow, chiefly because they lacked a convincing theoretical framework, but they enjoyed a major renaissance from about 1990, thanks to a number of workers, especially Grafen (1990a,b), who provided the missing theoretical backbone. Not all types of signals are costly, and there has been considerable confusion in the literature about how to classify different types of signals. Recent reviews by Maynard Smith and Harper (1995), Hurd (1995), and Reeve (1997) provide a clear path through the maze of different ideas.

Before talking about signaling in the context of parent-offspring conflict, I briefly describe ideas about signaling male quality in courtship, as this provides an important parallel. Consider the extreme ornamentation of many male birds that has puzzled biologists since the days of Darwin and Wallace. Could the long tail of a peacock or bird of paradise be an honest signal of the quality of the male? If so, what mechanism prevents the signaling system from breaking down and cheating arising? As hypothesized by Zahavi, the signaling system can be stable if the production of the signal is costly, but if the costs are greater for poor-quality males compared with high-quality males. In the simplest model, an ESS exists at which males of different quality signal with differing intensities. The female can assess the quality of a potential mate simply by noting the level of the signal. The signaling system is stable from the point of view of the transmitter because the marginal benefits of cheating and attracting more males are exactly counterbalanced by the marginal costs of producing the greater signal. The evolutionarily stable signaling level for males of different quality varies because males in poorer condition experience higher costs. Grafen (1990b) and Johnstone and Grafen (1992, 1993) have shown that the signaling system is stable to receiver costs, and to errors in the assessment of the signal, whereas Johnstone (1995a, 1996) has explored how signals using multiple cues may evolve.

Can a similar idea explain why communication between parents and young appears to be costly, and what would this tell us about the resolution of conflict over resource allocation? Consider the following candidate evolutionarily stable signaling system (fig. 6.2). Suppose (1) young that require no food (or the minimum amount of food) do not signal but that signaling (perhaps begging) increases monotonically with increasing need; (2) signaling is costly, perhaps because it requires the expenditure of energy or risks attracting predators—costs may or not be related to an individual's needs; (3) the benefits of obtaining food or other resources go up as need increases; (4) the parent uses the signaling level of the offspring to assess its need and then imposes its optimal resource allocation.

A quantitative model of this signaling system (Godfray 1991) can be shown to be locally stable and hence an ESS (I return below to questions of global stability and whether the local ESS can be reached). We can carry out a "verbal stability analysis" that mimics the mathematics. We need to show that the model is stable both from the point of the receiver, the parent, and the transmitter, the offspring. From the parent's point of view, the stability of the system is easy to understand. The parent obtains accurate information

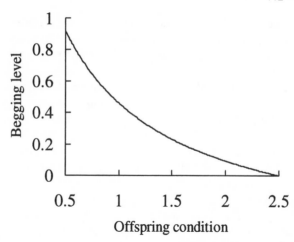

FIG. 6.2. Begging level and offspring condition. At the signaling ESS, offspring in poor condition (great need) signal at a high rate, whereas those in good condition (little need) signal at a low rate. Offspring in any condition that deviate from the ESS suffer a fitness penalty.

and is able to make an informed decision about resource allocation. Either to ignore the signal or to allocate other than the parental optimal amount of resources would obviously lead to a reduction in parental fitness. What about the offspring? What is to prevent the young from "pretending" to be needier than it actually is? Consider an individual in a particular state of need. The candidate ESS states that the offspring should signal at a set level that the parent can interpret to assess its food requirements. Let us suppose a mutant arises that causes its bearer, when in that state of need, to signal at a level slightly higher than that set by the ESS. Because it is begging at a level higher than it "ought," the parent interprets it as needier than it actually is and allocates extra food. This is thus a benefit of "cheating." There are, however, two costs. First, the extra food it gets is at the expense of its parent's future offspring; the cheating offspring suffers a reduction in nondescendant kin, the importance of which is scaled by the coefficient of relatedness. Second, the offspring suffers the direct cost of producing a higher level of signal. For the ESS to be stable, the benefits gained by exaggerating the signal must be exactly counterbalanced by the costs. Similarly, the reduction in costs resulting from producing a signal below that of the ESS must be exactly balanced by the loss in fitness of obtaining fewer resources.

Does the offspring or parent "win"? In one way, the parent wins. The resource allocation that occurs is at the parental optimum, though this is a function of offspring need. But the parent wins at a price. For the signaling system to be stable, the signal must be costly. And the costs of signaling reduce offspring fitness and hence parental fitness. If this resolution model is correct, it reconciles Trivers's identification of the presence of parent-off-spring conflict with Alexander's argument that the parent is in a position to impose its will. Resource allocation is at the parental optimum as Alexander predicted, but the fact that natural selection acts differently on genes expressed in parents and offspring—as Trivers recognized—requires a costly signaling system. This model of signaling of need between parents and young is closely related to Maynard Smith's (1991b) "Sir Philip Sydney game," which explores signaling of need between nondescendant relatives.

If young produce a signal that is accurately interpreted by their parents, do we have an honest signal of need with no room for cheating or misrepresentation? As Johnstone and Grafen (1993) have discussed in the context of signals of quality, there is still room for an element of deception. Suppose that two young of identical need differed in that one could produce a signal with lower cost that the other; further, assume the parent is unable to assess which offspring finds signaling cheaper. At the ESS, the parent allocates food according to the expected distribution of needs of offspring signaling at that level. Those that find signaling relatively inexpensive obtain more food than those who find it relatively expensive.

Is there any evidence that begging signals are expensive and that parents respond to increased levels of solicitation by providing more food? Haskell (1994) placed speakers in bluebird nests and played back the sounds of young begging. Playback increased the risks of predation (by blue jays) on nests near the ground, but not in nests in trees. In nests near the ground, the risks of predation increased with higher frequencies of begging calls. Many studies have found that increased begging results in greater resource allocation. Kilner (1997) has recently shown that canaries may signal need by flushing the insides of their mouths a bright red color. She ingeniously manipulated mouth color using cochineal dye and was able to increase the food received by the young.

As Rodriguez-Girones et al. (1996) and Bergstrom and Lachmann (1997) have recently stressed, the presence of a signaling system may lower a parent's fitness below what would occur if the offspring did not signal and if the parent doled out an average amount of food to each offspring. Might this mean that the local ESS described above is unlikely to be reached in nature? Before confronting this, I need to be more specific about the concept of offspring need. For each offspring, there will be a function relating the amount of food it gets on any particular day to its future lifetime fitness (though this function may be very difficult to measure in the field). Exactly what this function is will depend on two classes of factors. First, those that can be assessed with essentially no ambiguity by the parent: for example, the offspring's age, size, and other manifest indications of its general well-being. Second, there will be some factors that the parent cannot assess directly or that might be capable of misrepresentation. These second, cryptic components of need are the ones communicated by the costly ESS signaling system described above. One can imagine an initial state in which the young produced no special signal of need and the parent was selected to monitor its young to assess its resource requirements. Once this monitoring had evolved, the young (or rather genes expressed in the young) would come under selection to exploit parental assessment. In some circumstances, it is possible that parental assessment would contract to a core of unfalsifiable indicators, whereas in other circumstances, the trajectory of the signaling system would lead to the local ESS that involves costly signaling.

However, recent research has revealed hitherto unrealized complexities in signaling evolution. At the ESS I described above, offspring have a unique signal determined purely by their cryptic need. Lachmann and Bergstrom (1998) have elegantly shown that other "pooling" ESSs are possible under which certain classes of offspring all give the same signal, though the monotonic (increasing) relationship between the level of the signal and the offspring's need is retained. At the moment, theory does not provide a clear prediction as to which of these candidate ESSs will occur in nature. My expectation is that the eventual prediction will be for a costly signaling system, with begging intensity rising with need, but not necessarily smoothly: In particular, sets of offspring in unusually good condition may all give the same signal (technically, I expect the probability distribution of offspring needs, which does not enter the solution of my [Godfray 1991] models, to determine the precise ESS).

Looking ahead, to be of real use in generating hypotheses to test in the field, we probably require a more sophisticated model of parent-offspring interactions than the comparatively simple (at least biologically) models described so far. In particular, most models implicitly assume that parental care consists of a series of resource-allocation decisions that have independent effects on future offspring fitness. This is clearly a gross oversimplification as what you are fed today is likely to influence your food requirements tomorrow. Moreover, it excludes a variety of more subtle parent-offspring interactions, such as the possibility that a parent can learn about the behavior of its young or resort to punishing them. Thus far, virtually no models have attempted to study the dynamics of parent-offspring conflict (as opposed to the dynamics of parental care; see, e.g., Clark and Ydenberg 1990). An important exception is Clutton-Brock and Parker's (1995b) work on punishment. Suppose an offspring engages in a "selfish" act of behavior that reduces its parent's fitness. If this action only occurs once, then the parent must make the best of a bad job and respond in a way that maximizes its fitness. If the act can reoccur, however, the parent might respond by punishing the young in a way that reduces the fitness of both the offspring and the parent. The punishment works if the young is deterred from repeating the selfish act, or in other words, if the young of the species have evolved the capability to respond to parental chastisement. One can envisage young trying to counterpunish their parents, though as Alexander first pointed out, the greater capacity of the parents to dominate their young physically is an important consideration here.

Family Life

Although the last section was concerned almost exclusively with birds and mammals that exhibit an extended period of parental care, it considered the simple but rather unusual case of a single parent looking after a single young. Relaxing this basic model of family life leads to a number of new complications: conflict between the sexes and sibling conflict. I briefly discuss sibling conflict here.

Consider first sibling conflict in species without parental care. A hypothetical example has already been mentioned: larval insects competing for limited resources where individuals can preempt resources by feeding faster, but at the cost of a reduced assimilation efficiency. This is a specific example of exploitation competition among siblings, something that has been modeled on a number of occasions (e.g., Macnair and Parker 1979; Godfray and Parker 1992; Roitberg and Mangel 1993; Sjerps and Haccou 1994). The models predict competitive behavior that tends to reduce total brood fitness. The exact extent of this reduction depends on the detailed assumptions, for example whether the costs of an increase in competitive behavior are visited on the individual or on the whole brood. As discussed earlier, the parent cannot directly intervene in this type of sibling competition, though the nature of sibling competition may indirectly influence parental behavior such as choice of clutch size or oviposition site.

How is sibling competition modified by the presence of the parent? Consider birds in the nest begging food from their parent. At one extreme, parents may allow siblings to compete among themselves and feed the winner. In some birds, chicks jostle each other to be in pole position when the parent returns to the nest, and the parent appears to feed the first offspring it encounters (Rydén and Bengtsson 1980; McRae et al. 1993). In other bird species, the parent causes the young to hatch asynchronously so that chicks vary in size and older, larger individuals seem to dominate their younger siblings (Magrath 1990). In groups such as the herons and egrets, a very rigid pecking order exists (Mock 1987). In these species, it is likely that the parent is completely or largely abrogating resource allocation decisions to the young. Why might this evolve? Possibly it is advantageous for the parent to feed only the strongest chick, or possibly the benefits of imposing its own resource allocation fail to outweigh the costs in wasted time.

In the majority of cases, the parent probably takes an active role in resource allocation among siblings. Most theoretical studies of this problem have been battleground models, designed to predict how clutch or brood size influences the potential extent of parent-offspring conflict (Macnair and Parker 1979; Lazarus and Inglis 1986; Harper 1986; Godfray and Parker 1992; reviewed by Mock and Parker 1997). The models assume that the parent has a fixed response to offspring behavior and normally that all young are of equal competitive ability. No clear consensus about the relationship between clutch size and parent-offspring conflict emerges from a comparison of these models, and it appears that the results are quite sensitive to the natural history incorporated in the theory. Moreover, it is difficult to test these models against real data because there is no strong reason to suppose that the extent of the battleground of parent-offspring conflict is always reflected in the observed behavior of parent and young. Again, what is required to provide testable hypotheses are models of the resolution of parent-offspring conflict.

Of the resolution models discussed in the last section, Parker and Macnair's (1979) model and the signaling model (Godfray 1995b) have been extended to families greater than one. As before, the Parker and Macnair model predicts that, at the joint parental and offspring ESS, the resource allocation is intermediate between that optimal for the parent and the young. The signaling model, on the other hand, predicts resource allocation at the parental optimum but with offspring producing a costly signaling that lowers both their fitness and that of their parents. A further prediction made by the signaling model is that the level of the honest signal of need—for example, begging rate—produced by the offspring should depend not only on its own resource requirements but also on those of its brood mates. This suggests a way of testing the hypothesis: Manipulate the food requirements of all but one member of a brood by feeding them or depriving them of food and then measure the begging level of the remaining offspring. The theory predicts that, as long as some of the inclusive fitness costs of an individual getting more food are experienced via siblings in the same brood (as opposed to all costs being experienced as a reduction in future siblings), then feeding brood mates should reduce levels of signaling by the focal individual, and the reverse should occur when brood mates are deprived of food. Three such experiments have now been performed. Although the three are not exactly comparable, work on American robins (Smith and Montgomerie 1991) and yellow-headed blackbirds (Price et al. 1996) support the prediction, whereas experiments with starlings Cotton et al. (1996) do not. It should be stressed, however, that the theory is still exceedingly simplistic compared with the behavioral complexities of real animals, and better theory, tailored to specific systems, is required before hypotheses can be tested confidently in the field and laboratory.

A common feature of many bird and mammal species is brood reduction, the loss of one or more young through starvation or through infanticide or siblicide. The two most important explanations for brood reduction are that it is an adaptation that allows the parent to adjust the number of young it rears to the resources available each breeding season (Lack 1947, 1954), and that the parent produces extra eggs as insurance against some being infertile, which may or may not involve culling the excess when all hatch (Anderson 1990a,b; Forbes 1990). To what extent will selection acting on genes expressed in the parent, the surviving young, and victim differ in when they allow brood reduction to occur? O'Connor (1978) used simple kin-selection arguments to suggest that, as resources become scarcer, brood reduction is first favored by selection acting on genes expressed in the surviving young, second on genes expressed in the parent, and finally on genes expressed in the victim itself (i.e., selection for suicide). But kin-selection theory should be applied only in the case of weak, additive selection pressure (Grafen 1984), something that clearly does not apply to brood reduction (an individual cannot be killed twice). Explicitly genetic models do support O'Connor's conjecture, but with one proviso: The victim must be chosen independently of genotype, perhaps a runt (Godfray and Harper 1990).

O'Connor's work and more recent models (Mock and Parker 1986; Parker and Mock 1987; Dickens and Clark 1987) show that a conflict can exist over when brood reduction can occur, but its extent in nature and how it is resolved is still not clear. Where rearing extra young is impossible or very expensive, and supernumerary eggs are laid purely as an insurance against infertility, then the rapid dispatch of extra young may be strongly selected for in all parties. In species where this is thought to occur—for example, many raptors (Anderson 1990b)—there is usually a strong size hierarchy among siblings, and this has possibly evolved to facilitate efficient brood reduction (Hahn 1981; though there are many alternative hypotheses, see Magrath 1990). Where brood reduction has evolved to allow the parent to exploit a variable environment, there would seem to be greater scope for parent-offspring conflict. However, parents may respond to the loss of one their offspring not by the redistribution of food among the survivors but by decreasing effort to the current brood or even by desertion. Such behaviors would markedly decrease the benefits of brood reduction. Forbes (1993) has modeled such a scenario, based loosely on the behavior of many ardeids (herons, egrets, and allies). Parents are selected to put less effort into nests following brood reduction, and, as a result, the battleground—the conditions under which disagreement occurs over brood reduction—shrinks considerably.

Other Scenarios of Parent-Offspring Conflict

I have concentrated very heavily on parent-offspring conflict over resource share in birds and mammals with parental care. As described above, this is the classic problem that has received the most attention, both from theoreti-

Solitary   Gregarious

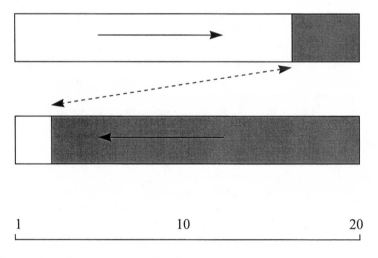

1   10   20

Clutch size

FIG. 6.3. Evolutionary hysteresis caused by parent-offspring conflict. Solitary parasitoid wasps generally have fighting mandibles, and only a single individual develops in each host. Gregarious larvae do not fight, and two to many develop on a host. The top bar shows what happens if host size increases over evolutionary time so that the parent is selected to produce a larger clutch size (see bottom scale). Selection acting only on the larvae causes larvae to lose the fighting mandibles when the parental optimum clutch size is quite large. The exact point of transition depends on the detailed assumptions of the model, but it is frequently between 10 and 20. If clutch size moves in the opposite direction, however, then the larval fighting behavior only spreads when the parental optimum drops to 2 to 4. There is thus a hysteresis, and the predicted state of the system depends on its evolutionary history. The hysteresis is largely the result of rare alleles for nonfighting behavior being at low fitness in a population composed largely of fighters.

cians and experimentalists. I finish by highlighting a few other areas in which selection acts in different ways on genes expressed in parents and young, and in which the theory of parent-offspring conflict has usefully been applied.

Outside vertebrates, parent-offspring conflict theory has been most extensively applied in the social Hymenoptera. Colonies normally consist of reproductive females (queens) and their nonreproductive daughters (workers), and there are ample opportunities for genetic conflict between the two. Trivers and Hare (1976) first pointed out that, because of the haplodiploid genetic system of Hymenoptera, queens favor an equal sex-allocation ratio, whereas workers favor reproductive females and males in the ratio 3:1. This and other issues are discussed by Keller and Reeve (chap. 8).

The social Hymenoptera derive from parasitoid Hymenoptera, which lay their eggs on or in the bodies of other insects that act as hosts for their developing larvae (Godfray 1994). Parasitoid wasps can be categorized as solitary or gregarious depending on whether one or more than one larva develops on the same host. Solitary larvae normally have large mandibles that they use to attack conspecific and other parasitoids laid on the same host; gregarious larvae lack these fighting mandibles. The number of eggs a female wasp lays on a host can be viewed as a problem of clutch size and analyzed using the standard optimality approach developed for clutch size in birds. Consider a wasp laying a solitary egg on a small host. Suppose that over evolutionary time, hosts increase in size (either because of selection acting on the host or because the parasitoid changes its host range). We would expect the parasitoid to increase its clutch size (though an increase in body size might be an alternative response). However, the parent could only realize a larger clutch size if selection acted simultaneously on genes expressed in the larvae to stop them from fighting and killing other individuals in the host. Models of this process suggest a wide battleground of parent-offspring conflict and also a hysteresis in the resolution; that is, the position of the transition between the solitary and gregarious states depends on whether the species is originally solitary or gregarious (Godfray 1987). The precise outcome depends on the detailed assumptions of the model, but a typical result is shown in figure 6.3. Hosts of a particular size may contain a solitary or a gregarious brood, depending on the evolutionary history of the species. Le Masurier (1987) surveyed the relationship between clutch size and host size for a genus of braconid wasp then called *Apanteles* but now split into many smaller genera. He indeed found that hosts of equivalent size could harbor solitary wasps or wasps with clutches of about 2–10. A reexamination of this data within a phylogenetic framework would be very interesting.

In angiosperm plants, the seed is nourished by endosperm tissue, which is genetically distinct from the parent plant and normally different from the seed itself. The endosperm contains genetic material from the parent plant and the pollen and may be diploid, though it is more usually triploid (and sometimes of even higher ploidy levels). If triploid, one chromosome set derives from the father and two from the mother, and the latter may or may not be genetically distinct. It is possible that genes expressed in the seed, the endosperm, and the parent plant all influence the resources invested in the seed and that natural selection will act in different ways on genes in these separate tissues. This has been confirmed by a variety of what are largely battleground rather than resolution models (e.g., Law and Cannings 1984; Queller 1984, 1989, 1994; Bulmer 1986; Uma Shaanker et al. 1988; Haig 1992a).

The resolution of parent-offspring conflict in plants over resource allocation to seeds is likely to involve competing biochemical signals, which may be easier to investigate than the complex behavioral interactions in families of birds and mammals. But the situation with plants may have a parallel in interactions between vertebrate parents and their eggs or embryos before birth. This is especially true of mammals that spend a long time in the womb. The importance of genetic conflicts in vertebrates before birth has been championed by Haig, who is also responsible for the recognition of a further area of potential genetic conflict: Selection may act differently on paternally and maternally derived genes (Haig and Westoby 1989; Haig 1992b, 1993b, 1996; Haig and Graham 1995). Maternally derived genes in the offspring have a genetic interest in the future well-being of the parent in contrast to paternally derived genes, which are effectively unrelated to the mother. The two types of genes are distinguishable because of genetic imprinting: differential methylation of genes in male and female gametes that influences whether they are expressed. Battleground models suggest that paternally derived genes would "prefer" greater resource investment in offspring than maternally derived genes. A fascinating insight into how this conflict has been resolved is revealed in mice whose eggs have been manipulated to carry either two paternally derived or maternally derived chromosome sets. In the first case, the placenta is abnormally large, and in the second case, abnormally small. A normal placenta requires a balanced chromosome set consisting of chromosomes from both parents. The molecular and physiological resolution of the genetic conflicts between genes expressed in the parent and on the two chromosome sets in the young is likely to be one of the most exciting areas in the study of parent-offspring conflict in the coming decade.

Acknowledgments

I wish to thank Austin Burt, Kate Lessells, and Geoff Parker for their helpful comments on this chapter.

7 Intragenomic Conflict

Andrew Pomiankowski

Organisms are designed by natural selection to transmit their genes to future generations. Most genes within an organism act in a cooperative manner because the common good benefits individual genes. Conflict arises at reproduction, however, because not all offspring inherit the same set of genes, and not all genes are transmitted in the same way. There is always room for some genes to exploit the common good even if this is detrimental to the rest of the genome.

Conflict within an organism, or *intragenomic conflict*, has two main causes. The first is that organisms are composed of multiple genetic entities: sex chromosomes, autosomes, organelles, transposable elements, plasmids, and a variety of intracellular symbiotic genes. These genes do not share the same interests because they have different modes of inheritance. For instance, organelle genes are normally passed on only by one sex and thus benefit from distorting the sex ratio toward that sex (Eberhard 1980). The second cause of conflict is sexual reproduction. Sexual mixing is preceded by meiosis, with half the nuclear genes segregating in each gamete. If meiosis is fair, each allele or chromosome segment is inherited by half the gametes. There is, however, strong selection for genes to subvert meiosis and gain a transmission advantage (Hamilton 1967).

Genes that cause intragenomic conflict have come to be called *selfish genetic elements* (or selfish DNA, ultraselfish genes, genetic parasites, etc.). Such elements enhance their own transmission to future generations while being either neutral or harmful to the fitness (survival or fertility) of the individuals that carry them. Harm can arise as a side effect of transmission distortion. For example, transposable elements increase their own fitness by inserting extra copies elsewhere in the genome. But as a side effect, insertions often cause mutations, and recombination between transposable elements in different locations produces unbalanced chromosomes (Charlesworth et al. 1994). Harm can also arise because selfish genetic elements actively destroy or disable competitors. For example, meiotic-drive genes cause the failure of gametes carrying the wild-type allele (Lyttle 1991), and a number of intracellular symbionts and organelles gain by killing male offspring or diverting resources to the female reproductive structures (Hanson 1992; Hurst and Majerus 1993).

These deleterious effects favor countermeasures. Most genomes contain suppressors that limit the activity of selfish genetic elements. Suppression can be so strong that it obscures evidence of intragenomic conflict. For example, suppressors of sex-linked meiotic drive are so prevalent in some populations of *Drosophila simulans* that drive was not even suspected to exist until interpopulation crosses were made (Mercot et al. 1995; Atlan et al. 1997). Selection on the selfish genetic element itself can also favor reduced harm when the reproductive interests of host and element overlap. For example, transposable elements within bacterial lineages have a number of mechanisms designed to inhibit transposition (Kleckner 1990). This makes sense for the transposable elements because most of the time their transmission to future generations is tied into their bacterial hosts' asexual reproduction.

Understanding the evolution of selfish genetic elements and their importance in causing intragenomic conflict does not require any novel concepts. The same logic used to understand social interactions between separate organisms applies to the evolution of cooperation and conflict between genes within an organism. In this chapter, these ideas are used to interpret a natural history of selfish genetic elements: transposable elements, cytoplasmic genes, meiotic drive, and post-segregational killers. The categories are not hard and fast or fully inclusive, but they serve as groupings of the major types of selfish genetic elements. The objective is to introduce the reader to the variety of forces giving rise to intragenomic conflicts and the way in which selfish genetic elements evolve.

Transposable Elements

Transposable elements (TEs) are a class of selfish genetic elements that occur in multiple copies, dispersed throughout the genome. They are common in eukaryotes and are found in some prokaryotes. At least 35% of the mammalian genome is composed of TE sequences (Yoder et al. 1997). There are two main classes of TEs, which differ in their mode of replicative transposition (fig. 7.1). Retroelements use reverse transcriptase to convert RNA transcripts back into DNA, which are then incorporated into a new chromosomal location (e.g., *Drosophila gypsy*, yeast Ty, and mammalian LINEs). The second class of elements transpose without RNA intermediates. Element DNA is excised from a replicating chromatid and then inserted at a new site (e.g., *Drosophila P*, *E. coli* IS*10*, and *mariner*, which is found in many species). Transposition of DNA elements is usually replicative because the double-strand damage caused by excision is repaired using the sister chromatid, which still carries a copy of the original element (Kleckner 1990; Gloor et al. 1991).

TEs exploit sexual reproduction to increase their transmission to future generations. Normal meiotic segregation means that chromosomal genes are passed, on average, to only half the gametes and resulting offspring. Mobile elements overcome this segregational loss by replicating within the genome.

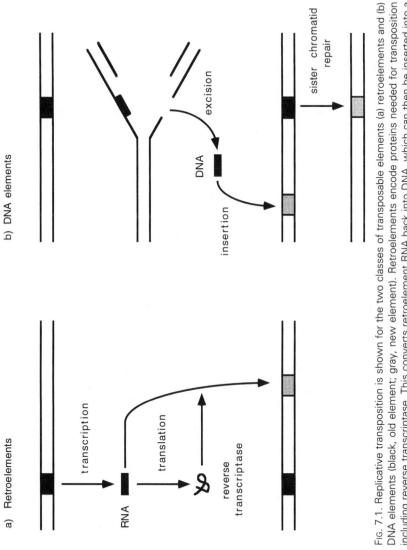

FIG. 7.1. Replicative transposition is shown for the two classes of transposable elements (a) retroelements and (b) DNA elements (black, old element; gray, new element). Retroelements encode proteins needed for transposition including reverse transcriptase. This converts retroelement RNA back into DNA, which can then be inserted into a new location in the host chromosomes. DNA elements also encode proteins involved in transposition. The molecular basis of transposition is less well understood. It is thought that excision of element DNA during DNA replication is followed by insertion elsewhere. The double-strand gap left by excision is filled by sister chromatid repair.

An element that has copies on several chromosomes increases the number of progeny it can infect. Thus, even if transposition is costly to the host, it can still be favored because TEs potentially double their fitness by having multiple copies scattered through the genome (Doolittle and Sapienza 1980; Orgel and Crick 1980).

The exploitation of sexual reproduction also underlies prokaryotic transposable elements. In bacteria, the predominant mode of reproduction is asexual. However, DNA is transferred between lineages through infective agents like plasmids and viruses. TEs hitch a ride by transposing into plasmids and viruses. They also gain by moving back into the bacterial DNA, as this ensures transmission through clonal reproduction (note that plasmids and viruses are regularly lost during cell division; Kleckner 1990; Summers 1996). TEs also exploit another sexual route, bacterial transformation, in which chromosomal fragments are released, picked up by other cells, and then incorporated into the recipient by homologous recombination (Levin 1988). Having dispersed copies throughout the bacterial genome increases the probability that the transforming DNA contains transposable elements.

TEs Are Deleterious to Their Hosts

Transposable elements are rightly thought of as selfish genetic elements because transposition is often harmful to the host (Charlesworth et al. 1994). This probably explains why TEs often occur at low frequencies at specific sites (Montgomery and Langley 1983; Charlesworth et al. 1992). Element insertions are known to be a major cause of spontaneous mutation in natural populations. In *Drosophila*, more than half of all spontaneous mutations are due to TE insertions (Yoder et al. 1997). Most insertions into coding sequences or nearby control regions disrupt normal gene activity and cause a decrease in host fitness. In addition, ectopic recombination between non-homologous copies causes the production of deleterious inversions, deletions, and duplications (see below).

Several properties of TEs have probably evolved to reduce the virulent effects of transposition. TEs do not usually transpose in somatic cells, as this harms host fitness without enhancing transmission to future generations. Second, some mobile elements are able to self-splice from primary RNA transcripts and are thus known as group I introns (Cech 1990). This greatly reduces the likely magnitude of their mutagenic effects. Finally, there is some evidence that TEs preferentially insert in regions that do not disrupt host gene activity. For instance, yeast Ty1 shows a nonrandom genomic distribution with most insertions into promoter rather than coding regions (Eibel and Philippsen 1984; Natsoulis et al. 1989). Likewise, yeast Ty3 is commonly inserted immediately upstream of tRNA genes (Kinsey and Sandmeyer 1991). It is thought that these patterns reflect insertion bias, but they could also partly be explained by retention bias of TE insertions where they cause little or no reduction in fitness.

These modifications only hide a few sources of intragenomic conflict caused by transposable elements. In sexual eukaryotes, elements that reduce their transposition rate will lose out to other, more active, elements. Clearly there must be forces acting to restrict the number of elements and their rate of transposition; otherwise, the host genome would have disintegrated under the accumulation of an ever expanding number of elements. I will delineate three views of the evolutionary outcome of this conflict. These differ in where counterselection is supposed to act: (1) against individuals with more elements; (2) for self-repression by TEs; or (3) for host repression of transposition.

If element number is controlled by selection against individuals with more elements, some force must increase exponentially with TE copy number. The view that has received the most attention is that negative selection is generated by ectopic recombination (Langley et al. 1988). Because elements occur at low frequencies at each insertion site (Montgomery and Langley 1983; Charlesworth et al. 1992), they generally lack a pairing partner at meiosis. This means that the unpaired DNA can undergo ectopic recombination with a homologous element at a different chromosomal site. If this occurs, it can cause the production of deleterious chromosomal rearrangements (inversions, deletions, and duplications). The frequency of ectopic recombination and thus the decrease in individual fitness are expected to increase sharply with copy number.

Evidence for this hypothesis is mixed. A number of indirect tests have been made. Ectopic recombination is expected to be more frequent in heterozygotes than in homozygotes, because heterozygotes lack a pairing partner. This was shown experimentally for *roo* elements in *Drosophila* (Montgomery et al. 1991), but heterozygosity had no effect on the rate of ectopic recombination in yeast (Petes and Hill 1988). A similar prediction is that inbred species can tolerate higher numbers of TEs, because homozygosity retards the rate of ectopic recombination. Recent data for *Lycopersicon* plant species support this hypothesis (Charlesworth and Charlesworth 1995). Inbred red-fruiting species have higher *Lyt1* copy number (3 species, 41–46 elements per genome) than outbred green-fruiting species (5 species, 2–8 elements per genome). This is, however, a weak comparison of two taxa (red vs. green fruiting) that are distinguished by many factors other than inbreeding.

The ectopic recombination hypothesis also predicts that TEs should be common in genomic regions with reduced recombination. Consistent with this, *Drosophila* TEs are abundant in β-heterochromatin (Miklos et al. 1988) and show a high rate of accumulation on the nonrecombining neo-Y chromosome of *D. miranda* (Steinemann and Steinemann 1992; Hochstenbach et al. 1994). *Drosophila* TEs are also more common among rare inversions (Sniegowski and Charlesworth 1994). In the most detailed study of individual *Drosophila* elements (*P*, *hobo*, *I*, *copia*, *mdg1*, *mdg3*, *412*, *297* and *roo*), however, none of the individual elements showed a negative relationship between element number per chromosome division and recombination rate, and there was no overall relationship between element number and recombination rate (Hoogland and Biémont 1996; Hoogland et al. 1997).

These equivocal results suggest that selection against ectopic recombination is not the only force limiting copy number. The alternative insertional mutagenesis hypothesis proposes that there is strong selection arising from mutations caused by TE insertions. As the rate of mutation increases with copy number, there will be a limit to the number of TEs that can be tolerated at the individual level.

Again, tests of this hypothesis are indirect. It predicts that TEs will accumulate where insertions are less likely to have detrimental effects on host fitness. Most of the patterns reported above fit this hypothesis. For example, β-heterochromatin and Y chromosomes are genetically inert regions, so that insertions in these regions are unlikely to be deleterious. Likewise, rare inversions are rarely homozygous, making insertions with recessive effects effectively neutral.

The insertional mutagenesis hypothesis makes one prediction that distinguishes it from the ectopic recombination hypothesis: There should be fewer TEs on the X chromosome (compared to the autosomes) because X-linked mutants are fully expressed in hemizygous males and are thus under stronger selection. The prediction is based on the reasonable assumption that most mutants are recessive and are considerably more deleterious in a hemizygous (or homozygous) state. A pattern of fewer TEs on the X in *Drosophila* is not predicted by the ectopic recombination hypothesis because X and autosomal genes have similar rates of recombination. Although several early studies failed to show any X deficiency, a recent, more extensive survey across several populations of *D. melanogaster* and *D. simulans* established that several TEs have relatively few copies on the X chromosome (Vieira and Biémont 1996). It is still unclear, however, whether the cost of deleterious mutation at the individual level is sufficient to halt increases in TE copy number.

CONTROL OF TE NUMBER

As in other parasites, there is a trade-off between TE transmission to offspring (increased copy number) and virulence (deleterious effects on the host) (Anderson and May 1982; Ewald 1983; May and Anderson 1983). TEs need to replicate themselves to raise the chances that they are transmitted to future generations, but at the same time, they want to limit their numbers to avoid harm to their current host or future offspring.

Another trade-off faced by the TEs is competition between different element genotypes within a host (Bremmerman and Pickering 1983). Elements with high transposition rates do better within a host, but again this advantage needs to be balanced against the harm done to the host. The expected degree of competition (and hence virulence) will be inversely proportion to average genetic relatedness of the elements within an individual (Frank 1996). For transposable elements, relatedness will be determined by the degree of mixing caused by meiosis and sexual reproduction, the mutability inherent in TE replication, and any horizontal transmission between hosts.

Sometimes the balance of these trade-offs can favor very low rates of transposition. This is particularly obvious for bacterial TEs in which the predominant mode of host reproduction is asexual and transfer between lineages (via plasmids, viruses, or transforming DNA) is rare. This means that relatedness between elements within a lineage is high, and there is less benefit to within-host competition. This situation favors elements with low rates of transposition that can persist within lineages without causing much damage to host fitness.

Evidence of self-regulation can be seen in a number of bacterial TEs. For example, both the promoter and transposase binding sites of the *Escherichia coli* element IS*10* have specific GATC sequences that attract adenine methylation, which blocks their activity (Kleckner 1990). Transposition only occurs from the hemimethylated fork of a replicating chromosome. Removing methylation increases transposition a hundredfold. In addition, IS*10* produces an antisense RNA inhibitor that binds transposase (Kleckner 1990). Inhibition multiplies exponentially with copy number. These two mechanisms ensure rates as low as one transposition event every 1,000-cell generations and a low limit to the number of elements per genome. These limit the deleterious effect of IS*10* on host fitness.

A similar system underlies *P*-element repression in *Drosophila*. *P* elements code for two proteins, a transposase and a transcription inhibitor, the latter produced by alternative splicing of transposase transcripts (Lemaitre et al. 1993). As inhibition is maternally inherited, extensive transposition occurs when *P*-carrying males are crossed to females without *P* elements (so-called dysgenic crosses; Engels 1989). This conditional transposition makes sense because *P* elements multiply on naive backgrounds, where the benefits of transposition are high, but remain inactive at high copy number, where the costs of transposition are high.

Another form of repression of *P* elements occurs through the invasion of nonautonomous elements (Engels 1989). These do not produce transposase but are still able to transpose. They arise as partial deletions of complete elements and use transposase produced by other fully competent elements. Nonautonomous elements are the most common elements in many *Drosophila* populations. Brookfield (1991) has argued that selection on individuals favors those with nonautonomous elements because they lower the overall transposition rate and thereby reduce the rate of deleterious mutation.

An additional possibility is that nonautonomous elements are favored within individuals. It is not obvious how this might apply to *P* elements, but the situation appears to apply to retroelements (Nee and Maynard Smith 1990). In this case, complete TEs produce transcripts that are either used as mRNA to produce transposase or act as RNA templates for transposition. In contrast, nonautonomous TEs, which lack coding sequences, only produce RNA templates for transposition. These nonautonomous templates will outnumber complete templates and hence will be more likely to be transposed. In the long term, this can result in dominance by nonautonomous elements with a reduction in overall transposition rates. Once mutants occur in the remaining complete elements, transposition rates will fall to zero.

Finally, there is some evidence of host control of TE transposition. The best example is *flamenco*, a gene that controls *gypsy* transposition in *D. melanogaster*. Usually *D. melanogaster* strains contain fewer than five elements, plus some inactive elements in the pericentromeric heterochromatin (Prud'homme et al. 1995). However, some strains contain high copy numbers (>20) and evidence of raised mutability because of high rates of transposition. This is due to the *flamenco* mutation. Progeny of mothers homozygous for the *flamenco* mutation have highly elevated rates of transposition of *gypsy*. It appears that *flamenco* is a supressor of *gypsy*.

Unlike this specific control, methylation of cytosine residues has been suggested as a general defense against repetitive genomic parasites like TEs (Bestor and Tycko 1996). This has been most clearly established in *Neurospora* fungi where RIP (repeat induced point mutation) acts as a general defense against TEs (Selker 1990). The RIP system works by detecting and methylating duplicated sequences, leading to their transcriptional silencing. In the longer term, the high mutability of methylated cytosine residues ($C \longrightarrow T$ transitions) leads to permanent, mutational inactivation of genes encoded by duplicated sequences. The effectiveness of RIP probably explains why TEs are rare among filamentous fungi.

Similar systems for silencing repetitive sequences operate in flowering plants (RIGS, repeat-induced silencing) and other fungi (MIP, methylation induced pre-meiotically) (Yoder et al. 1997). Methylation probably serves the same role in mammals, where most cytosine methylation is associated with transposable element sequences. Even though about one-third of the mammalian genome is composed of TE sequences, only a small number of transposition events occur per generation.

TEs BENEFICIAL TO HOSTS

The evolution of TE copy number and transposition rate are short-term outcomes of intragenomic conflict between elements and the hosts in which they reside. In the longer term, phylogenies show that TEs are frequently lost and also undergo extensive horizontal transfer (Engels 1992; Robertson

1993). This implies that TEs regularly invade new lineages but eventually lose out. It is not obvious what forces are important in causing TE loss from lineages.

As with other parasites, it is possible that TEs evolve toward a more symbiotic relationship with their hosts. There are many claims that specific TE insertions have been selected because of their contribution to the control of gene expression or because they have facilitated gene duplication (Britten 1996). Circumstantial evidence in support of this hypothesis comes from the distribution of some TE insertions. For instance, the androgen sensitivity of the C4 mouse protein is due to the insertion of a retroelement (Stavenhagen and Robbins 1988); control of human amylase protein expression in the parotid gland is due to another retroelement (Ting et al. 1992); and database searches for the plant retrotransposon *Hopscotch* have uncovered a large number (21) of ancient, degenerate insertions in close proximity to coding sequences (White et al. 1994).

Insertions favored by selection ought to go to fixation, to have mutated so they no longer transpose, and to have undergone subsequent modifications to fulfill their new control role (though this may make it harder to identify them). In addition to a role in controlling gene expression, it appears that the ability to transpose has been taken over for other cellular functions. Possible examples are the V(D)J recombination system of the vertebrate immune system (van Gent et al. 1996), mating-type switching in fungi, and the maintenance of telomeres in eukaryotic chromosomes (Zeyl and Bell 1996). It seems likely, however, that the utility of TE insertions is merely a rare side effect rather than the reason for their existence. Most insertions are probably disadvantageous, end up losing their ability to transpose, and then slowly dissolve into "junk" DNA.

Cytoplasmic Genes

Typically genes in the eukaryotic nucleus undergo sexual mixing followed by equal transmission through both sexes. In contrast, cytoplasmic genes are generally not mixed by sexual reproduction and are predominantly or exclusively transmitted through one sex. In animals and angiosperms, the female is usually the transmitting sex, whereas in gymnosperms, both sexes fulfill this role. This is true both for genes in organelles (mitochondria and chloroplasts) and genes in other intracellular symbionts (e.g., bacteria and protists).

The different modes of nuclear and cytoplasmic inheritance mean that natural selection acts differently on the two types of DNA (Eberhard 1980; Cosmides and Tooby 1981). Selection on nuclear genes favors transmission through both sexes, whereas selection on cytoplasmic genes only favors transmission through females (or males, if they are the single transmitting sex). This results in intragenomic conflict, which once again has sexual reproduction at its root. If reproduction in eukaryotes was exclusively asexual, both nucleus and cytoplasm would share the same interest over inheritance.

In this section, a number of nuclear-cytoplasmic conflicts will be discussed: cytoplasmic incompatibility, male killers, feminization, and parthenogenesis. The phenomena are diverse, as are the agents that cause them (e.g., organelles, bacteria, protozoa). All involve manipulations of the host reproductive system that are deleterious to the host but promote transmission of cytoplasmic genes. Often these distortions are self-limiting, but sometimes they lead to the evolution of host counter adaptations.

CYTOPLASMIC INCOMPATIBILITY

Cytoplasmic incompatibility (CI) is a developmental defect that involves the mortality of zygotes that are not infected with a maternally transmitted bacterial symbiont (Rousset and Raymond 1991; fig. 7.2). No or fewer viable offspring are produced by crosses between infected males and uninfected females. Yet reciprocal crosses, between uninfected males and infected females, and crosses in which both sexes are infected or uninfected, produce normal broods. In haplodiploid organisms such as the wasp *Nasonia*, a slightly different phenotype is produced. Eggs of uninfected females that are fertilized by infected sperm do not die. Instead, there is early paternal genome loss in incompatible crosses and development of male-only (haploid) broods (Breeuwer and Werren 1990).

In all cases studied so far, CI is caused by *Wolbachia* bacteria (Werren 1997). These bacteria can be visualized microscopically in the ovaries and testes of infected individuals. Individuals can be cured by antibiotic or heat treatments that wipe out *Wolbachia* infections. The mechanism by which *Wolbachia* causes CI is not known, but embryonic mortality (and paternal genome loss in haplodiploids) is known to involve early defects in the fertilized egg that cause failure of the first mitotic division (Werren 1997). Current models envisage a two-component system, with one bacterial gene modifying sperm (mod$^+$) and the other rescuing eggs (res$^+$). If eggs lack the rescue gene product (res$^-$), then normal mitotic division is disrupted when they are fertilized by modified sperm (mod$^+$). However, if sperm lack the bacterial gene product (mod$^-$), normal development follows whether or not eggs carry the rescue gene product. A possible mechanism is that modified sperm carry a toxin that can only be deactivated and rescued by antitoxin in the egg.

The paradox about *Wolbachia* is how it gains though killing offspring of uninfected females. The explanation appears to be a form of kin-selected altruism (Hurst 1991a; Rousset and Raymond 1991). Bacteria like *Wolba-*

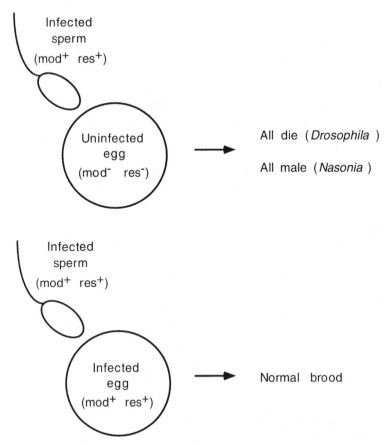

Fɪɢ. 7.2. Cytoplasmic incompatibility shown for crosses with males infected with *Wolbachia* bacteria. Crosses with uninfected males produce normal broods irrespective of the female infection status.

chia are vertically transmitted through the maternal lineage. As far as the bacteria are concerned, males are an evolutionary dead end. Therefore, *Wolbachia* do not lose by killing the offspring of crosses between infected males and uninfected females, as they are not transmitted by the infected male. In fact, killing is advantageous if it reduces local competition with other offspring sired by infected females. These carry clonal relatives of the *Wolbachia* present in infected males.

This kin-selection argument concerns a selfish act (killing offspring of uninfected females) that does not benefit the actor (*Wolbachia* in an infected male) but benefits relatives of the actor (*Wolbachia* in nearby offspring of infected females). At its simplest, this argument merely requires that the coefficient of relatedness is greater than zero in the neighborhood benefiting from reduced competition (Frank 1997). This depends on a number of factors like dispersal rates and transmission probability of *Wolbachia* from mother to offspring (normally less than 100%), as these determine relatedness.

The same logic underlies cytoplasmic incompatibility in haplodiploids like *Nasonia* (Breeuwer and Werren 1990). Sperm carrying *Wolbachia* cause paternal chromosome loss when they fertilize uninfected eggs, converting diploid female offspring into hapoid males. This masculinization confers a kin-selected advantage because of the life history of *Nasonia*, which is a parasitoid wasp. When a host is parasitized by an infected and uninfected female, some of the uninfected female's daughters will be fertilized by infected males (females are fertilized within the host puparium before dispersal). In the next generation, these daughters will produce only male offspring. Because males are nondispersing (they are wingless and never leave the host in which they are born), the absence of female offspring will reduce competition for nearby females that carry the infection, thereby increasing the frequency of *Wolbachia* in the transmitting sex.

A different aspect of the dynamics has been stressed by Hoffman and Turelli (1997). Even if *Wolbachia* have no local effect on competition (i.e., no kin-selected benefits), uninfected females will be at a disadvantage as a certain fraction mate with infected males and lay inviable eggs. The strength of this disadvantage is clearly frequency dependent. The effect is weak if few males are infected, whereas uninfected females will suffer greatly if most males are infected. Whether *Wolbachia* successfully invades a new population will then depend on the cost of carrying the infection. If female survival or egg laying is impaired by *Wolbachia* infection, there will be a threshold frequency below which selection acts against the infection and above which the infection spreads. This threshold will obviously be weakened if there is population substructure and kin-selected benefits.

Several studies have demonstrated reductions in female fecundity by comparing infected flies with those cured by antibiotic treatment, for example, U.S. populations of *D. simulans* (Hoffman et al. 1990) and *Tribolium confusum* (Wade and Chang 1995). Fitness loss is not observed in all cases, however. Infection appears to have no detectable effect on female fitness in populations of *D. simulans* from the Seychelles (Poinsot and Mercot 1997) or in *D. melanogaster* from Australia (Hoffman et al. 1994), but these are measurements of laboratory fitness. *Wolbachia* may be less benign under natural conditions.

Evolutionary pressure probably underlies this variation in *Wolbachia* virulence. In particular, there is likely to be a trade-off in the density of *Wolbachia* causing increased male incompatibility and increased female transmission probability but at the cost of reduced female fecundity (Frank 1997). High virulence (loss in female fecundity) may be advantageous in U.S. populations of *D. simulans* where *Wolbachia* has only recently spread and where many individuals are uninfected (Werren 1997). But in the older Seychelles and Australian infections where *Wolbachia* has reached high frequency, lower virulence may be favored because infected males tend to encounter infected females. The advantage of incompatibility is thus self-limiting.

This has led to the hypothesis that CI invasion is unstable (Hurst and McVean 1996). CI invasion is predicted to be followed by the evolution of *Wolbachia* strains that lack the ability to cause CI but are resistant (mod$^-$ res$^+$). Recent work has shown that some *Wolbachia* have exactly this ability to resist CI without being able to modify sperm (Bourtzis et al. 1998). Hurst and McVean (1996) assume that these *Wolbachia* spread because they do not cause such severe fitness loss in females. In turn, these strains are likely to be lost once fully competent *Wolbachia* disappear, as there is no longer any reason to be resistant. If this is correct, *Wolbachia* infections only persist in the long term through regular horizontal transmission (Hurst and McVean 1996).

This view is undoubtedly too simplistic. It is unclear why the loss of sperm modification (mod$^-$) should have a pleiotropic benefit on female fitness. In addition, the assumption of discrete *Wolbachia* types is probably also unrealistic. Both modification and resistance are likely to show continuous variation, especially if the strength of these traits depends on bacterial density. Selection will then act on the trade-off between female virulence on the one hand and transmission rates to males and females on the other. The evolutionary outcome will depend on a number of factors like population structure, local relatedness, and host coevolution (Frank 1997). Just as in other examples of host-parasite coevolution, persistent infection and even limit cycles are possible outcomes; epidemic invasion followed by loss is not inevitable.

Another possibility is that new types of *Wolbachia* regularly evolve and invade within a single species. There is evidence for this in *D. simulans*, where there are several strains of *Wolbachia* with different incompatibility types, some of which coexist in single populations (Clancy and Hoffman 1996). Likewise, incompatibility in mosquitoes is extremely complex, with strains showing variable degrees of incompatibility (Curtis 1992). Finally, it might even be possible for *Wolbachia* to persist through the evolution of mutualism with its host. Benign strains of *Wolbachia* (mod$^-$ res$^-$) have been detected in several populations (Clancy and Hoffman 1996; Hoffman et al. 1996). Perhaps these have beneficial effects on host fitness?

MALE KILLERS

Fisher (1930) argued that the optimal investment sex ratio is 1:1. This is the case for autosomal genes because they are inherited equally through the two sexes. Because cytoplasmic genes are not passed from father to offspring, these genes favor a female-biased sex ratio. Hence, there is intragenomic conflict between nuclear and cytoplasmic genes over the sex ratio (Hamilton 1967). Because the sex ratio is about 1:1 in most populations, it appears at first glance that nuclear genes have control over the sex ratio. However, the existence of sex-ratio distorters is hard to detect if they occur at low frequency or if their action is repressed by host modifiers. More detailed studies have revealed that sex-ratio distortion is common and is caused by a variety of elements acting in several different ways.

One of the clearest examples of distortion of an even ratio of sex allocation is cytoplasmic male sterility (CMS) in plants. CMS is caused by mitochondrial genes that disrupt pollen development (Hanson 1992). CMS is widespread among angiosperms (over 140 species; Laser and Lersten 1972), where mitochondria generally show uniparental maternal inheritance. At the population level, CMS results in gynodioecy, a mixture of hermaphroditic and female (male sterile) individuals.

The mitochondria causing CMS are selfish genetic elements because they destroy normal male function and thereby remove one of the standard routes by which nuclear genes are transmitted to future generations. CMS causes the death of mitochondria in the tissues that give rise to the male sexual structures. As with cytoplasmic incompatibility, this is not harmful to the mitochondria because they are not transmitted in male pollen anyway. CMS-causing mitochondria benefit by the release of resources from male function being redirected into increased egg and seed production (van Damme and van Delden 1984). Because eggs contain clonal relatives of the mitochondria found in pollen cells, CMS mitochondria enjoy greater rates of transmission by disrupting male function. Thus, CMS-causing mitochondria are favored through kin selection.

In principle, the spread of CMS could cause extinction through the loss of male fertile plants. It is difficult to know whether extinction has occurred, and if it has, at what frequency on a local or specieswide scale. However, extinction is unlikely in the presence of nuclear restorer loci that repress the action of male-killing mitochondria. Restorers are known in most well-studied cases. Selection in favor of restorers becomes increasingly strong as CMS spreads and the population sex ratio becomes more female-biased. Selection will favor restorers of male fertility even if they have deleterious side-effects on host fitness. But as restorers spread, the population sex ratio returns toward 1:1, thereby diminishing selection in favor of restorers.

A natural coevolutionary outcome is that both CMS mitochondria and nuclear restorers remain polymorphic if restorers carry a cost (Frank 1994). Restorers can be fixed if costs are low. This appears to be the case in many species. In these cases, CMS was uncovered through interspecific crosses (Laser and Lersten 1972).

Because CMS has agricultural importance, its genetic basis has received extensive study, revealing that many different mitochondrial mutations can give rise to CMS (Saumitou-Laprade et al. 1994). The four best-studied cases all involve chimeric genes created by repeated within chromosome recombination events producing novel gene expression (T-*urf13* in maize, S-*pcf* in Petunia, *pvs* in the common bean, and B-*atp6* in rice; Saumitou-Laprade et al. 1994). Why new genes are required for CMS is unclear, except that this will not disrupt existing gene functions.

None of the CMS genes sequenced from different species show homology; they are all derived from different precursors. This suggests that there are many ways in which mitochondria can potentially cause male sterility. All that is required is a mitochondrial gene with male-specific expression that causes the developmental failure of male tissues. The variety of genes also suggests that CMS has a limited evolutionary life span. Presumably host counteradaptations soon evolve to repress particular CMS mechanisms. Once CMS is repressed, the genes causing it no longer gain an advantage and may merely encumber host fitness, so that, in the long term, they will be lost.

The analogue of CMS in animals are male killers. Cytoplasmic agents causing male killing are widespread in arthropods; for example, son-killer in *Nasonia vitripennis* (Werren et al. 1986); sex ratio in the *D. willistoni* group (Malagolowkin and Carvalho 1961); and male killer in the two-spot ladybird *Adalia bipunctata* (Hurst and Majerus 1993). Male killing usually occurs early in development (Hurst 1991b). It is not caused by mitochondria but by a variety of bacteria that reside in the cytoplasm or intercellular matrix of infected individuals (Hurst et al. 1997); for example γ-Proteobacteria in *N. vitripennis*, a Spiroplasma in *D. willistoni*, a different Spiroplasma and a *Rickettsia* in *A. bipunctata* (fig. 7.3). Unlike CMS, male killing can be cured by antibiotics and is often sensitive to heat treatment (Hurst 1993a).

Male killers are clearly selfish genetic elements because they reduce host fitness by killing male offspring. Male killers may also have other deleterious effects; for example, they reduce oviposition rate and longevity in infected female ladybirds (Hurst et al. 1994). Selection will tend to alleviate these detrimental effects, because the success of symbionts depends on the lifetime reproductive success of their maternal hosts. However, transmission rates to offspring undoubtedly depend on bacterial density, which inevitably has some cost to host fitness.

Because males are an evolutionary dead end, male suicide is not costly for cytoplasmic genes. The most obvious benefit of this action is a kin-selected advantage to clonal relatives in sisters. This fits well with observations in the two-spot ladybird because female siblings gain resources by cannibalizing their dead brothers (Hurst and Majerus 1993; Hurst et al. 1997). Female consumption of dead eggs increases survival probability. Because ladybird eggs are laid in clumps, the benefits of male killing will be local and preferentially transmitted to hosts carrying related bacteria. Male killing may also

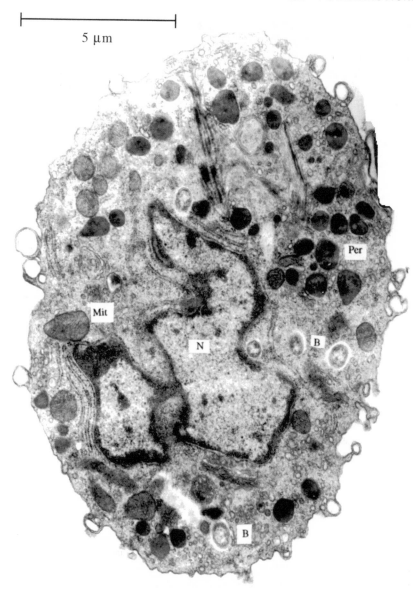

FIG. 7.3. Electron micrograph of a blood cell of a two-spot ladybird, *Adalia bipunctata*, infected with male-killing *Rickettsia* bacteria (B). Other cell structures shown are the nucleus (N), peroxisome (Per) and mitochondria (Mit). (Reprinted with permission from Greg Hurst and Academic Press.)

reduce local competition for resources, but clear evidence for this is lacking.

Another proposed benefit is that male killing reduces inbreeding and thereby raises host reproductive success (Werren 1987). Male killing seems an extremely wasteful means of avoiding inbreeding for the host genome. From the viewpoint of cytoplasmic genes, male killing is not costly because males are the nontransmitting sex. As yet, there is no evidence in species with male killers that enforced outbreeding causes significant increases in female offspring fitness (Hurst and Majerus 1993).

Another possibility is that male killing is advantageous because it causes horizontal transmission. This appears to explain mosquito male killing caused by microsporidian protozoa (Hurst 1991b). Microsporidians cause male suicide late in development, in the last larval instar. The larval cuticle ruptures, releasing haploid spores that infect a copepod intermediary host. In turn, the copepod releases spores capable of reinfecting mosquito larvae (Sweeney et al. 1988). Alternatively, the spores released by male killing directly infect other mosquito larvae (Bechnel and Sweeney 1990). Killing is restricted to males because adult males (unlike adult females) are incapable of vertical transmission of the parasite. Presumably, killing has evolved to occur in the last larval instar because this maximizes the opportunity for spore replication before the host is killed (Hurst 1991b).

PARTHENOGENESIS AND FEMINIZATION

Two other forms of sex-ratio distortion are caused by cytoplasmic agents: parthenogenesis and feminization. Both benefit cytoplasmic genes because females are the sole transmitting sex. Bacteria-induced parthenogenesis was discovered after antibiotic and heat-treatment experiments on asexual strains of *Trichogamma* wasps were shown to restore male production (Stouthamer et al. 1990). The bacteria that cause parthenogenesis are from the *Wolbachia* clade and are closely related to those causing cytoplasmic incompatibility (Stouthammer et al. 1993). *Wolbachia*-induced parthenogenesis has now been identified in five different wasp genera. The phylogenetic pattern of *Wolbachia* indicates multiple origins and frequent horizontal transmission (Schilthuizen and Stouthamer 1997).

It is unclear why *Wolbachia*-induced parthenogenesis is associated with the Hymenoptera. The obvious reason is that Hymenoptera normally produce haploid males by parthenogenesis. This might enable female parthenogenesis for two reasons. First, sperm fertilization is not needed for normal development, and second, the preexistence of male haploidy removes deleterious recessives. The latter is important for female parthenogenesis because this results from the failure of the first mitotic division, which creates homozygosity as well as restoring diploidy (Stouthamer and Kazmer 1994). An alternative explanation is study bias; once one example of parthenogenesis was found in the Hymenoptera, there was a good incentive to search for more. It is still too early to say whether *Wolbachia*-induced parthenogenesis will be found in other insect groups.

In the wild, some populations of *Trichogamma* contain sexual as well as parthenogenetic individuals. It is unclear why sexual forms persist. One reason is the imperfect transmission of *Wolbachia* from mother to offspring. Parthenogenetic females infected with *Wolbachia* also suffer from reduced lifetime fecundity compared with sexual females derived from them by antibiotic treatment (Stouthammer and Luck 1993). The co-occurrence of males and parthenogenetic females has the interesting consequence that asexual strains are not genetically isolated. Parthenogenetic females mate with males, and fertilized eggs develop normally as females (Stouthamer and Kazmer 1994). This makes it easy for the asexual population to absorb adaptive changes that might occur more quickly in the sexual population.

Wolbachia is also one of several cytoplasmic agents—the others being microsporidia and paramixydia protozoa—that are known to cause feminization in crustaceans (Rousset et al. 1992; Stouthamer et al. 1993; Rigaud 1997). The best-known case is the woodlouse *Armadillidium vulgare*, in which feminization is caused by *Wolbachia*. Sex determination in uninfected *A. vulgare* is controlled by sex chromosomes. Females are heterogametic ZW, and males are homogametic ZZ, but infected males develop as females. *Wolbachia* achieves this sex change by suppressing the development of the male androgenic gland in late-instar larvae. This gland is needed to secrete hormones that stimulate testes formation and the subsequent development of male sexual characters. Presumably the reason why feminizing symbionts are common in crustacea is because sex determination occurs late in development and can be easily disrupted (though the mechanisms are not well understood).

In some populations, *Wolbachia* has taken over as the major sex-determining mechanism (Rigaud 1997). These populations have lost the W sex chromosome, and all individuals are ZZ. Uninfected individuals develop as males; infected individuals develop as females. Because *Wolbachia* infection is passed from mothers to 80–90% of their offspring, these populations are heavily female biased.

In other populations, the intragenomic conflict caused by feminizing bacteria has selected for nuclear resistance genes in the *Armadillidium* host. Two types of resistance genes are known, one that suppresses the feminizing action of *Wolbachia*, and one that reduces the transmission rate of bacteria (Juchault et al. 1992; Rigaud and Juchault 1993). Both types of resistance gene increase the frequency of males, the rarer sex with higher reproductive value.

In this and the preceding section, we have considered three causes of cytoplasmic sex-ratio distortion: male killing, parthenogenesis, and feminization. Other mechanisms undoubtedly exist. One possibility is increased rates of fertilization in haplodiploids, resulting in female-biased broods. This may underlie *msr*, a maternally inherited trait in the wasp *Nasonia vitripennis* (Skinner 1992). Females carrying *msr* produce more female offspring than normal females. A plausible hypothesis is that *msr* is caused by a cytoplasmic factor. However, the causal agent remains unknown. Another possibility is that cytoplasmic genes favor meiotic segregation of the W chromosome in species with heterogametic (ZW) females, or favor fertilization by X-bearing sperm in species with heterogametic (XY) males, in both cases resulting in female-biased broods. Some tentative evidence exists, but further investigation is required to establish whether cytoplasmic genes are involved (Hurst et al. 1997).

Meiotic Drive

Meiosis is a form of reduction division. It causes diploid cells ($2n$) to produce multiple haploid (n) gametes. In higher organisms, there is full cooperation between pairs of alleles during the diploid phase. Meiosis marks the breakup of this partnership and appears to be designed to divide resources fairly between the partners (Haig and Grafen 1991; Haig 1993a).

Normally, meiosis ensures that each allele or chromosome segment is inherited by 50% of the gametes. However, equal division creates a tension in the cell. Each allele would do better (up to twice as well) if it could be transmitted to more than half the gametes. This strong selection has resulted in a class of "meiotic drive" genes that subvert meiosis and gain a disproportionate representation among gametes (Hamilton 1967).

Meiotic-drive genes are classic selfish genetic elements. By distorting meiosis, they increase their own rate of transmission, but at a cost to the rest of the genome (either to viability or fertility). Not only is there intragenomic conflict between the distorting allele and its homologue, but there is also conflict with the rest of the genome. This has led to a variety of countermeasures to stop drive.

Despite this intragenomic conflict being the few against the many, meiotic drive has been discovered throughout nature. In addition to the intensively studied cases in *Drosophila* (*SD*) and mice (*t*), drive occurs in a variety of other insects, mammals, fish, fungi, and plants (Hurst and Pomiankowski 1991; Jones 1991; Lyttle 1991). Far from being one of nature's oddities, segregation distorters appear to be a frequently erupting problem for many meiotic systems.

a) *t*-locus

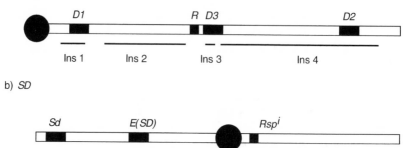

b) *SD*

FIG. 7.4. Schematic diagram (not to scale) of (a) the mouse *t*-locus and (b) *Drosophila melanogaster SD* drive chromosomes. The *t*-locus is a 100-Mb region located close to the centromere (black circle) on chromosome 17. There are three drive loci (D1, D2, and D3) and a *t* complex responder locus (R). The *t*-complex is characterized by several inversions (Ins 1–4). The *SD* region lies close to the centromere of chromosome two. On the left arm is the *Sd* drive locus and the nearby *E(SD)*, enhancer of drive. On the right arm is *Rsp^i* the insensitive responder. Recombination between these loci is suppressed as *E(SD)* and *Rsp^i* lie within the centromeric heterochromatin and most *SD* chromosomes carry an inversion farther along the right arm. Based on Silver et al. (1992), Lyttle (1991), and Lindsey and Zimm (1992).

AUTOSOMAL DRIVE

The two best-characterized systems of meiotic drive are *SD* in *D. melanogaster* and *t* in the mouse. Both are autosomal and were discovered by serendipity (autosomal drive has no obvious phenotypic effect). In both cases, males heterozygous for drive transmit the drive chromosome to 90–99% of their offspring. In *SD*, this occurs because + bearing spermatids fail to mature (Crow 1979), whereas in *t* this occurs because + bearing gametes mature but have abnormal sperm motility and sperm-egg interaction (Olds-Clarke and Johnson 1993; Johnson et al. 1995).

Both drive systems involve two factors (Lyttle 1991). Segregation distortion is caused by the action of a driver on a responder site. In *D. melanogaster*, the drive chromosome alleles are known as *Sd* and *Rsp^i*, respectively (fig. 7.4). If an *Sd* allele is present, gametes inheriting *Rsp^i* show a segregation advantage over those carrying the sensitive version (i.e., wild type) of the responder Rsp^+. Segregation is not distorted in the absence of the drive allele or if both chromosomes carry the same *Rsp* type. These two genes are closely linked on chromosome 2 on either side of the centromere. In addition, this region has an inversion and *Sd* maps close to the centromeric heterochromatin. All these features greatly reduce the probability of recombination between the drive and responder alleles. Recombination suppression is clearly necessary for segregation distortion. Without it, *Sd* would often be separated from *Rsp^i*, causing *Sd* to attack itself and *Rsp^i* to lose its association with the drive agent.

A very similar genetic system characterizes the *t* locus in mice (Lyttle 1991). The *t* locus is located on chromosome 17. At least three distorter loci (*D1, D2, D3*) are closely linked to the responder (*R*) within a region characterized by multiple inversions. Again, recombination between drivers and the responder is severely restricted.

At a mechanistic level, however, *SD* and *t* are different. The *Sd* drive locus consists of a duplication that produces 4-kb *Sd*-specific transcripts (McLean et al. 1994). The *Sd* responder locus maps to an array of repetitive, satellite DNA. Insensitive alleles have a low copy number (<200 repeats), whereas wild-type alleles have a high copy number (750–2,500 repeats; Wu et al. 1988). This suggests that the drive gene product is a toxin that is attracted to the repeated sequence of the responder locus (Lyttle 1991). Such a model fits with the finding that the strength of drive is proportional to satellite copy number (Lyttle 1991). In contrast, the two *t* drive loci that have been analyzed are both nonfunctional. *D3* is a deletion, and *D1* is either a deletion or a non-informative coding sequence (Lyon 1992; Braidotti and Barlow 1997). The structure of the *t* responder locus is not yet known. This suggests that *t* differs from *SD* in that drive is caused by low concentrations of drive gene products to which the insensitive responder is more resistant.

The variation of mechanisms is probably typical of meiotic-drive systems. Any gene can potentially benefit from segregation distortion. All that is needed for segregation distortion is some way of marking self and something causing a disadvantage to non-self-segregants. This necessarily involves two loci, the driver and its target, which must be kept in close linkage. The latter requirement places a strong constraint on the appearance of autosomal distortion.

SD and *t* are polymorphic in natural populations. It is difficult to know whether this is typical for autosomal drivers or whether they often go to fixation. If fixation occurs, it will leave little obvious trace (i.e., no segregation distortion). In addition, once a driver reached fixation, there would be nothing to maintain its driving ability. A possible example of a fixed autosomal drive is gamete eliminator in rice (Sano 1990). This meiotic-drive gene appears to be fixed in the African rice species *Oryza glaberrina*, where it has no obvious phenotypic consequence. It was uncovered by crosses to another cultivated species, *O. sativa* from Asia, in which it causes meiotic drive.

There are several reasons why *SD* and *t* drivers have not gone to fixation. In both systems, homozygotes have low fitness; *t* homozygotes are sterile because of the absence of gene products from the amorph *D1* and *D3* genes. The products of these genes are necessary for normal spermatogenesis. In addition, in *SD* and *t*, the drive alleles are closely linked to deleterious recessive alleles that cause strong homozygous viability loss (Lyttle 1991). Many of these deleterious recessives may be recent additions (because of the lack of recombination), but some may have been originally linked to the *SD* region and contributed to stopping fixation.

Another reason why *SD* has not spread to fixation is that it naturally gives rise to resistance. Chromosomes that are immune to drive arise as rare recombinants that lack the drive allele but are insensitive at the responder locus ($+,Rsp^i$). These chromosomes do not cause drive but cannot be driven against.

Though immune chromosomes have invaded and occur at high frequencies in natural populations of *D. melanogaster*, they have not gone to fixation (Lyttle 1991). This implies that immunity must carry a cost. Immunity spreads when *SD* is common but becomes disadvantageous when drive is rare. This results in a three-way polymorphism of drive, immunity, and wild-type chromosomes (Charlesworth and Hartl 1978). The cost to immunity has been experimentally investigated by comparing the fitness of chromosomes with different *Rsp* satellite copy numbers (Wu et al. 1989). As predicted, individuals with immune chromosomes (20 repeats) were less fit than those carrying wild-type chromosomes (700 repeats). This study is not, however, a definitive proof of costly immunity. Variation in satellite copy number was created by a deletion that also removed other sections of DNA. It is possible that deletion of these additional regions of DNA contributed to the loss of fitness found in immune chromosomes (Moschetti et al. 1996).

There are no similar *t* haplotypes that are immune to drive. This is puzzling because immune recombinants can be generated in the lab (i.e., $+,Rsp^i$). Chromosomes with this genetic combination have been experimentally investigated and have been found to cause negative drive in favor of the wild-type chromosome in immune–wild-type heterozygotes, resulting in only 20% of viable gametes carrying the immune chromosome (Lyon 1991). This might explain the absence of immune haplotypes from natural populations of mice.

There are several other modifiers of drive. Closely linked to the *SD* locus is a linked enhancer of drive, *E(SD)* (Lyttle 1991; fig. 7.4). Enhancers are like the drive locus in their requirement of close linkage to the target *Rsp* locus. In addition, natural populations of *D. melanogaster* carry suppressors of drive on the X and autosomes other than chromosome 2 (where *SD* is located). Selection for these suppressors does not arise from drive per se. It must arise because drive is associated with fitness loss at the level of the organism. One cause of host fitness loss is the reduction in the number of viable sperm, which probably reduces male fertility. Although this has not been investigated in *SD*, lack of sperm has been shown to reduce male fertilization success under conditions of sperm competition in *D. pseudoobscura* males subject to sex chromosome drive (Wu 1983a,b).

X AND Y DRIVE

It seems that sex chromosome drive is more common than autosome-linked drive (Hurst and Pomiankowski 1991). Sex-linked drive also appears to be more persistent on an evolutionary timescale (see below). Part of the reason for this pattern can be attributed to reporting bias. Sex-linked drive is much easier to recognize than autosomal drive because it causes sex-ratio distortion, but the main reason that sex-linked drive is common is the lack of recombination between X and Y chromosomes (Hurst and Pomiankowski 1991; Lyttle 1991). This allows any distinguishing locus on the Y to act as a target for an X-linked driver (and vice versa). Hence, it should not be surprising to find cases of X versus O drive in species that lack a Y chromosome (e.g., aphids; Hurst and Pomiankowski 1991). Here the absence of an X product acts as a marker of the sex chromosomes contained by a gamete.

Sex-chromosome drive is probably more common than currently realized. If drive is expressed, it is likely to be obvious (i.e., as a biased brood sex ratio). Coevolution between the driver and the rest of the genome is, however, likely to lead to strong suppression of sex-linked drive. As with autosome-linked drive, there is selection for suppressors on the homologous chromosome that suffers from drive because they are not transmitted to viable gametes. In addition, there is strong selection for autosomal suppressors because X and Y drive create populationwide sex-ratio distortion. These autosomal suppressors are favored because they will be inherited more frequently by the rare sex.

Many suppressors are linked to the driven chromosome as well as to autosomes. *Drosophila simulans* harbors *SR*, an X-linked meiotic drive. Indo-Pacific island and African populations of *D. simulans* contain autosomal and Y-linked suppressors of *SR*, and suppression is at high frequency so that drive is rarely observed (Mercot et al. 1995; Atlan et al. 1997). Similar suppressors of *SR* have been identified in other *Drosophila* species that are Y-linked (Voelker 1972; de Carvalho et al. 1997) or autosome-linked (Varandas et al. 1996). Likewise, several non-*Drosophila* species—for example mosquitoes (Wood and Newton 1991) and stalk-eyed flies (Presgraves et al. 1997)—also have sex-linked and autosome suppressors of sex-linked drive.

The best demonstration that sex-ratio distortion generates suppressors is an ingenious experiment by Lyttle (1977, 1979). He investigated a novel *D. melanogaster* Y chromosome to which part of the *SD* autosome had been translocated. In this new location, *SD* still caused drive, but now the Y^{SD} chromosome showed segregation distortion (i.e., against the X chromosome). Cage populations with the new driver soon showed extreme sex-ratio bias, but they rapidly developed suppressors that returned these populations to a more even sex ratio. Suppression was polygenic and mapped both to X and autosomes.

A further, but more bizarre, example of drive suppression is the *Stellate* phenotype of XO *D. melanogaster* males (Livak 1990). Occasionally male zygotes lose the Y chromosome and develop as XO males. These individuals are sterile because of overproduction of protein from the multicopy *Stellate* gene, which is linked to the X chromosome. *Stellate* is normally suppressed by another multicopy gene, *Su(Ste)*, linked to the Y chromosome. It appears that *Stellate* coevolved with *Su(Ste)* so that both genes now have high copy number. Recently it has been proposed that *Stellate* is a meiotic-drive gene and *Su(Ste)* is its suppressor (Hurst 1992a, 1996b). Removal of all suppression in XO males causes complete gamete breakdown, whereas loss of some *Su(Ste)* elements causes drive (Hurst 1996b).

As yet, little is known about the phylogenetic distribution of drivers. In many cases they are unique and are not found in closely related species. For example, *SD* is found in *D. melanogaster* but in no closely related species (Lyttle 1991), and the *SD* region harbors less genetic polymorphism than the same region on the wild-type chromosome, which is indicative of a recent origin and sweep to high frequency (Palopoli and Wu 1996). The *t* locus is more widely spread among *Mus* species, but its origin has been estimated to be only 3 million years ago (Morita et al. 1992). This suggests that drivers are often lost after a short period of time. However, there are several exceptions among sex-linked drivers. The *SR* driver has been found in five species of the *D. quinaria* group (Jaenike 1996). Likewise, Y-linked drive in mosquitoes (Wood and Newton 1991), X-linked drive in *Diopsid* stalk-eyed flies (Presgraves et al. 1997), and spore killers in fungi (Turner and Perkins 1991) all have widespread distributions in these genera. It is not known whether these distributions are due to single or multiple origins. A single origin seems more likely and suggests that drive can persist through evolutionary time despite suppression by the host genome. In none of these cases do we have a good estimate for the age of the drive system.

B Chromosomes

B chromosomes are dispensable extra chromosomes that are widespread in animals and plants (Bell and Burt 1990; Jones 1991). They are another type of selfish genetic element that make no direct contribution to the morphology and behavior of individuals carrying them. In most cases, as the number of Bs increases, there is a measurable reduction in host viability and fertility. These deleterious effects probably stem from the increase that Bs make to DNA content, which elevates the time taken for cell division.

To offset this disadvantage, Bs have a variety of mechanisms equivalent to meiotic drive. For example (Bell and Burt 1990), they double their numbers in germ tissue before meiosis in *Crepis capillaris* (premeiotic drive), preferentially move to the egg nucleus rather than being eliminated in a polar body in *Lilium callosum* (female meiotic drive), or double their number in the first pollen mitosis before fertilization in rye (postmeiotic drive). Unlike the examples of meiotic drive discussed above, Bs do not cause drive against another chromosome, and their mode of action can be premeiotic, meiotic, or postmeiotic. But the net effect is a similar increase in the rate of transmission to future generations.

Once again, suppressors of B drive, linked to the normal nuclear chromosomes, are known to occur (Jones 1991). These are selected because of the deleterious effects of Bs on host fitness. In populations of the grasshopper *Eyprepocnemis plorans*, resistance genes appear to be at sufficiently high frequencies that there is no evidence of B drive (Herrera et al. 1996). However, crosses between populations with Bs and those without them cause B drive to reappear, revealing that resistance is only common in populations that have suffered from B infections in the past (Herrera et al. 1996). Crosses between different populations with B chromosomes suggest that new B variants evolve that can evade existing resistance mechanisms (López-León et al. 1993).

Probably the most interesting B chromosome is *PSR*. This chromosome gains a transmission advantage by masculinizing females in the parasitoid wasp *Nasonia vitripennis* (Nur et al. 1988). Because *Nasonia* is a haplodiploid, fertilized eggs develop as females. When eggs are fertilized by a sperm containing *PSR*, the paternal genome with the exception of *PSR* becomes highly condensed and is eliminated at the first mitotic division. The result is a haploid, male individual. This is advantageous to *PSR* because it has much higher transmission through the male lineage. In oogenesis, *PSR* tends to be lost and segregates into very few eggs, but in spermatogenesis, *PSR* is transmitted to nearly all sperm.

Postsegregation Killers

Meiotic-drive genes act before the production of offspring. Postsegregation killers achieve the same end by acting after offspring are produced. These elements cause the death of non-carrying offspring. They are clearly selfish genetic elements because they greatly reduce host fitness, by up to 50%. As with cytoplasmic male killers, the benefit of postsegregation killing probably lies in the reduction of competition faced by carriers. However, experimental investigation remains to be undertaken.

In this section I discuss the few examples of postsegregational killing in beetles and mice that have recently been uncovered. A parallel example from prokaryotes will also be mentioned. The best example of a eukaryotic postsegregational killer is *Medea*, which is found in several species of flour beetles (Beeman et al. 1992). Females heterozygous for the *Medea* gene $(Ml+)$ show high early brood mortality of offspring that fail to inherit the

Medea gene (*M/M* and *M/+* live, *+/+* die). Early death is prevented by inheritance of *Medea* from either parent. Neither wildtype (*+/+*) nor homozygous (*M/M*) mothers show high early offspring mortality, even when they are mated to heterozygous males.

A similar example has recently been discovered in the red fire ant *Solenopsis invicta* (Keller and Ross 1998). Red fire ants show a widespread polymorphism at the *Gp-9* locus, with a large excess of heterozygous *Bb* queens. This excess arises because *Bb* workers detect and kill *BB* homozygous queens. Recognition appears to be by odor or contact chemical cues as workers rubbed with *BB* queens were attacked, whereas workers rubbed with *Bb* queens were not (Keller and Ross 1998). This system differs from *Medea*, as killing is achieved through workers rather than by lack of inheritance of the selfish genetic element. One might predict that the advantage of the *b* allele should cause it to spread to fixation. But as with many selfish genetic elements, the homozygous *bb* state is highly deleterious, so the system has remained polymorphic.

Other possible examples of postsegregational killing in eukaryotes are *Scat* (Hurst 1993b) and *MUT* (Weichenhan et al. 1996) in mice. Again, higher brood mortality is only seen in the offspring of mothers heterozygous for these genes. In both cases, offspring failing to inherit the selfish genetic element have higher death rates. Baby mice lacking the *Scat* gene suffer from severe combined anemia and thrombocytopenia and die in early life (Peters and Barker 1993). In *MUT*, there is a 1:1 ratio of four-cell blastocysts in heterozygous mothers mated to wild-type males, but by the early embryonic stage, there is a relatively large loss of *+/+* compared to *+/MUT* embryos (Weichenhan et al. 1996).

Medea, Scat, and *MUT* can be explained by a two-factor system similar to that observed in cytoplasmic incompatibility and meiotic drive. This assumes that a "toxin" is produced in the eggs of mothers carrying *Medea*, which must be suppressed by antitoxin expressed in zygotes. Zygotes failing to inherit *Medea* lack the antitoxin and are killed by the maternally inherited toxin. There is no distinct evidence for this model, but it is interesting that *Scat* is found near a centromere and *MUT* is in an inversion, both regions of low recombination. A similar two-factor system may underlie *Gp-9* in the red fire ant. Even though recognition and killing both map to the same marker locus (*Gp-9*), these two traits may well be coded for by separate genes held together by close linkage or an inversion. Presumably paternal-effect lethality is also possible but unlikely because far fewer gene products are inherited in sperm.

The benefit to a gene causing severe damage or death of noncarrier siblings presumably arises through reduced competition. *Medea, Scat,* and *MUT* carrying embryos and young gain because sibling rivals are eliminated in early life and thus do not compete for maternal or other limited resources. Likewise, the *Gp-9* b allele in worker ants gains by killing *BB* queens that will compete for reproductive opportunities with carrier *Bb* queens. In each case, the action of the selfish genetic element is beneficial if it increases the survival or reproductive rate of carriers, but such a benefit remains to be demonstrated in natural systems.

RESTRICTION-MODIFICATION

Bacteria are different from eukaryotes because sex takes place without cytoplasmic mixing. This limits the possibility of intragenomic conflict between different types of DNA present in bacterial cells. However, genes in the bacterial genome and co-resident plasmids do not have identical modes of transmission. When a bacterial cell divides, the bacterial genome replicates with one copy segregating to each daughter cell (Maynard Smith and Szathmáry 1995). Plasmids are not always transmitted to all daughter cells. First, plasmids do not show coupling of DNA replication and cell division, nor do they have fail-safe ways of ensuring segregation in both daughter cells (Summers 1996). Second, bacteria are often transformed by plasmids from other cells, which can result in the resident plasmid being outcompeted and lost from daughter cells.

It has recently been proposed that bacterial restriction-modification (R-M) systems are examples of postsegregational killing that have evolved to stop plasmid loss (Naito et al. 1995). R-M systems consist of two components (fig. 7.5) (Buckle and Krüger 1993). Restriction enzymes recognize and cut highly specific DNA sequences. For example, *Pae R7* induces double-strand cuts in 5'-CTCGAG-3' sites. Modification enzymes recognize the same sites and protect them by methylation of adenine and cytosine residues. Methylation modification of bacterial DNA stops it from being attacked by restriction enzymes produced by its own plasmids, which otherwise would cut the DNA and cause cell death.

When daughter cells fail to inherit plasmids, they are quickly killed (fig. 7.5). Because restriction enzymes have a longer half-life than modification enzymes, some recognition sites soon fail to be methylated and are cleaved by the remaining restriction enzymes. Likewise, a cell whose resident plasmid is displaced by another plasmid lacking the same R-M genes will also be killed. The net effect is stable maintenance of R-M plasmids through the elimination of plasmid-free bacteria.

The power of R-M plasmids has been experimentally investigated (Naito et al. 1995). Resident bacteria carrying competent plasmids (r^+m^+) were found to be resistant to invasion by plasmids lacking these genes (r^-m^-). Even though the incoming plasmid could infect bacteria cells and eliminate the resident plasmid, all such colonies died out. But when resident plasmids lacked the restriction enzyme (r^-m^+), the introduced plasmid did much better and were able to proliferate. Further experiments using thermosensitive R-M plasmids showed that these patterns could be explained by restriction

a) Bacteria with R-M plasmid

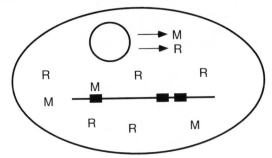

b) Plasmid loss

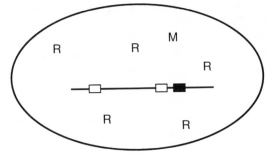

c) Chromosome breaks

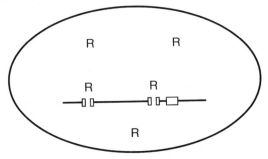

FIG. 7.5. (a) Bacterial R-M plasmids produce modification enzymes (M) that methylate specific nucleotide sequences (filled boxes) on the bacterial and plasmid genomes, and restriction enzymes (R) that recognize the same sequence. (b) Modification enzymes have a shorter half-life than restriction enzymes, leading to loss of methylation (empty boxes) if the plasmid is lost. (c) This is followed by cutting of DNA by restriction enzymes and cell death.

enzyme killing of cells that had lost competent plasmids. Under permissive temperatures, bacteria with competent plasmids (r^+m^+) grew as well as control strains. But once at nonpermissive temperatures that caused plasmid loss, bacteria that had contained competent plasmids (r^+m^+) stopped growing and died out, whereas controls were unaffected.

The exact benefits of R-M host killing in nature have not been quantified. The death of bacteria that lose their plasmids (or have them replaced by other incompatible plasmids) reduces competition. Presumably, the benefit of this action is local and increases the fitness of clonal relatives that carry the R-M plasmid.

More than 50 different R-M systems have been characterized, each with distinct recognition specificity (Buckle and Krüger 1993). The lethal effect of plasmid loss helps explain how high diversity and specificity have arisen (Kusano et al. 1995). If two R-M systems recognize different DNA sequences, loss of either plasmid type will cause host killing. But if two systems partially overlap in their recognition sequences, loss of one will be compensated by the other, and host death will not occur. Thus, selection favors a multiplicity of R-M plasmids, each with a unique recognition site.

R-M plasmids are good examples of selfish genetic elements. Host killing is advantageous to the plasmid but has no obvious benefit to the hosts (Naito et al. 1995). This view has only recently gained acceptance. Originally R-M systems were thought to be beneficial to their hosts. It was proposed that they had a function in defending the bacterial cell from viral attack (Korona and Levin 1993), with their diversity evolving through host-parasite coevolution (Frank 1994). But laboratory studies do not give much support to this interpretation (Korona and Levin 1993). Defense against phage by R-M plasmids is very short-lived. The main way that bacterial populations respond to viral attack is by quickly evolving mutants that prevent phage adsorption and penetration. The minor advantage conferred by R-M plasmids is unlikely to explain their persistence or diversity in bacterial populations.

Similar selfish behavior underlies another type of plasmid-controlled suicidal behavior, the *Hok-Sok* system (Gerdes et al. 1986). There is, however, less controversy over the evolutionary origin of this system because it serves no role in defense against viral attack. It also is a two-factor system: The *Hok* gene produces a lethal toxin, and the *Sok* gene produces its antidote. The *Sok* gene product has been identified as antisense RNA that binds *Hok* mRNA (Thisted et al. 1994). This stops *Hok* mRNA from being translated and suppresses its lethal effect. *Sok* antisense RNA is very unstable and is quickly lost from plasmid-free bacteria. Like the action of R-M systems, *Hok-Sok* action mediates plasmid stability by killing segregants lacking the plasmid. Once more one must assume that the benefits of killing are manifested as a reduction in local competition that increases the fitness of bacteria carrying the *Hok-Sok* plasmid.

Conclusion

In this final section I draw some general observations about the causes and evolution of intragenomic conflicts.

CAUSES OF INTRAGENOMIC CONFLICT

This chapter has discussed how selfish genetic elements contribute to intragenomic conflicts. These elements exploit the reproductive process to enhance their transmission to future generations. At the same time, they are detrimental to the fitness of their hosts. This harm or virulence is the cause of intragenomic conflict with other genes in the host genome.

There are two main types of intragenomic conflict. The first is caused by cell divisions that reduce the genetic content of the cell. For instance, many conflicts are associated with meiosis, which halves cell ploidy. This creates the opportunity for drive alleles (e.g., *t*, *SD*, and *SR*) and postsegregational killers (e.g., *Medea*) that disable or destroy noncarrier gametes or noncarrier zygotes. It also creates the opportunity for transposable elements that gain by making dispersed copies throughout the genome; these copies have a better than even chance of being carried in each gamete.

The benefit to these selfish genetic elements is that they increase their transmission rate to future generations. The mechanism of their action is highly variable, however. Postsegregational killers and male meiotic drivers generally benefit through reduced competition (for resources or fertilization, respectively), whereas transposable elements and female meiotic drivers (i.e., avoidance of segregation into the polar bodies) benefit by increasing the fraction of gametes that carry the selfish element. Conflict is caused because these elements reduce the fitness of the rest of the genome. There are a variety of causes: loss of gametes, death of offspring, reduced host viability, distorted sex ratios, increased mutation rates, and deleterious chromosomal rearrangements. It is therefore unsurprising to find that the rest of the genome contains numerous modifiers that suppress the activity of these selfish elements.

The second main type of intragenomic conflict is caused by the unisexual inheritance of cytoplasmic genes, usually solely through the female lineage. For cytoplasmic genes, males are an evolutionary dead end. Therefore, cytoplasmic genes in organelles and intracellular symbionts benefit by converting males into females (feminization and parthenogenesis), destroying males (male killers or cytoplasmic male sterility), or using males to benefit carrier females (cytoplasmic incompatibility). Benefit is passed on to females that carry clonal relatives of the cytoplasmic genes found in males by making female-biased broods or reducing competition for females. Clearly these actions are deleterious for genes in the nucleus, because these are transmitted by males as well as females. Again, it is unsurprising to find evidence of suppressor genes located in the nucleus that inhibit the behavior of selfish cytoplasmic genes.

THE FORCES INVOLVED

The principal forces involved in intragenomic conflicts are the same as those governing social interactions between separate individuals (Keller and Reeve, chap. 8; Kitchen and Packer, chap. 9). Kin selection is of great importance because it underpins many of the examples of suicidal killing, like cytoplasmic male killers, cytoplasmic male sterility, and cytoplasmic incompatibility. In all these cases, male death benefits clonal relatives of the cytoplasmic agent in neighboring females (or female-differentiated tissue). Because the intracellular symbionts that cause these effects reproduce asexually, the coefficient of relatedness in the individuals that benefit is likely to be high.

A second important factor is the trade-off between transmission and virulence. Because the reproductive interest of selfish genetic elements are in part tied up with those of the host, it is advantageous for elements to minimize their cost to host fitness. This is seen in some very obvious adaptations; for instance, TEs do not usually transpose in somatic tissues because this does not enhance their transmission. Where the reproductive interests of host and selfish genetic elements overlap, the elements tend to evolve lower virulence, even though this also means lower transmission rates. This explains the low element activity of some bacterial TEs. Similar lower virulence and transmission rates may also be associated with some *Wolbachia* infections that cause cytoplasmic incompatibility (CI). Once *Wolbachia* reaches high frequency, there is little opportunity for males to cause CI, because most females are already infected. Therefore, selection increasingly favors *Wolbachia* that do not harm host fitness.

Finally, most elements causing intragenomic conflict are subject to suppression. This is a form of "policing" (Frank 1995) by other genes that lose out because of the action of the selfish genetic elements. Suppression is common in drive systems both on the homologous chromosome attacked by the driver and on other chromosomes not directly involved (e.g., autosomes suppress sex-chromosome drive). Various types of suppressors are also found for TEs. Less is known about suppressors of cytoplasmic distortion.

THE OUTCOME

The population dynamics of selfish genetic elements has been an area of intense study (e.g., Lyttle 1991; Charlesworth et al. 1994; Frank 1997). There are several possible outcomes. One is element fixation. Examples are rare, partly because fixation is predicted to cause extinction (X or Y drive;

Hamilton 1967) or lead to the disappearance of any obvious phenotype (e.g., autosome drive or *Wolbachia* causing cytoplasmic incompatibility). Mainly fixation is prevented because selfish genetic elements have deleterious effects once they become common, and their spread is limited by the evolution of suppressors.

This will lead to the evolution of polymorphism. For example, *SD* meiotic-drive genes are polymorphic because *SD* homozygotes are inviable and *SD* cannot drive against insensitive (*I*) chromosomes. Thus, *SD* invades wild type (because of drive), *I* invades *SD* (because *SD* homozygotes are inviable), and wild type invades *I* (because resistance is costly). The result is a three-way polymorphism (Lyttle 1991). Similar forces probably limit the invasion of *Wolbachia* bacteria that cause cytoplasmic incompatibility. Selection in favor of *Wolbachia* weakens once most individuals become infected and exploitable, and uninfected hosts become less frequent. In turn, this favors resistant strains of *Wolbachia* that are insensitive to cytoplasmic incompatibility but are less costly to their hosts (Hurst and McVean 1996).

In the long term, coevolution of the host and its selfish genetic elements may follow a course equivalent to host-parasite coevolution (Herre, chap. 11), with the element continually developing new ways of evading host defenses. However, the phylogeny of many selfish genetic elements suggests that persistence is limited unless horizontal transmission is possible. Autosomal drive genes appear to be limited to a single species or to a few related species (*SD* to *D. melanogaster*, *t* to *Mus*). X-linked drivers appear to be more widely distributed both among *Drosophila* species (Hurst and Pomiankowski 1991; Jaenike 1996) and *Cyrtodiopsis* stalk-eyed flies (Presgraves et al. 1997). But it is unclear how old these groups are or whether they represent single or multiple origins (genetic markers for X-linked drive have not yet been developed). Broad-scale phylogenies have been developed for *Wolbachia*. These show small-scale cladogenesis of *Wolbachia* with its host, plus frequent horizontal transmission between arthropod taxa (Werren 1997). Many TE phylogenies also show evidence of weak persistence within a clade coupled to frequent horizontal transmission (Engels 1992; Robertson 1993), although some show evidence of long-term persistence (Eickbush and Eickbush 1995).

Acknowledgments

I am grateful for corrections and comments provided by John Brookfield, Greg Hurst, Laurence Hurst, Laurent Keller, and several anonymous referees. This work started while I was a Fellow of the Wissenschaftskolleg zu Berlin. I am grateful for the support of a Research Fellowship from the Royal Society.

8 Dynamics of Conflicts within Insect Societies

Laurent Keller and H. Kern Reeve

Social insects fascinate because of their extreme cooperation and social cohesion. They indeed provide some of the most remarkable examples of altruistic behavior, with a worker caste whose individuals forgo their own reproduction to enhance reproduction of the queen. The level of such worker self-sacrifice can be extreme, as exemplified by the evolution of "kamikaze" weapons such as detachable stings and exploding abdomens used in defense of the colony (Wilson 1971). Workers also collectively exhibit highly organized, sophisticated behavior that is adaptively fine-tuned to ecological conditions. For example, workers of some ants react to the presence of workers from other colonies, and the heightened risk of conflict, by increasing the production of soldiers that are specialized in colony defense (Passera et al. 1996). The honeybee waggle-dance communication system provides yet another example of the sophistication of the cooperative behavior of social insect workers (see Seeley 1995). Such examples of group harmony and cooperation have given rise to the concept that colonies are harmonious fortress-factories in which individual-level selection is muted, with the result that colony-level selection reigns. In other words, the colony often appears to behave as a "superorganism" operating as a functionally integrated unit (Wheeler 1928). In this vein, Seeley (1997) has described the elegant group-level adaptations of honeybee societies.

The concept of a superorganism as the only unit at which natural selection operates has been challenged, however, both on theoretical grounds and by the observation that life within the colony is not always as harmonious as it may first appear. The more we come to understand the dynamics of life in the society, the more we realize that the colony is far from a superorganism emerging from a nexus of self-sacrificing altruism (Ratnieks and Reeve 1992; Bourke and Franks 1995; Crozier and Pamilo 1996; Bourke 1998). Social life may involve conflicts of genetic self-interest, resulting in tactics of coercion, manipulation, and even deadly aggression. These conflicts arise because colony members should favor individuals that are more closely related to maximize their inclusive fitnesses (Hamilton 1964b). Because the pattern of relatedness to a set of individuals differs for different colony members within genetically heterogeneous insect societies, and because the colony has finite amounts of resource to allocate to its reproductive propagules, the stage is set for a multitude of potential conflicts (Hamilton 1964a). It has also become increasingly clear, however, that threats of counter-

Levels-of-selection approach

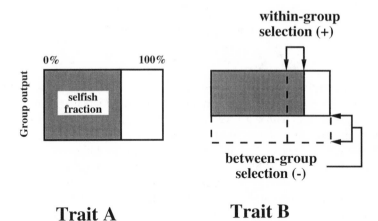

FIG. 8.1. Levels-of-selection (trait-group selection) versus broad-sense individual selection approaches. The levels-of-selection approach divides the fitness of broad-sense individual selection models (i.e., offspring number) into two components, the selfish fraction of group reproduction, and the total group output. Shown are two traits, A and B, the latter of which would (relative to A) experience positive within-group selection, but negative between-group selection. The broad-sense individual fitness is just the area of the shaded rectangles. Inclusive fitness equals the sum of these areas, weighted by relatedness.

strategies can suppress such manipulations by punishing them (as in policing; Ratnieks 1988) or rewarding their termination (as in bribing; Reeve and Keller 1997). These counterstrategies appear to limit the amount of intra-group warfare that is actually observed so that the actual conflict often appears less than the potential conflict (Ratnieks and Reeve 1992). Understanding exactly how and to what degree actual conflict is suppressed is the key to understanding the extent to which social insect colonies can be viewed as adaptively organized group-level units (Seeley 1997).

To study the evolutionary outcomes of within-colony conflicts, it is helpful to use a multilevel selection approach. Although genes are the entities that are ultimately transmitted over generations, it is important to keep in mind that genes are packaged in organisms, organisms in groups, and groups in populations, and that selection theoretically may act at any of these levels. For instance, some genes may be expected to be selected against at the within-colony level (if they decrease the proportional personal reproductive output of their carriers relative to other group members) but selected for at the among-colony level (if they increase the colony's overall reproductive output). Selection acting within and between colonies can be analyzed by using a multilevel analysis of selection, but it is important to remember that models of colony-level selection can always be translated into generalized inclusive fitness models (Dugatkin and Reeve 1994) (fig. 8.1). The multi-level approach is useful because it is well-designed for the analysis of how socially mediated mechanisms that restrain within-group selfishness may evolve and remain stable (Reeve and Keller 1997).

The aim of this chapter is to outline the major areas of potential conflicts among colony members and discuss how these potential conflicts are resolved. Resolutions (i.e., evolutionarily stable outcomes) can range from high levels of actual conflict to the complete absence of actual conflict, the latter empowering colony-level selection as the main force shaping colony-level phenotypes. As we shall see, the degree of peaceful cohesion of insect societies appears sensitively determined by the benefits of group membership and the benefits and costs of selfish manipulation and social suppression. We will review what we know about the potential and actual conflicts occurring within insect colonies. The three main areas of conflicts discussed are (1) the reproductive division of labor, (2) the relative investment in the sexes of reproductives, and (3) within-colony nepotism.

Conflicts over Who Reproduces

The most pervasive potential conflict within insect societies arises over who gets to reproduce. In many species, individuals engage in fierce fights to establish dominance hierarchies or to reverse reproductive roles within colonies, and it is not uncommon for these fights to end with severe injuries or the death of some of the protagonists (e.g., Keller et al. 1989; Heinze 1993; Keller and Ross 1993; Bourke 1994b). There is a clear tendency for aggressive interactions to be more common in species with low or no caste dimorphism than in species where the reproductive division of labor is associated with a strong morphological specialization of the castes. As the presence of morphological castes also influences the nature of within-colony conflict and the proximate factors regulating reproduction within the colony, we separately consider species with and without morphological castes.

Species without Morphological Castes

Many species of social insects (e.g., paper wasps, hover wasps, allodapine bees, sweat bees, and some Ponerine ants) are characterized by the absence or the presence of only a low level of dimorphism between the queen and worker castes (Wilson 1971; Keller and Vargo 1993; Peeters 1993; Bourke 1997). In these species, all group members have the ability to mate and reproduce, but aggressive interactions (including ritualized dominance displays) generally lead to one or a few individuals achieving most of the colony reproduction

(Heinze et al. 1994). There is, however, tremendous intra- and interspecific variation in the number of individuals taking part in reproduction and the variance in their relative contribution. In some societies, all individuals reproduce equally (low-skew societies), whereas in others, a single individual completely monopolizes reproduction (high-skew societies). A central issue in the study of social insect societies is therefore identifying the role of ecological, genetic, and social factors influencing the apportionment of reproduction among group members. This has been made possible by the development of optimal-skew models that allow consideration of these factors in a single explanatory framework (Emlen 1982; Vehrencamp 1983; Reeve 1991, 1998b; Reeve and Ratnieks 1993; Keller and Reeve 1994b).

Current models of evolutionarily stable reproductive skew share the critical assumption that the dominants control reproduction of the subordinates. If the dominants benefit from retention of the subordinates, it may pay the dominants to leave some reproduction to subordinates as inducements for these subordinates to remain in the society and cooperate peacefully rather than to leave or fight for exclusive control of the group's resources. (These inducements are called *staying* and *peace* incentives, respectively.) The skew models delineate the theoretical circumstances under which the dominant will yield reproduction to the subordinates and also predict the magnitude of reproduction forfeited.

The optimal-skew models predict that partitioning of reproduction in a society will be affected by four parameters: (1) the expected success of a subordinate that reproduces solitarily; (2) the group's overall productivity if the subordinate cooperates; (3) the genetic relatedness among group members; and (4) the probability that a subordinate would win a fatal fight with a dominant. Skew will increase under four conditions: (1) when group productivity increases, because enhanced group productivity reduces the attractiveness of the leaving and fighting options for the subordinate; (2) when ecological constraint on independent breeding increases, because subordinates can expect only small payoffs for leaving if ecological conditions are harsh; (3) when fighting ability of the subordinate is low, because subordinates with lower fighting ability will be less tempted to engage the dominant in a lethal fight for complete reproductive rights; and (4) when the relatedness between a dominant and a given subordinate increases, because subordinates that are more closely related to dominants automatically receive larger indirect (kin-selective) benefits for cooperating peacefully with dominants, hence they require smaller direct reproductive inducements for such cooperation. Reeve and Keller (1995) also showed that the asymmetry in relatedness occurring in mother-daughter associations versus sibling associations should tend to increase the degree of skew in the former (see also Emlen 1996; Reeve and Keller 1996). The effects of these parameters on reproductive skew are illustrated in figure 8.2.

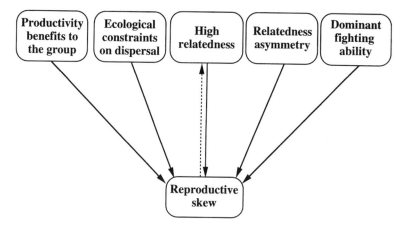

FIG. 8.2. Factors that increase the stable level of reproductive skew (under the optimal-skew models). Each arrow indicates that the factor causes a rise in the factor to which that arrow is pointing. (From models of Vehrencamp 1983; Reeve and Ratnieks 1993; Reeve and Keller 1995). Increased skew can also result in increased relatedness when individuals are recruited from within the group (dashed arrow on the figure), because increased skew decreases the effective number of breeders and therefore increases the average relatedness of individuals produced (Heinze 1995).

The optimal skew-models involve a dominant's "bidding" for the services of a subordinate helper against the payoff afforded by independent breeding in a fixed environment. A more recent model considers what happens when dominants in different colonies bid against each other for the subordinate's services (Reeve 1998b). In this case, the skew is expected to be lower than in the optimal-skew models because dominants should be willing to yield reproduction up to the point where they just barely gain from the subordinate's presence. Two somewhat surprising predictions of this bidding-game model are that relatedness does not affect skew and that skew decreases with increasing group output, unlike in the optimal-skew model. This model should apply to cases in which subordinates can cheaply sample multiple colonies before joining one, and thus is most likely to apply to (1) species in which there are relatively few potentially joining subordinate helpers compared to dominant residents and (2) species whose nests are present at high densities and are not cryptic (Reeve 1998b).

Recent empirical studies support the predictions of optimal-skew models (Creel and Waser 1991; Bourke and Heinze 1994; Keane et al. 1994; Bourke and Franks 1995; Emlen 1995, 1997; Heinze 1995; Reeve and Keller 1995; Bourke et al. 1997; Jamieson 1997), although the bidding-game model may apply to some ants and burying beetles (Scott 1994; Evans 1996; discussed in Reeve 1998b). Moreover, Reeve et al. (1998) also found some evidence for the main assumption of optimal-skew models and bidding games, namely that the dominants voluntarily forfeit direct reproduction to prevent subordinates from leaving the group or initiating a fight to monopolize the group resources. They analyzed an alternative model's predictions that reproduction by subordinates results because it is too costly for dominants to prevent subordinates from reproducing, and not because dominants are yielding some reproduction to subordinates as inducements to stay and cooperate peacefully. One important prediction of these "incomplete control" or "tug-of-war" models is that reproductive skew should either *decrease* with or be insensitive to the genetic relatedness between the subordinate and the dominant (at least when the genetic relatedness between them is symmetric). This occurs because, as relatedness increases, either both the dominant and the subordinate are predicted to reduce their efforts to enhance their own shares of the direct reproduction (resulting in little net change in skew), or only the dominant is predicted to decrease its effort (resulting in a reduction in skew). These predictions differ dramatically from that of optimal-skew models, which predict a positive association between reproductive skew and relatedness.

Data from both vertebrate and hymenopteran societies appear to support the optimal-skew models, as they reveal a consistently negative relationship between relatedness and a subordinate's reproductive share (Reeve et al. 1998). Many of the species groups studied, however, were composed of parents and offspring, that is, there was an asymmetry between dominant and subordinate in their relatednesses to each other's offspring. In such cases, the incomplete-control models, like optimal-skew models for two-person groups, predict maximum skew in favor of the dominant parent. Fortunately, it is still possible to discriminate between the optimal-skew and incomplete-control models in the case of parent-offspring groups because they make different predictions for sufficiently large groups. The incomplete-control model predicts maximum skew regardless of parent-offspring group size. In contrast, optimal-skew models often predict incomplete skew if the parent-offspring group size is larger than two (Reeve et al. 1998). The few data available on reproductive skew in symmetric relatedness groups strongly favor the optimal-skew models, but many more studies of reproductive partitioning are required to assess the general applicability of optimal-skew versus incomplete-control models.

One very counterintuitive prediction of the optimal-skew models is that intragroup aggression should often be higher in groups of close relatives than in groups of nonrelatives. In the former, high relatedness leads to high skew, and high skew leads to higher payoffs for the aggressive testing of dominants by subordinates and thus advertisement of fighting ability by dominants. In contrast, the incomplete-control models always predict decreasing intragroup aggression as relatedness increases. Again, the available evidence for associations of co-nesting social insect queens overwhelmingly favors the optimal-skew models (Bourke and Heinze 1994; Keller and Reeve 1994b; Reeve et al. 1998).

An important implication of skew models is that there is a social contract between the dominant and the subordinate over the allocation of direct reproduction. That is, for cooperative association to remain stable, it is necessary that individuals have the ability to detect reproductive cheating by other group members (i.e., attempts to obtain more than a member's reproductive share). This assumption was tested in *Polistes* paper wasps (Reeve and Nonacs 1992). Simulation of cheating by the dominant through experimental removal of subordinate's eggs in foundress colonies elicited enhanced aggression by the subordinate toward the dominant queen, especially if the subordinate was similar in size to the dominant. There was no such response by the subordinate if only worker-destined eggs or pupae were removed. These results indicate that subordinates monitor their amount of reproduction and aggressively retaliate if this amount falls below some minimal value. Strassmann (1993) proposed the alternative hypothesis that egg removal elicited the subordinate's aggression because the empty brood cells indicated a weak queen vulnerable to being overthrown, but Reeve and Nonacs (1993) showed that queens who replaced the removed eggs at high rates received significantly more aggression than did queens that replaced them at low rates, in accordance with the social-contract hypothesis but not the weak-queen hypothesis. The ability of animals to assess carefully each other's contribution to alloparental care and other nonreproductive tasks within the groups, as well as to monitor direct reproduction of other group members, is probably a common phenomenon in both vertebrate and invertebrate societies (Clutton-Brock and Parker 1995b), although it remains undocumented for most social taxa.

A recent extension of optimal-skew theory is the prediction that dominants and subordinates will always be favored to give reproduction to each other ("bribe" each other) to suppress destructive, selfish acts, that is, acts that reduce overall group output while increasing the actor's share of the reproduction (Reeve and Keller 1997). This has important implications for the issues discussed in this book because bribing provides a mechanism by which dominants and subordinates can suppress the selfish acts that would otherwise decrease overall group productivity. Interestingly, the magnitudes of bribes are predicted to decrease with increasing relatedness (Reeve and Keller 1997). The reason for this is that both optimal-skew and bribe models are examples of *transactional* partnerships in which dominants yield reproduction to subordinates in order to obtain some benefit (retaining a subordinate and preventing it from engaging in a selfish act, respectively; Reeve et al. 1998). For both types of models, the magnitude of reproductive incentive necessary to induce a subordinate to engage in some behavior beneficial to the dominant will always decrease with increasing relatedness between them.

Using Hamilton's rule and a multilevel selection approach, Reeve and Keller (1997) further showed that reproductive bribing is more likely to be favored over social policing (i.e., the suppression of selfish acts by other

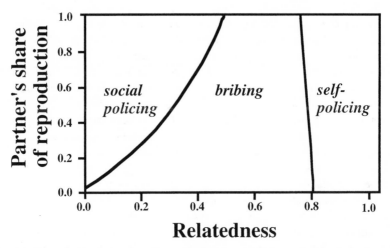

FIG. 8.3. Example illustrating the effects of relatedness and the partner's share of reproduction on whether reproductive bribing will be favored over social policing and self-policing. Information on the other relevant parameters (e.g., group-level cost of policing; loss of group productivity that a selfish act would entail; and benefits of group life) affecting skew are given in Reeve and Keller (1997). Adapted from Reeve and Keller (1997; with permission of the University of Chicago Press).

group members through aggressive acts and coercion [punishment] or through sabotage destroying the benefits of selfishness) (1) when the group cost of policing is high; (2) when the benefits of being in a group are not particularly high; (3) when the increment in personal reproductive share resulting from a selfish act is moderate; and (4) when intragroup relatedness is high. In short, bribing is most expected between genetic relatives and when the threatened acts are mildly selfish but quite destructive. The condition favoring bribing versus social policing and "self-policing" (i.e., when the individual itself benefits from not acting selfishly because of the harmful effects to kin) are illustrated in figure 8.3. The possibility of reproductive bribing has not been examined in any social insect, but the preliminary support for optimal-skew theory, which is based on a parallel logic, suggests that it may well occur. The required evidence is of the following form: When a potential breeder is induced to refrain from a cooperative act—for example, foraging—it should be observed to receive a smaller share of the reproduction than if it performs such an act. Any boost in reproductive share achieved as the result of performance of a cooperative act should occur without aggression by any party, suggesting that a subtle bribing transaction has taken place.

Although the possibility of reproductive bribing remains to be investigated empirically, there is recent evidence that the degree of self-policing is modulated by the benefit of being in a group. Reeve and Nonacs (1997) presented a general model of how the optimal level of aggression by a group member should vary in different social contexts characterized by different degrees to which group members are reproductively valuable to each other. They assumed that aggression increases the aggressor's share of the group's expected total reproductive output (positive within-group selection), but at the same time decreases the magnitude of the overall reproductive output (negative between-group selection; e.g., fig. 8.1). Under such conditions—which are likely to apply in nature—they showed that the optimal level of aggression toward a recipient will decrease with increasing value of the recipient in terms of effect on overall group productivity. Thus, an aggressor is more likely to self-police when there is a larger benefit to being in a group. In a series of field experiments, Reeve and Nonacs provided support for this prediction by manipulating the value of nestmates in colonies of paper wasps. They found that aggression decreased when the value of the cofoundress was increased by reducing the size of the future worker force, whereas the experimental decrease in the value of the cofoundress resulted in increased aggression.

SPECIES WITH MORPHOLOGICAL CASTES: CASTE DETERMINATION

Many social Hymenoptera—including the honeybee, bumblebees, vespine wasps and virtually all ant species—are characterized by the existence of specialized morphological castes, with queens reproducing and workers assuming other tasks such as foraging and brood care (Wilson 1971; Wheeler 1991; Keller and Vargo 1993; Peeters 1993; Bourke 1997). Partitioning of reproduction in species with morphological castes is therefore primarily defined by caste membership. Understanding the factors influencing reproductive roles in these species requires understanding the ultimate causes that direct or force individuals to commit themselves developmentally or physiologically to becoming workers.

It had long been believed that queens in large colonies were able to manipulate brood development chemically and force female larvae to develop into workers rather than female sexuals. That queens produce chemical substances (pheromones) preventing the differentiation of female brood into sexuals has been demonstrated for several species (Wilson 1971; Fletcher and Ross 1985; Vargo and Passera 1991). Traditionally, this inhibition has been taken as evidence of queen manipulation of the brood against the brood's genetic interests. This view has been challenged, however, by Seeley (1985) and Keller and Nonacs (1993), who argued that the queen pheromone may in fact act as a signal to which workers respond in ways to increase their inclusive fitness. Based on a pheromonal signal, each immature individual may decide whether to develop into a queen or a worker, with the choice depending on the relative benefits of becoming a queen or a worker (e.g., there may be a lower incentive to become a worker in older colonies because the larger the colony is, the lower may be the benefit of each additional worker in terms of increasing colony survival and productivity).

In this view, the queen pheromone may be a relatively low-cost cooperative signal. The possibility of low-cost but honest signaling between relatives has now been given a firm theoretical foundation by a synthesis of kin-selection theory with the modern evolutionary theory of honest communication (Reeve 1997). The synthetic theory shows that observed levels of signaling, and thus observed signal costs, will generally decrease with increasing relatedness between the signaler and the receiver. As relatedness increases to 1.0, signal costs should approach 0. Indeed, within organisms, where the relatedness among cells is virtually 1.0, hormones can be viewed as minimal-cost signals obviously enhancing the joint fitness of the genes within these cells. In close analogy, queen pheromones might be thought of as colony-level hormones.

The marriage of kin-selection theory and honest-communication theory also has demonstrated that evolutionary stability of honest signaling among relatives does not require that signals be differentially costly to signalers with higher values of the signaled attribute (a central assumption of Zahavi's "handicap" model [1977b] of competitive signaling; Grafen 1990a). All that is required for stable honesty among relatives is that signalers with higher values of the signaled attribute benefit more from a given receiver's assessed value of that attribute. This clearly is plausible for a "queen present" pheromonal signal. That is, a queen who is present will benefit more from the workers' appropriate response to this signal than would a worker who falsely produced the same signal in a queenless colony. For example, the false signal might cause a selectively disadvantageous delay in the rearing of a replacement queen that is more fecund than the worker.

An alternative hypothesis for the function of queen pheromonal signaling (more in line with a Zahavian model) is that the signal communicates the queen's level of fecundity (say, to prevent workers from replacing the queen with a more fecund queen). In this case, the queen pheromone may have to be costly to produce for it to be an honest signal. Moreover, it would have to be costlier to produce for less-fecund queens so that the latter would not falsely signal high fecundity. These conditions might be satisfied if, for example, highly fecund queens receive more food from workers and thus are better able to bear the cost of producing high levels of pheromone than less fecund queens (A. Bourke pers. comm.).

The factors that may induce development of individuals into a worker morph are basically the same as those favoring high skew (Keller and Reeve 1994b). It is critical to determine the lifetime consequences of developing into a worker versus a breeder morph and not simply the instantaneous breeding opportunities available for subordinates (Shellman-Reeve 1997).

Hence, the decision to develop into a worker morph will be favored when the lifetime probability of successful dispersal and breeding is low and when the reproductive output of kin is greatly enhanced by the assistance of workers with a specialized morphology. In other words, individuals should develop into workers only when (1) there is a low probability of becoming a successful breeder during one's lifetime; (2) the relatedness among group members is high; (3) individuals with a worker morph significantly increase colony productivity; and (4) workers are likely to remain throughout their lives with kin that can benefit from their help. In many species of the lower termites, for example, workers can facultatively transform into widely dispersing alates when the log containing the nest runs out of nutrients, illustrating the converse of (3) and (4) (Shellman-Reeve 1997).

This view of caste determination stresses that the factors determining caste will be sensitive indicators of the costs and benefits of pursuing the alternative reproductive strategies represented by the alternative caste decisions. For example, larval nutrition clearly affects the chances of successfully competing with breeders for direct reproduction within the natal colony and, thus, is understandably an important determinant of whether the worker or breeder strategy is pursued (Wheeler 1986). It may not pay everybody in the same situation to do the same thing. On the basis of an analysis of caste decisions among termite juveniles, Shellman-Reeve (1997), proposed that the benefit of pursuing a given reproductive strategy is often negatively frequency-dependent; that is, the greater the proportion of individuals choosing to develop into workers, the less the increment in colony output from a single worker. The evolutionary equilibrium would be a mixed evolutionarily stable strategy, with some individuals choosing to become workers and others breeders. In the Hymenoptera, the larvae are entirely nutritionally dependent on tending adults, which may manipulate the nutrition to ensure that nearly all of the larvae are selected to become workers during the colony growth phase (Keller and Reeve 1994b). In species like termites, however, the juveniles can forage for themselves, and there is likely to be a more even mixture of workers and breeders that have similar expected inclusive fitnesses at the moment the caste decision is made, with their precise ratio reflecting the relative payoffs of staying and helping versus dispersing and breeding (Shellman-Reeve 1997).

Species with Morphological Castes: Worker Reproduction

Although workers have reduced reproductive abilities, they have retained the ability to produce male offspring from unfertilized eggs in many hymenopteran species with morphological castes (Bourke 1988). This may lead to conflicts over male parentage. In a colony with a single, once-mated queen, workers are more closely related to their own sons ($r = 0.5$) and the sons of their full sisters (0.375) than they are to their brothers (0.25). Thus, each worker would benefit from monopolization of male reproduction. If a given worker cannot monopolize male reproduction, however, this worker should still favor male parentage by other workers (full sisters) rather than by the queen (their mother). This leads to a queen-worker conflict, as a queen should prefer to produce sons (0.5) rather than let her daughters produce males to which she is less related (0.25).

Interestingly, the nature of conflict over male parentage is influenced by the number of times a queen has mated. An increased number of matings decreases the average relatedness among workers and thus the relatedness between workers and worker-produced males. The relatedness of workers toward males is $0.125 + 0.25/n$, where n is the number of mates per queen. Hence, if the queen has mated with more than two unrelated males (and uses the sperm equally), workers are, on average, more related to queen-produced eggs than to worker-produced eggs. Under such conditions, workers may thus benefit from actively preventing each other's reproduction (Starr 1984; Woyciechowski and Łomnicki 1987; Ratnieks 1988)

The honeybee provides a nice example of such worker policing. Queens mate up to 20 times (Estoup 1994), and only 0.1% of the adult males produced derive from workers (Visscher 1989). Workers ensure the virtual absence of worker reproduction through (1) physical aggression against workers with developed ovaries (Visscher and Dukas 1995), and (2) selective destruction of worker-laid eggs (Ratnieks and Visscher 1989). In other words, worker policing effectively regulates male parentage, and it provides a mechanism ensuring a relative harmony within the colony despite the potential for kin conflicts (Ratnieks and Reeve 1992).

Sex-Ratio War

Conflicts over reproduction involve not only parentage of reproductive offspring but also the sex of these offspring. In eusocial Hymenoptera such as ants, wasps, and bees, diploid females develop from fertilized eggs and haploid males from unfertilized eggs. As a result, queens are equally related to their sons and daughters, whereas workers are more related to their sisters than to their brothers (Trivers and Hare 1976). These asymmetries in relatedness suggest that queens should favor an equal investment in both sexes, whereas workers should favor higher investment in females than in males (Trivers and Hare 1976; Charnov 1978; Nonacs 1986a; Pamilo 1991a). Hence, a sex-ratio conflict arises between queens and workers because workers may enhance their inclusive fitness by altering colony sex ratios in their favor and thereby acting against the interests of the queen.

The first tests of sex-ratio theory in social Hymenoptera have been done by determining sex-ratio investment at the population level. If workers control the sex ratio, one expects a 3:1 population sex-ratio investment in monogyne (single-queen) species when queens are singly mated. In polygyne (multiple-queen) ant colonies, the expected sex ratio should be closer to 1:1, even under worker control, when nestmate queens are related (Nonacs 1986b; Bourke and Franks 1995; Crozier and Pamilo 1996). This occurs because the presence of extra queens in a colony reduces the workers' relatedness asymmetry to nestmate females and males, as fewer females will be full sisters (Nonacs 1986a; Frank 1987; Boomsma 1993). Empirical studies showed the average sex-ratio investment across 40 species of monogyne ants is significantly female biased (0.63, expressed as the proportion of investment in females), whereas polygyne ants (25 species) have a sex-ratio investment of 0.44, which is not significantly different from 0.5 (e.g., Trivers and Hare 1976; Nonacs 1986a; Boomsma 1989; Pamilo 1990; Bourke and Franks 1995; Crozier and Pamilo 1996). Additional evidence for worker control (or at least partial control) of sex ratio comes from the study of sex ratios in monogyne slave-making ants. These are social parasites whose workers steal pupae from nests of other species, which, after maturation, raise all the slave-makers' brood. Because slave-maker workers lack the ability to control sex allocation (they do not raise the brood), the queen optimum is expected to prevail, leading to a 1:1 investment ratio. Sex-ratio data are available for three species of slave-makers, and the observed investment ratio (0.48) does indeed not differ significantly from 0.5, the value expected if queens control sex ratios (Trivers and Hare 1976; Bourke and Franks 1995).

Although these data support the view that workers have at least partial control over sex ratio in ants, some caution is needed because several sources of error are bound to occur with these broad interspecific comparisons. First, there are important interspecific differences in sexual dimorphism leading to a nonlinear relationship between cost of production of male and female sexuals and their relative dry weight (the measure that has been used to assess relative cost of production of males and females) (Boomsma 1989; Boomsma et al. 1995). Second, queens of many species mate more than once (Page 1986; Keller and Reeve 1994a; Boomsma and Ratnieks 1996), and this is an additional uncontrolled factor affecting relatedness asymmetry. Third, many ant species use budding as a mode of reproduction, which entails complications in measurement of the cost of production of females (because workers that accompany queens need to be considered investment in the female function and because budding may lead to local resource competition with nearby relatives; Pamilo 1990, 1991a; Nonacs 1993a). Finally, local mate competition (Hamilton 1967) may occur in several species where mating occurs in or near the nest (see Tsuji and Yamauchi 1994; Bourke and Franks 1995).

These sources of error are likely to have important consequences for the comparison between monogyne and polygyne species because these two classes of species frequently differ in their breeding system, life history, and mode of colony reproduction (Keller 1993a,b, 1995, 1997; Ross and Keller 1995; Chapuisat and Keller 1999). Thus, polygyne species are generally characterized by a lower queen-to-male dimorphism (Keller and Passera 1989; Stille 1996), and young queens tend to have fewer fat reserves than their monogyne counterparts (Keller and Passera 1989). This may lead to biases when using relative dry weight of males and females as an estimate of the relative cost of males and females in these two categories of species (Boomsma et al. 1995). Moreover, in polygyne species, young queens frequently return to their mother colony after mating (Keller 1995), and in some species, matings occur in the nest (Keller and Passera 1993; Passera 1994). These strategies are frequently associated with budding as a mode of reproduction (Rosengren and Pamilo 1983; Keller 1991; Bourke and Franks 1995), which results in local mate competition and the necessity of considering part of the workers produced as part of the investment in the female function (Pamilo 1991b). Finally, it has also been realized that monogyny and polygyny do not constitute discrete categories. Interspecific variation in queen number is better considered on a continuum, with species varying in the proportion of polygyne colonies and the average number of queens per colony. Only a few species always have colonies with single or multiple queens, making classification of species into just two discrete categories problematic (Nonacs 1986a).

Interestingly, at about the same time that the shortcomings of using interspecific comparisons to test queen-worker conflict were fully realized, Boomsma and Grafen (1990, 1991) developed the theory of split sex ratios. Their theoretical work was partly fueled by empirical studies showing that colonies of many social insect species produce mostly or only individuals of one sex (Pamilo and Rosengren 1983; Nonacs 1986a). Because relatedness asymmetry decreases with queen mating frequency and the number of related queens per nest, theory predicts that workers should rear mainly or only females in colonies with relatedness asymmetries above the population average and mainly males in colonies with relatedness asymmetries below the average.

Their theory was received with some skepticism (and still is by some colleagues) as it assumed that workers should be able to assess the number of mates their mother mated with. It might seem at first glance that their theory assumes too high a sophistication in both the ability to assess colony kin structure and the ability to bias sex-ratio investment patterns based on this assessment. However, empirical studies soon provided striking support for split sex-ratio theory (Queller and Strassman 1998). Evidence for relatedness-induced split sex ratios comes from halictid bees, in which relatedness asymmetry varies according to whether workers raise full siblings or offspring of a full sister (Boomsma 1991; Mueller 1991; Mueller et al. 1994; Packer and Owen 1994), ants and wasps with variable queen number (Herbers 1990; Queller et al. 1993; Chan and Bourke 1994; Evans 1995); and populations of two monogyne ants in which queens mate either singly or multiply (Sundström 1994; Sundström et al. 1996). In some species, however, sex-ratio variation among colonies does not seem be associated with differences in relatedness asymmetry (Pamilo and Seppä 1994; Vargo 1996).

An assumption of Boomsma and Grafen's (1990, 1991) theory is that workers in a colony can assess their relatedness asymmetry. This does not require worker ability to discriminate among different kin within the colony, but workers must be able to judge whether they are in a colony with relatedness asymmetry above or below the population average. This may be possible, for example, if workers are able to assess the genetic variability of the brood. This idea was tested by Evans (1995), who experimentally added unrelated larvae to colonies of the polygyne species *Myrmica tahoensis*. As expected, if workers use brood genetic variability as a cue to assess colony relatedness asymmetry, colonies into which foreign brood had been added reared more male-biased broods.

The ability of workers to manipulate colony sex ratio in their favor raises the question of *how* they do so. The queen controls the primary sex ratio by the proportion of haploid (male) and diploid (female) eggs she lays, but workers may subsequently adjust the sex ratios in their own interest by selective rearing of the brood. Consequently, selective male elimination is expected in the singly-mated class of colonies (high relatedness asymmetry), whereas no change or preferential male rearing is expected in the multiply-mated class of colonies (low relatedness asymmetry) (Boomsma and Grafen 1990, 1991). Sundström et al. (1996) used a recently developed technique for sexing eggs (Aron et al. 1994, 1995; Keller et al. 1996a,b) to compare primary (egg) sex ratios and secondary (adult) sex ratios between colonies headed by singly- and multiply-mated queens in a monogyne population of the ant *Formica exsecta*. They found that queens contributed a similar fraction of haploid eggs in both types of colonies. However, workers in colonies with a singly-mated queen, but not those in colonies with a multiply-mated queen, altered the queen-laid egg sex ratio by preferentially eliminating males to raise new queens. These results indicate that worker-queen conflict is manifest only in colonies headed by a singly-mated queen, as predicted by split sex-ratio theory. By eliminating males in these colonies, workers preferentially raise the sex that yields the largest marginal fitness return per unit of investment (Boomsma and Grafen 1990, 1991) thereby enhancing their inclusive fitness.

The manipulation of colony sex ratio through male elimination must entail decreased colony productivity, because the resources invested in rearing males until the moment of elimination are partly or completely lost. All else being equal, one would thus predict that workers should eliminate males as early possible in development to minimize the cost of sex-ratio manipulation. Surprisingly, this is not the case. By following the sex ratio of the brood throughout development, Chapuisat et al. (1997) found that workers eliminated a significant proportion of their brothers, and possibly all of them, at a late developmental stage, when they were close to pupation. In fact, workers destroyed males that had grown beyond the size threshold where workers should replace them with extra females to increase their inclusive fitness. Why are males still eliminated at such a late developmental stage?

Chapuisat et al. (1997) suggested that late elimination of males may stem from the necessity to eliminate part of the brood because of resource limitation (e.g., in years that are less productive than average). In this case, workers would still maximize their inclusive fitness by selectively eliminating the sex to which they are less related compared to the population average, that is, males in colonies headed by singly-mated queens. Although this explanation may account for late male elimination, it does not explain why workers did not replace supernumerary males with extra females earlier. Two mutually nonexclusive hypotheses are possible. First, early sex-ratio biasing may entail a high cost if workers make frequent errors when determining the sex of young brood. A high error rate early in development may arise if the mother queen and males conceal the sex of the brood in order to resist worker manipulation (Nonacs and Carlin 1990; Nonacs 1993b). Workers may thus reduce the cost of errors by postponing elimination until males and females can be reliably distinguished. Second, queens may prevent early sex-ratio biasing by limiting the number of female eggs. If there are not enough extra female eggs to replace young male larvae, workers may gain no benefit from early elimination of males.

These recent studies on ant sex ratio show that overt expression of the queen-worker conflict depends on the genetic structure of the colony. In colonies where queens are multiply mated, there is no conflict of interest because both parties benefit from producing males. Because the population-wide investment ratio is female biased (Boomsma and Grafen 1990, 1991; Sundström 1994; Sundström et al. 1996), males are therefore also more valuable for the mother queens. These studies also indicate that conflicts may be expressed even if the colony incurs costs in the form of the elimination of kin. The cost of male elimination in colonies headed by singly-mated queens remains to be investigated. It is also unknown whether some females are eliminated in colonies headed by multiply-mated queens or, alternatively, whether a greater proportion of females are channeled into the worker developmental pathway.

Conditional manipulation of colony sex ratio by workers in response to differences in relatedness asymmetry has interesting consequences for the evolution of queen mating frequency. Because the populationwide sex-ratio investment is female biased under worker control, queens producing sons have a higher fitness than queens producing female sexuals. Thus, multiply-mated queens will tend to have higher fitness than singly-mated queens because their colony will specialize in male production. This fitness advantage may generate selection for multiple mating (Queller 1993; Ratnieks and Boomsma 1995), a phenomenon that is relatively common in social Hymenoptera (Page 1986; Keller and Reeve 1994a; Boomsma and Ratnieks 1996). However, other hypotheses have been proposed for multiple mating—for example, that it promotes adaptive genetic diversity within colonies (Crozier and Page 1985; Sherman et al. 1988; Keller and Reeve 1994a; Liersch and Schmid-Hempel 1998; Schmid-Hempel 1998)—but the critical discriminating experiments have yet to be performed.

A further twist to the sex-ratio story is that sex-ratio biasing by workers may also affect male mating strategies. Because males develop from unfertilized eggs, males have no genetic interest in the sons of their prospective colony. Thus, worker manipulation of sex ratio toward females in colonies headed by singly-mated queens decreases queen reproductive success but increases that of the male mate(s). The number of matings by the queen therefore has a differential effect on male and queen reproductive success, because multiple mating increases the reproductive success of queens and decreases that of males (Boomsma 1996).

Boomsma (1996) thus pointed out that a worker-induced split sex ratio may lead to an unusual mating system with eager females and choosy males, because shared paternity leads to a tremendous decrease in inclusive fitness for males. It remains to be investigated whether specific adaptations have evolved for males to maximize the relatedness asymmetry in their prospective colony. Boomsma suggested that this may occur by several means. For example, after mating, males could deposit chemical substances on the queens to signal that they are mated. Another mechanism would be for the second male to transfer only a limited number of sperm so as to decrease the relatedness asymmetry to an extent that would not be detectable by workers.

Finally, an intriguing finding suggests that intracolonial conflict over sex ratio and differential investment in the sexes by workers can extend beyond the larval stage. Starks and Poe (1997) have recently found that when foraging workers of the social paper wasp *Polistes dominulus* return to mature, late-season colonies, the resident workers react by forcibly "stuffing" their brothers into empty brood cells until the food brought by the forager has been distributed to female adult nestmates and developing larvae. This temporary imprisonment of the males appears to ensure that resources critical to successful overwintering by the workers' reproductive sisters are preferentially channeled to these sisters and away from the less closely related brothers. This provides yet another example of behavior in the name of genetic self-interest. In *P. dominulus*, just as in the ant *F. exsecta*, workers actively disfavor their less related brothers in favor of their more related sisters to pass on more copies of their genes to the next generation.

Nepotism within Families

Colonies of social insects consist of a number of genetically distinct lineages when queens mate with multiple males or when colonies contain several queens. For example, colonies headed by one multiply-mated queen consist of full sisters ($r = 0.75$) and half sisters ($r = 0.25$) when the queen's mates are unrelated to her and each other (as is usually the case). Similarly, the relatedness between individual workers may vary between 0.75 and 0 within colonies headed by multiple queens, depending on the relatedness among queens. In colonies consisting of several genetically distinct lineages, workers could benefit by behaving nepotistically, favoring the most-related individuals. Earlier studies of honeybees suggested that workers favored full sisters over half sisters in interactions with other workers, including swarming and queen brood rearing (e.g., Getz and Smith 1983; Page et al. 1989). However, these studies have been justifiably criticized on a number of grounds including artificiality of colony composition, possible bias because of the genetic markers used, and faulty statistical analyses (Oldroyd et al. 1990; Frumhoff 1991; Breed et al. 1994). New empirical studies using molecular markers also failed to demonstrate differential family composition during swarming (Kryger and Moritz 1997).

Studies of other species of social insects have also failed to detect nepotism within colonies. In several ant and wasp species, queens may cooperate to initiate new colonies, but not in a nepotistic way. Queens of the wasp *P. annularis* do not prefer to cooperate with close relatives when initiating new colonies (Queller et al. 1990). Cooperation between founding queens is also common in the fire ant (*Solenopsis invicta*), but queens start fighting soon after the eclosion of the first workers, ending with the death of all but one queen. The first eclosed workers also may take part in the fights, but they do not favor their mother nor increase her probability of survival (Balas and Adams 1996; Bernasconi and Keller 1996; Bernasconi et al. 1997). Finally, mature colonies of some ant and wasp species are headed by several queens, but the workers fail to favor sisters over workers from other matrilines or to behave nepotistically toward their mother compared with other queens (Carlin et al. 1993; Snyder 1993; DeHeer and Ross 1997). Similarly, in multiple-queen colonies of the ant *Leptothorax acervorum*, queens eat eggs, but they do not eat their own eggs less frequently than those laid by their nestmates (Bourke 1994a).

Three nonmutually exclusive, general explanations for the apparent lack of nepotism within insect societies have been proposed (reviewed in Keller

1997). The first explanation is that selection has favored universal treatment of colony members because differential treatment of kin classes incurs costs that outweigh the benefits to nepotistic individuals (Reeve 1989; Carlin et al. 1993). When all individuals receiving help are related (to a variable degree) to the altruist, the benefits resulting from the increased fitness of more related individuals is mitigated by the decrease in fitness of the less-related individuals. For example, Ratnieks and Reeve (1992) considered the case of an insect colony headed by a multiply-mated queen in which workers may help either full or half sisters. They suggested that differential treatment of colony members would inevitably lead to losses in colony efficiency, for example, because time is wasted in assessment of patriline status. Therefore, if nepotism is associated with a decrease in colony efficiency, all patrilines may experience a net decrease in inclusive fitness, and kin discrimination might be selectively disfavored. In general, smaller differences in relatedness between colony members should tend to select against nepotism, because smaller difference in relatedness between altruists and classes of individuals receiving help result in smaller potential benefits for nepotistic behaviors.

The second explanation posits that kin-biased behaviors are disfavored because the high frequency of recognition errors so dilutes the benefit of nepotism that this benefit no longer exceeds the cost. Because no recognition system is perfect, the decision of an individual to behave nepotistically depends on the probability of correctly identifying desirable and undesirable recipients and the benefits versus costs of correct and incorrect assessment (Reeve 1989; Sherman et al. 1997). The efficiency of any type of kin-recognition system relies on the types of cues available. Because they share the same environment, workers in an insect colony cannot use environmental cues to distinguish genetic lineages; they would need to use genetically specified cues.

Recognition mediated by genetic cues might be unstable and error-prone. Theoretical studies indicate that allelic diversity of recognition should decrease over time because more frequent alleles will be continually favored until fixation (Crozier 1988; Ratnieks 1991). Thus, other selective forces such as gamete compatibility or disease resistance apparently must operate to maintain genetic diversity of recognition cues (Crozier 1988; Ratnieks 1991). Consistent with this idea, odors induced by the major histocompatibility complex (MHC) are used to recognize relatives in several vertebrates, and extremely high allelic variability at MHC loci is maintained because of their role in immune defense and, most importantly, the avoidance of kin matings (Potts and Wakeland 1993). Meiosis also decreases the efficiency of recognition by genetically specified cues because it leads to variable combinations of segregating alleles in family members (Getz 1981; Lacy and Sherman 1983; Gamboa et al. 1986; Waldman 1987).

Finally, a third hypothesis is that some colony members (e.g., workers, subordinate queens) benefit from actively scrambling recognition labels. The workers most likely to benefit from scrambling cues are those that belong to classes of close relatives constituting only a small fraction of the total pool of interactants (Reeve 1998b). This is because accurate signaling of kin status would produce some increased help from (rare) more closely related individuals but reduce the benefits provided by the (most common) less-related individuals. Colony members may also benefit by reducing or eliminating information about kinship within the group when nepotism entails reduction in colony productivity (Reeve 1998b).

Scrambling of recognition labels could be achieved by transferring odors between colony members, as indeed has been observed in many ant species (Hölldobler and Wilson 1990). It is not yet clear, however, whether the role of chemical cue transfer is to remove information about kinship within colonies or to provide a better system of recognition between members of different colonies. A direct prediction of the scrambling hypothesis is that odor transfer should be absent or less frequent in colonies where there is usually a single matriline and patriline (colonies with one singly-mated queen) than it is in colonies with different genetic lineages, because no nepotism is predicted to occur in the former (Keller 1997).

A recent study showed that subfamilies (offspring from different fathers) tend to have different cuticular hydrocarbon profiles in the honeybee (Arnold et al. 1996). As cuticular hydrocarbons are probably the chemical labels used by workers to discriminate nestmates from non-nestmates, this raises the possibility that such labels can be used for within-colony discrimination. However, Arnold et al.'s study also showed considerable overlap in the chemical profile of workers from different patrilines, suggesting that these chemical labels would at best provide a moderately efficient system of recognition for within-colony discrimination.

In conclusion, whether or not intracolonial nepotism will be favored ultimately depends on (1) the maximum efficiency of the recognition system, and (2) the minimum recognition efficiency required for nepotism to provide a net benefit. Figure 8.4 provides a schematic representation of the combined effects of factors favoring and disfavoring the evolution and maintenance of nepotism.

Conclusion

The advent of "selfish-gene" thinking has led to the recognition that there are many potential sources of genetic conflicts within insect colonies. In some cases, empirical studies have supported the view that these potential conflicts translate into actual conflicts, whereas in other cases, mechanisms seem to have evolved in order to prevent conflicts from being expressed.

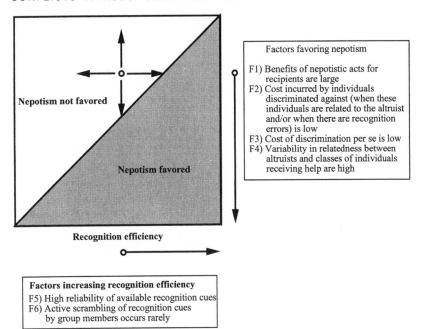

Factors favoring nepotism

F1) Benefits of nepotistic acts for recipients are large
F2) Cost incurred by individuals discriminated against (when these individuals are related to the altruist and/or when there are recognition errors) is low
F3) Cost of discrimination per se is low
F4) Variability in relatedness between altruists and classes of individuals receiving help are high

Nepotism not favored

Nepotism favored

Recognition efficiency

Factors increasing recognition efficiency
F5) High reliability of available recognition cues
F6) Active scrambling of recognition cues by group members occurs rarely

FIG. 8.4. Factors favoring/disfavoring nepotism. The actual efficiency of a recognition system depends on the nature of recognition labels available (F5) and whether active scrambling of these labels by group members occurs (F6). The level of recognition efficiency required for nepotism to be favored is set by four main factors (F1–F4), with nepotism being less likely to evolve if discrimination is a costly process (F3) and if the differences in relatedness between altruists and classes of individuals potentially receiving help are small (F4). By contrast, nepotism will be favored even when recognition efficiency is relatively low when the four following conditions are met: important benefits for recipients of nepotistic acts (F1); low costs for individuals discriminated against (F2); low cost of discrimination between kin classes for the altruist (F3); and large differences in relatedness between altruists and potential recipients of help (F4).

Queen-worker conflict over sex ratio and the potential conflict between nepotists from genetically distinct lineages, respectively, appear to provide the best examples of these two possible outcomes.

Conflict of interest over reproduction is a potential conflict that sometimes, but not always, translates into actual conflict. Aggression is common in species lacking morphological castes, yet, even in species with the highest level of aggression, individuals interact in a cooperative manner most of the time. This may be due to mechanisms that evolved to suppress selfish and destructive behavior manifesting intragroup conflicts. These mechanisms include policing and/or bribing by other group members. The study of intragroup conflict in insect and other animal societies is still in its infancy, but the recent development of theories of conflict suppression provides a rich source of predictions that are amenable to experimental testing. For example, one prediction is that reproductive bribing should be widespread, particularly among genetic relatives. This prediction is testable with careful observations of potentially quite subtle behavioral interactions representing transfers of reproduction within groups (Reeve and Keller 1997).

There are clear analogies between the dynamics of conflicts within insect societies and conflicts that occur at other levels of biological organization, which are discussed in the other chapters of this book. For example, one cannot consider the potential conflict among genetic lineages within social insect colonies without being struck by the close analogies with intragenomic conflicts. The potential benefits of workers' scrambling recognition cues in order to increase their inclusive fitness are much the same as the benefits received by organisms that evolve mechanisms preventing or reducing intragenomic conflicts (reviewed in Keller 1997; Reeve 1998b). Thus, a driving allele that increases its probability of being passed into the gamete by destroying nondistorting alleles on the homologous chromosome will generally inflict a cost on the whole organism because fewer gametes are produced. Hence, a modifier gene that inactivates a driving allele will be favored (Pomiankowski, chap. 7). Haig and Grafen (1991) suggested that recombination might have been selected to decrease the possibility that closely linked genes would cooperate to bias their transmission during meiosis. In the same manner, scrambling of recognition cues, social policing, and bribing may benefit all workers in a social insect colony if these processes increase colony productivity. Optimal-skew theory can even be applied to an organism when it is viewed as a group of cooperating cells or chromosomes. (1) Because only a small proportion of cells generate gametes, skew is very close to 1, just as predicted by optimal skew theory because relatedness among cells approaches 1. (2) By contrast, at meiosis, unrelated chromosomes (in outbreeding species) pair up and participate in an intricate molecular tango that ensures equal representation in gametes ("fair meiosis"), just as there is equitable reproduction in many societies of unrelated organisms. Thus, at both the organismal and intragenomic levels, the same principles may determine how conflicts are resolved and when the collective interest of the group wins over the selfish interest of each group member.

Finally, the study of within-colony conflict (and conflicts at other levels of biological organization) and its resolution has important consequences for our understanding of evolution. As pointed out by Leigh (chap. 2), genetic conflicts enable one to look for footprints of the role of natural selection in evolution. In the same vein, the study of the outcome of sex-ratio conflicts between queens and workers and the recent demonstration that workers manipulate colony sex ratios according to differences in relatedness asymmetry certainly provide strong evidence of the correctness of kin-selection theory.

It is somewhat paradoxical that the outcome of within-colony conflicts provides one of the best demonstrations of kin-selection theory, a theory that was first proposed to explain the evolution of extreme cooperation!

It is meaningful to ask how the balance of cooperation and conflict should tend to change within a given biological unit (organisms, societies) over long periods of evolutionary time. Will scrambling, bribing, policing mechanisms, and their equivalents eventually remove all traces of within-unit competition, so that harmony will continually crystallize upward from lower to higher levels of biological organization? There is a simple, but compelling, reason why they will not. When bribing and policing remove the incentives for a particular selfish manipulation, they increase group output. This increase in output automatically increases the temptation to engage in other kinds of selfish manipulations, because the payoffs for getting a bigger piece of the now bigger reproductive pie increase (Reeve and Keller 1997). Thus, as one conflict is resolved, others inevitably will open up, just as plugging up the proverbial hole in the dike creates new holes via increased water pressure! Perhaps this is why conflict, and the mechanisms thwarting it, persist within individual organisms (and a fortiori within animal societies).

Acknowledgments

We thank K. Boomsma, A. Bourke, W. Brown, M. Chapuisat, P. Christe, E. Fjerdingstad, J. Goudet, E. Leigh, K. Ross, and D. Shoemaker for useful comments on the manuscript. We were supported by grants from the Swiss and U.S. National Science Foundations.

9

Complexity in Vertebrate Societies

Dawn M. Kitchen and Craig Packer

Individuals are the preeminent vehicles for selection. Yet individuals consist of a collection of genes, the physical elements of evolution. Does the transition from selfish gene to individual genome provide lessons for higher-order phenomena, in particular, the relationship between the individual and the social group? Maynard Smith and Szathmáry (1995) argued that there have only been two general contexts in which such transitions may have occurred. First, eusocial species display an extreme degree of cooperation and division of labor, but they are in effect "extended genotypes," in which the genetic self-interests of each group member are highly similar. Thus, sufficient kinship permits specialization similar to the cells of a single superorganism (Seeley 1997; see also Keller and Reeve, chap. 8). Second, human cognition and language permits rational planning and a unique capacity for cultural transmission, enabling the development of an elaborate social organization beyond the reach of most organisms (see Maynard Smith, chap. 10).

Social evolution is often considered to have reached only these two remarkable pinnacles; everything else appears to be stranded on lower ground. But is the topography of vertebrate sociality really so uniform? Detailed studies of animal societies often reveal hints of true complexity, sometimes even giving the appearance of group-level coordination or a well-organized division of labor. Do these provide evidence of a higher level of selection that has superseded individual selection? Or has individual selection alone produced these novel levels of organization?

Although there is theoretical evidence that group selection can operate under certain strict criteria, fitness differences among groups will only rarely supplant the effects of differential individual fitness within the population (Williams 1966a; Dawkins 1976). But can important evolutionary pressures be revealed by measuring fitness effects at the group level? Proponents of neo-group selection advocate that group-level adaptations have produced "emerging properties" that can only be understood from their effects on the relative fitness of different groups (e.g., Wilson 1997a,b; Sober and Wilson 1998). They also urge the necessity of viewing the group as an entity in its own right rather than reducing everything to the sum of individuals. Strict individual selectionists, however, suggest that such complexity can best be understood by building from the simplest unit. If it is advantageous for individuals to form groups, individual selection will lead to adaptations that maximize personal fitness within the group and thus produce the complexities observed in nature. For example, certain physical structures may transcend the contribution by any single individual (e.g., termite mounds, weaver bird nests, acorn woodpecker granaries), but these are epiphenomena that result from individuals working for themselves and their kin rather than the result of a preplanned blueprint (Maynard Smith and Szathmáry 1995).

One way to view these contrasting approaches would be to apply to vertebrate societies criteria similar to those that Maynard Smith and Szathmáry (1995) have suggested for the evolution of human societies: Rousseau's social contract versus Adam Smith's free market. According to Rousseau, society is designed to maximize benefits to the society itself, whereas Smith contended that society results from the behavior of individuals working for their personal benefit.

Our task for this volume was to determine how the most complex and apparently coordinated vertebrate social behaviors conform to this dichotomy of societal evolution. Many vertebrate societies are centered around cooperative breeding, and this pathway apparently leads toward eusociality in the same manner as naked mole-rats, termites, and hymenoptera (Wilson 1971; Lacey and Sherman 1991; Sherman et al. 1995). Because strict eusociality is rare among vertebrates (see Alexander et al. 1991), and the behaviors associated with cooperative breeding systems have been the subject of several recent reviews (Stacey and Koenig 1990; Emlen 1995, 1997; Solomon and French 1997), we have chosen instead to examine alternative examples of emergent properties in vertebrates. Predator avoidance, food acquisition, and resource competition all provide clear examples of complex, group-level behavior. Can simple rules of individual behavior account for these phenomena? Or do we see signs of a superorganism with near-perfect coordination, altruism, specialized division of labor, and advanced group-level decision making beyond the sum of its parts? In each case, we characterize the complexity of the behavior, providing a plausible scenario for the evolution of the trait and exploring its maintenance at the observed level of complexity. Where possible, we discuss factors that may have prevented each trait from attaining an even higher level of complexity.

Predator Avoidance

Predation has been an important force in the evolution of group living, primarily through the dilution effect (Treisman 1975). Individuals benefit from dilution whenever a predator can only capture one prey at a time and a group of n individuals is attacked less than n times as often as a solitary. But group living affords numerous other antipredator advantages beyond the sheer safety of numbers, including the confusion effect and corporate vigilance.

Predator-avoidance behavior often shows a superficial resemblance to higher cognition or a coordinated division of labor. We discuss these phenomena within the context of evasion and vigilance.

COORDINATED EVASION

A flock of dunlins flies low over Puget Sound; Mt. Rainier glows in the afternoon light. The dunlins fly along in a loose-knit swarm until a merlin swoops down from the sky, and the dunlins make a sudden sharp turn, bunching together and moving as one. A few stragglers stand out from the crowd, and the merlin catches its dinner.

Such coordinated evasive maneuvers can appear choreographed, giving the impression of a supraindividual intelligence. Indeed, the idea that cognition involves an extensive group-level process such that the group literally has a mind of its own (Wilson 1997b) has been around for decades (e.g., Selous 1931). This impression may tell us more about the neurological attributes of the observers than of the birds themselves, however. Slow-motion film analysis of the dunlins' tightly executed maneuvers reveals a precise sequence of individual decisions. When first faced with an external threat, one to three birds react by banking toward the rest of the flock, and the remainder respond one by one, forming a synchronized "maneuver wave" (Potts 1984). This results from a pattern of neighbor following neighbor, not a direct response to the predator, and the individual birds react so fast that the sum of their actions appears to have a life of its own. Yet no higher level of organization is involved.

The proximate mechanism of this behavior, however, does not diminish its complicated group-level effect. The response time between dunlin flock neighbors decreases along the front of the "wave" and becomes, on average, three times faster than the average startle response of a lone individual. This suggests that each individual synchronized its movement to coincide with the approaching maneuver wave propagating across the flock, giving rise to a "chorus line" (Potts 1984). A similar phenomenon is found in schooling fish, where close proximity allows each individual to detect rapid pressure waves with their otolith and lateral line organs (Gray and Denton 1991). Individuals synchronize their movements with those of their neighbors (Pitcher and Parrish 1993), resulting in remarkably coordinated predator-evasion maneuvers (fig. 9.1).

The chorus line is possible because the animals' absolute speed of response is so quick that each group member can wait its turn and still escape the approaching predator. Otherwise, they should follow a direct-response rule in which each group member flees as soon as it detects the predator. Neighbor-rules maintain integrity within the flock during evasion, and order-

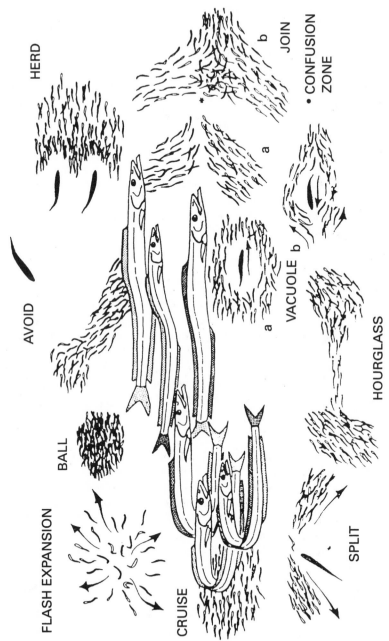

FLASH EXPANSION

BALL

HERD

AVOID

JOIN

CONFUSION ZONE

VACUOLE b

a

CRUISE

SPLIT

HOURGLASS

Fig. 9.1. Schematic representations of evasion maneuvers by schools of sand eels while pursued by hunting mackerel. Taken from Pitcher and Parrish (1993).

liness is individually advantageous because it reduces the risk of collision and maintains benefits from the dilution effect (Heppner 1997; Parrish and Turchin 1997). The chorus line will only be successful, however, in species in which the maneuver wave can be reliably detected (Lima 1995; Lima and Zollner 1996) and the absolute reaction time is very rapid in comparison to the speed of the approaching predator.

What would a group-selected evasion tactic look like? In a world of like-minded individualists, each animal's only goal is to escape, but in a world where group interests supersede the needs of individual group members, the birds should try to minimize the risk that *anyone* is ever captured. Thus, individuals should maintain a precise location within the flock because novel movements could decrease the efficiency of the group's coordinated pattern (Schilt and Norris 1997). It is a common observation, however, that individuals attempt to move toward the center of the flock when threatened by a predator (Hamilton 1971; Romey 1997), and random movement patterns are continually generated by individuals within the group (Heppner 1997).

COORDINATED VIGILANCE

In Tsavo National Park, a family of dwarf mongooses searches the grass for insects and spiders. The alpha male perches atop a nearby termite mound and scans the sky. As the chirpings of his family recede, he notices a subordinate female. A quick and efficient forager, she sits perched on another termite mound, digesting and watching the sky. The alpha male rejoins the pack, and everyone remains absorbed in a search for food. Startled by the shrill cry from the watchful female, the family safely joins the sentinel in her termite mound just before a pale chanting-goshawk completes its futile plunge.

Extreme examples of vigilance behavior involve an apparent division of labor with individuals taking turns as "sentinels." Several well-documented cases in birds and mammals indicate a highly coordinated system with at least one, but often only one, sentinel on duty most of the time (e.g., Gaston 1977; Rasa 1977, 1989a; Moran 1984; Ferguson 1987; McGowan and Woolfenden 1989; Hailman et al. 1994). In Florida scrub jays, the occurrence of a single sentinel was more frequent than expected by chance (McGowan and Woolfenden 1989), and, in a study of captive meerkats, at least one individual played sentinel over 95% of the time (Moran 1984). The meerkats' coverage dropped to 70% immediately after the death of one group member—a decrease comparable to the time it normally spent on duty—though the survivors soon adjusted their guarding time to cover the gap. Substitutions between group members also seem to be well synchronized in these systems. In the scrub jays, one sentinel typically relieves another within the same minute (McGowan and Woolfenden 1989), whereas sentinel exchange in dwarf mongooses involves a consistent sequence of individuals (Rasa 1977, 1989a).

Vigilance has traditionally been considered to involve some form of altruism: Animals must forego foraging time to scan for predators, and vigilant individuals may inevitably alert their companions of an impending attack (Pulliam 1973; Pulliam et al. 1982; Parker and Hammerstein 1985). Thus, the coordination observed in these sentinel systems implies a degree of active cooperation, indicating a group-level emergent property. More recent approaches to predator-detection behavior, however, have all emphasized an inherent *individual* advantage to being the first to spot the predator (e.g., FitzGibbon 1989, 1994; Packer and Abrams 1990; McNamara and Houston 1992; Caro et al. 1995; Godin and Davis 1995; Bednekoff and Lima unpub.).

Although their elevated positions might make them seem more exposed, sentinels probably gain direct benefits from their behavior. Sentinels refrain from all other activities besides vigilance (e.g., Moran 1984; Wickler 1985; Hailman et al. 1994); their posts provide an improved view of the surrounding area (Rasa 1986, 1989a; Hailman et al. 1994); and they may be situated close to shelter (Rasa 1989a, but see Rasa 1987). Thus, active vigilance may pay more than continued foraging once an individual has reached its gut capacity, and any asynchrony in feeding requirements among group members would give the appearance of a coordinated division of labor. If this is the case, rather than look for evidence of cooperation, we should look for factors that lead to feeding asynchrony.

Animals with different nutritional requirements or foraging efficiencies will be expected to approach satiation at different times (Gaston 1977; Bednekoff 1997). Perhaps the first sentinel of the day was the last individual to have fed the night before or is the most efficient forager in the group. The staggered energy reserves of individuals would affect the relative payoff of vigilance behavior and the time at which they switch to this task. Some individuals may be more efficient foragers (e.g., adults vs. juveniles) or may have lower energy requirements (e.g., males vs. pregnant or lactating females), thus explaining why adults are more often sentinels than juveniles (e.g., Rasa 1977; Hailman et al. 1994), and why males are sentinels more often than females (e.g., Rasa 1977, 1989a; Moran 1984; Hailman et al. 1994). Consistent differences between individuals could also explain why the sequence of exchange between specific group members occurs with such regularity (Rasa 1977, 1989a).

However, an organized sentinel system cannot result from asynchronous requirements unless foragers benefit from the vigilance of the sentinel and sentinels can resume foraging when it is their best option to do so (Bednekoff 1997). Each animal's decision depends on what everyone else is doing: An individual can devote relatively more time to foraging (rather than vigilance) as long as there is at least one sentinel and the sentinel's response to an approaching predator is reliable and easily detected by other group members.

Why do sentinels signal an advancing predator? An alarm call might be the inevitable by-product of a startle response, a signal that encourages the predator to attack a different individual (e.g., FitzGibbon 1989; Caro et al. 1995), or a means of manipulating other group members (Charnov and Krebs 1975). Besides protecting close kin (Hamilton 1964a,b; Sherman 1977), animals may receive a strong benefit from the dilution effect and therefore benefit from protecting necessary companions (Lima 1989).

Assuming that sentinels enjoy even a minor improvement in predator detection compared to foragers, Bednekoff (1997) has shown that a coordinated sentinel system can arise purely through individual advantage. Bednekoff's model incorporates three factors: (1) At each time-step, individuals choose between foraging and vigilance, depending on their own energetic reserves; (2) individuals switch tasks at each time-step according to the foraging/vigilance behavior of other group members; (3) predator-detection information is effectively transferred from the sentinels to the foragers.

Consider first a group in which everyone forages, and no one is vigilant. Once one individual has attained an adequate energy reserve, it can benefit more from behaving as a sentinel rather than continuing to forage. The remaining group members now enjoy considerably less risk of predation so that they only benefit from acting as sentinels if their energy reserves are sufficiently high (fig. 9.2). Once the lone sentinel's reserves have fallen below the threshold, it resumes foraging, but its role is filled by any other group member with adequate reserves to become the lone sentinel. As a result, coordinated sentinel behavior can confer a selfish advantage as long as individual priorities can be satisfied asynchronously (Bednekoff 1997).

Sentinel behavior shows an irregular taxonomic distribution. What ecological differences might account for the presence of a coordinated vigilance system in one species and its absence in a related species or in a different population? Sentinels might be more likely in a population or species in which (1) elevated lookout posts provide a genuine predator-detection advantage (Bednekoff 1997); (2) food supplies are adequate (Gaston 1977; Bednekoff 1997); and (3) predator densities are high (McGowan and Woolfenden 1989).

Dwarf mongooses in the Serengeti are an example of the latter point, as they do not display the complicated and coordinated sentinel behavior seen in the Tsavo population. Predation pressure is much higher in Tsavo than in the Serengeti (Rasa 1986, 1989b; S. Creel pers. comm.), and the behavior of one particular raptor species apparently determines the payoffs from coordi-

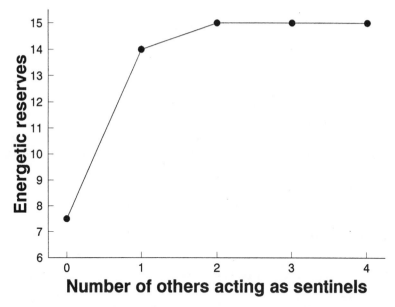

FIG. 9.2. Bednekoff's (1997; with permission of the University of Chicago Press) model of sentinel behavior. Bednekoff considered several different scenarios, but we only present the outcome when sentinels are the only animals to give an alarm call. Graph illustrates the optimal decision rule for a member of a group of five, given the number of other group members that are acting as sentinels. If an individual's current energetic reserves exceed the line, then it should become a sentinel, otherwise it should forage. The maximum possible energetic reserve is 15 in this model.

nated guarding. In Tsavo, the pale chanting-goshawk uses the cover of trees and bushes to ambush from behind the traveling pack and exerts a potential predation pressure almost ten times higher than any other species (Rasa 1983, 1989b). This apparently explains why the mongoose sentinel focuses its vigilance 180° away from the foraging direction of the pack (Rasa 1989a,b). In the Serengeti, however, pale chanting-goshawk do not attack dwarf mongooses, and the mongooses never emit alarm calls in their presence (S. Creel pers. comm.).

Sentinel behavior seems to have arisen from a simple system of mutual benefit: well-fed animals bide their time looking out for predators, while the remaining group members exploit their companion's vigilance. Individuals benefit from specializing in one behavior at a time but are not locked into a particular caste as in eusocial insects. Although dwarf mongooses show some of the most elaborate forms of sentinel behavior, this system is highly facultative, so the sentinel system is unlikely to provide a platform on which group-level adaptations could be built.

Cooperative and Collaborative Hunting

A pride of lions sleeps near a waterhole as a reedbuck advances to drink. Stretched out on her side, one of the lions spots her prey, waits until it lowers its head, then rolls onto her chest, every muscle tense. The reedbuck looks up; the lioness stays frozen. The reedbuck walks toward the water, and the lioness starts her slow careful stalk, using every shred of cover. Head up. Freeze. After 10 minutes she is within 15 m; the rest of her pride sleeps soundly. The reedbuck starts to drink; the female pounces and bowls it over. Woken by the scuffle, her pridemates run to join her. She snarls and attempts to swat them away while keeping her jaws clamped firmly on the reedbuck's throat. Undeterred, her hungry companions rip open the prey's abdomen and eat most of the entrails and muscle. Finally, the hunter relinquishes her grip and stands panting in the heat of the day—the last member of the group to feed.

Group hunting frequently captures the popular imagination, and anecdotes abound of highly organized hunters working to seize a large dangerous prey. However, recent theoretical work has refocused attention on the conflicting costs and benefits for each individual in a hunting group, and empirical evidence is accumulating that shows truly *collaborative* hunting is far less widespread than previously supposed (Busse 1978; Packer and Ruttan 1988; Scheel and Packer 1991). Nevertheless, numerous species clearly do cooperate in specific situations. How can we account for this diversity, and to what extent does cooperative hunting indicate some higher level of cognition (or some other form of emerging complexity)?

How do we define cooperation in this context? At its simplest, group hunting can be said to be cooperative as long as two or more individuals simultaneously pursue the same prey animal. Cooperation will evolve as long as each individual gains a higher payoff by participating in a group hunt rather than by scavenging from a companion's kill (as in the above vignette). Beyond this mere simultaneity of prey pursuit, group members may actively coordinate the hunt by modifying their behavior according to the tactics of their companions.

SIMULTANEOUS/COOPERATIVE HUNTING

When a prey item is large enough to feed several foragers and the effort of prey capture incurs an inevitable cost, the advantage of joining an ongoing hunt depends on the extent to which an additional individual can improve its companions' chances of successful prey capture (Packer and Ruttan 1988). If one individual can capture the prey by itself, the contribution of a second hunter may be too low to overcome the costs of hunting, and the second animal's best option would be to "cheat" and hence wait on the sidelines until the prey has been captured. If prey capture is difficult, however, each additional hunter may make an important contribution, and simultaneous hunting can evolve even in the absence of kinship or long-term relationships.

Using data from a variety of animal species, Packer and Ruttan (1988) tested these predictions indirectly, using data on group-size specific hunting success as evidence of cooperation. In most species in which individual hunting success was high, groups performed no better than solitaries, but in species in which solitaries suffered poor hunting success, larger groups performed at rates that were consistent with a simple model of simultaneous cooperation (in which the success rate of a group of n individuals, h_n, is given by $h_n = 1 - (1 - h_1)^n$, with $h_1 = $ success of a solitary, and thus $1 - h_1 = $ failure of a solitary, and $(1 - h_1)^n = $ the chance that everyone fails simultaneously).

Subsequent studies have confirmed that groups generally do cooperate when individual hunting success is very low (e.g., Eklöv 1992; Fanshawe and FitzGibbon 1993). The most detailed data come from studies of African lions. In Serengeti National Park, individual lions often refrain from group hunts (i.e., cheat), but refrain least often during hunts of prey species that are most difficult to capture (e.g., Cape buffalo; Scheel and Packer 1991). In Etosha National Park, individual hunting success is far lower than elsewhere, and these lions show a much greater degree of cooperation (also see below). The hunting success and grouping patterns of the Serengeti and Etosha lions as well as data from a third study in Uganda (Van Orsdol 1981) are compared in figure 9.3.

The Etosha lions showed the greatest improvement in hunting success with increasing group size (fig. 9.3a). Only in Etosha were pairs more than twice as successful as solitaries, and trios were more than three times as successful (fig. 9.3b). As might be expected, the Etosha lions spent considerably more time in groups than did the other populations (fig. 9.3c). The root cause of these differences appears to be the harsh conditions of Etosha, where prey abundance is very low and the lions are forced to specialize on prey species that are very difficult to capture (East 1984).

Before considering more complex forms of group hunting, we want to make two points. First, it is surprising how often "cooperative" hunting seems to consist merely of several individuals hunting simultaneously. Additional hunters rarely contribute to the success rate of larger groups more than is predicted by the simple multiplicative model of h_n as outlined above. Second, even when animals have actively cooperated to capture a large prey, their feeding behavior is often competitive and disorganized (e.g., Kruuk

A. Hunting success of each group size

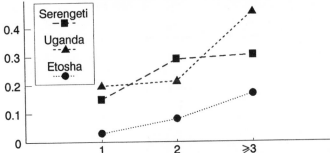

B. Group success divided by solitary success

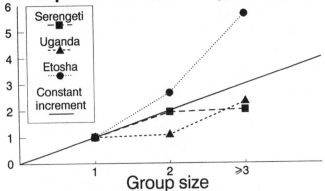

C. Time spent in groups of each size

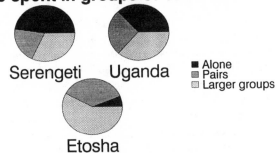

FIG. 9.3. Hunting success and grouping behavior in three different populations of African lions: Serengeti, Tanzania (Schaller 1972; Packer et al. 1990); Queen Elizabeth Park, Uganda (Van Orsdol 1981); and Etosha, Namibia (Stander 1992a,b). (A) Group-size specific hunting success. (B) Relative hunting success for each group size. (C) Proportion of time lions spent alone, in pairs, and in larger groups.

1972; Schaller 1972). Thus, their cooperation is context specific and does not lead to an overall increase in social complexity.

Coordinated/Collaborative Hunting

Hunting partners sometimes respond to each other's behavior, either recruiting additional companions in anticipation of a hunt or coordinating themselves during the hunt itself.

Active Recruitment of Hunting Partners

Zebra stallions vigorously defend their families from spotted hyenas, requiring the hyena to hunt them in large packs. Human observers can reliably predict when spotted hyenas are about to hunt zebra even if other prey species are nearby and no zebra are in sight (Kruuk 1972; Holekamp et al. in press; L. Frank pers. comm.). Before embarking on a zebra hunt, the hyenas assemble at "pep rallies," where they perform greeting ceremonies, scent marking, defecation, and social sniffing. Then one to two females often lead the group on long treks, ignoring easier-to-catch prey (such as wildebeest) in their path. Once they reach a zebra herd, they engage in a simultaneous (but not coordinated) hunt.

Although hyenas may appear to be making a group-level decision, these pep rallies arise from the fact that individuals in large hunting groups can expect to obtain a greater reward per capita by selecting zebra rather than wildebeest. Many more hyenas scavenge from a carcass than participate in prey capture, and zebra weigh around 40% more than wildebeest. Thus, even though hyenas form significantly larger groups when hunting zebra rather than wildebeest, feeding group size is comparable for both prey species. Hence, in areas of higher hyena density (and limited hunting opportunities each day), individuals gain higher payoffs by forming hunting parties large enough to catch the larger prey (Kruuk 1972; Holekamp et al. in press).

Again, it is significant that even though hyenas actively encourage the formation of large groups, their coordinated behavior breaks down once they pursue a specific prey animal. Mutual benefit encourages mutual participation but not any clear-cut division of labor.

Division of Labor During Group Hunts

Stander (1992a,b) described a remarkably stereotyped system of hunting behavior in the Etosha lions. These animals mostly hunted a single prey species, springbok, which was so small that the lions needed to capture several prey each day. Hunting in a homogeneous habitat, pride members fanned out to surround their prey, and certain individuals consistently approached from the left, others from the right, while the remainder approached directly. These "wings" and "centers" showed clear preferences and generally only altered their position if their group composition changed on a particular day. Most interestingly, hunting success appeared to depend on whether the lions were able to hunt in their preferred positions, suggesting that this division of labor conferred the sort of benefits that might lead to true specialization (Maynard Smith 1978; Oster and Wilson 1978).

In the Serengeti, complex hunting strategies are occasionally observed (e.g., Schaller 1972), but because these lions live in a more heterogeneous landscape and typically hunt several different species each day, their hunting techniques are more haphazard, and they often fail even to hunt simultaneously (Scheel and Packer 1991). The Etosha lions, on the other hand, probably needed to coordinate their cooperation owing to the low success rates of solitary hunters, and the sheer repetition of stalking a single species of prey may have enabled each lion to learn a specific tactic that reflected an associated skill: Heavier females were more likely to be "centers"; lighter females to be "wings." These Etosha prides lived in an extremely harsh environment and were unable to rear cubs during the study period (P. Stander pers. comm.), so it is noteworthy that their highly developed hunting strategies, having arisen in extremis, may have been invisible to natural selection.

Another example of coordinated hunting is provided by the chimpanzees of Tai National Park, Ivory Coast, West Africa (Boesch and Boesch 1989). These animals typically show complementary actions such as driving, blocking escape paths, and encirclement, although no data are available on the consistency of individual behavior. The Tai chimps "collaborated" in 68% of group hunts compared to less than 20% in two Tanzanian populations. As in the lion comparisons, the Tai chimps show lower success rates when hunting solitarily, and they more typically hunt in groups than chimps in the other populations (even recruiting distant companions to the hunt). Boesch (1994a,b) speculated that the greater degree of collaboration at Tai ultimately arises from ecological factors: The taller and thicker forest structure requires the chimps to work together to capture their arboreal prey. Boesch and Boesch (1989) also suggested that because chimps in this population are more consistently gregarious than their Tanzanian counterparts, they have greater opportunity to learn complex hunting tactics.

However, another interpretation of the Tai chimps' behavior is that these animals are less tolerant of scavenging by "bystanders" or "latecomers" (Boesch 1994b), and thus the individual who actually captures the prey gains a significant advantage. This leads to an important question. Is an apparent division of labor the result of each individual's maximizing its personal chances of prey capture (Busse 1978), or do group members coordinate themselves to maximize the success rate of the entire group? Are individuals ever willing to reduce their personal chances of prey capture in order to maximize the success rate of their companions? This may vary from species to species. In chimpanzees, the successful hunter achieves the lion's share, but in lions (as typified by our vignette), the killer typically continues throttling the prey long after its pridemates have started feeding.

Group-Group Competition

It has stopped raining. The young chimpanzee shakes himself off and resumes feeding. He occasionally scratches his chest and spits out a palm nut, which crashes through the fronds to the ground below, punctuating the dripping hiss of the forest. He is all alone in this narrow valley; his usual companions are somewhere off to the south. Then all hell breaks loose. Four males from a neighboring community charge toward the adolescent; he tries to escape, but two of the neighbors tackle him and hold him down. The other two take turns biting, kicking, and stomping on his neck, his back, his legs. They leave him broken and bleeding, dying on the forest floor, and return to their own territory, where they climb a large fig tree and set off a chorus of hooting and screaming.

Intergroup competition is widespread in vertebrates: Group territoriality occurs in carnivores (Kruuk and MacDonald 1985), rodents (Lacey and Sherman 1991), primates (Cheney 1987), birds (Davies and Houston 1981; Brown 1987; Black and Owen 1989), and fish (Clifton 1989). If groups fight to the death or for exclusive access to key resources, the self-interests of each group member will coincide with the interests of the entire group, as has been documented in coalitions of male lions (Grinnell et al. 1995). These coalitions compete intensively for access to female prides, and they are typically only able to maintain residence long enough to father one or two cohorts of offspring. Unrelated partners will form lifelong relationships and cooperate wholeheartedly even in situations when their behavior cannot be monitored by their partner. However, this behavior is driven by the existence of larger groups of cooperative kin (which forces solitaries to team up in order to remain competitive) and is limited by the increasing degree of within-group competition in larger groups (the larger kin groups will tolerate reproductive skew, whereas unrelated companions will not) (Packer et al. 1991).

Here we explore whether group-group competition has led to complex forms of cooperative behavior within each species. We then describe cases in which groups behave like individuals in higher-order levels of competition.

THE COMPLEXITY OF "US" AGAINST "THEM"

The remarkable patroling behavior and organized gang warfare of male chimpanzees (Goodall et al. 1979; Goodall 1986) provides the best-known example of complexity in this context. Males seem to decide in advance when to make a foray to a territorial boundary. They seek out and stalk their quarry, usually a lone member of the neighboring community. The marauding males remain unusually silent and stealthy until launching a sudden attack, and those males who hold down the opponent make it easier for their companions to administer a coup de grâce. Of the few attacks that have been directly observed, specific individuals do not consistently show the same tactic (although certain individuals appear to be more actively aggressive). The caution with which they set out to maim or kill their opponent probably serves to minimize the risk of injury to their companions as well as to themselves. This is a dangerous task and is clearly an emergent property of group living: The entire group must keep themselves healthy in order to overpower even a single opponent.

Even in the case of group territoriality, however, individual costs and benefits may not always coincide, and intergroup conflicts may often involve a considerable degree of within-group decision making. When a large group greatly outnumbers its opponents, a lone defector might be able to gain the resource without paying any costs of territorial defense. Female lions compete against their neighbors for access to land, and larger groups dominate smaller ones (McComb et al. 1994). Recent studies show, however, that not all individuals pull their weight when confronted by strangers. During playback experiments, certain animals routinely hang back during their approach to the territorial invaders (Heinsohn and Packer 1995). Others (nicknamed "friends in need") participate when their assistance would be most likely to influence the odds of winning the encounter. Still others ("fair-weather friends") participate most often when their group safely outnumbers the opposing group.

No theoretical model currently exists that can account for such a diversity of individual strategies during intergroup encounters, but these results emphasize the fact that individuals probably weigh personal costs and benefits before deciding to participate in a territorial dispute.

Nevertheless, some form of group selection will be expected to operate in this context. Successful groups may annihilate unsuccessful groups, and traits that promote individual survival will also promote survival of the entire group. Intergroup conflict therefore involves relatively little opposition between individual advantage and the good of the group. Because the outcome of these conflicts may depend on the emerging properties of group-level competition, the question remains whether a group-oriented perspective is necessary to explain the evolution of these properties or if they can best be understood by emphasizing individual fitness. Perhaps an individual-fitness approach similar to Bednekoff's models of sentinel behavior would reveal the factors necessary to produce a division of labor and other characteristics of group warfare.

HIGHER LEVELS OF INTERGROUP COMPETITION

Resource competition is often restricted to disputes between members of the same group or struggles between neighboring groups. But individuals from separate groups may sometimes cooperate in competition against a common enemy (e.g., rock pipits, in which adjacent territory holders cooperate to evict potential newcomers; Elfström 1997), and entire groups may temporarily coalesce to form second- or third-order alliances. In fact, the spatially discrete social units of several vertebrate species actually consist of collections of matrilineal alliances, and thus more species show a multitiered social system than is generally recognized. These higher-order alliances are reminiscent of tribal societies in humans and involve some of the most intelligent mammalian species, including primates, carnivores, elephants, and cetaceans.

Social complexity is increasingly seen as the driving force in the evolution of intelligence, with a large brain size being required to track multiple social relationships (Byrne and Whitten 1988; Barton 1996). Boehm (1992, 1997) argued that the quality of these relationships has also been important in the evolution of our own species, specifically in permitting the development of enforced egalitarianism. Thus, the despotic tendencies of more powerful members of human society are countered by cultural traditions that create a powerful force for consensus and minimize the fitness outcome of such phenotypic variation. As outlined below, however, most multitiered vertebrate societies probably exist precisely because they are *not* egalitarian: Each individual is ordered according to dominance rank and participates in ever higher levels of social organization in a nested series of group-ordered dominance relationships.

Primate troops and hyena clans show a complex social organization based on family-level alliances (Chapais 1992; Frank et al. 1995). Mothers support their daughters in disputes against other families so that members of the same matriline enjoy adjacent rank in an overall hierarchy. But adjacently ranked families will cooperate against subordinate matrilines to maintain the status quo. Experiments in Japanese macaques (Chapais et al. 1991) showed that a second-ranking matriline benefits by allying itself with the top-ranked matriline in order to prevent the formation of a "bridging alliance" between the first- and third-ranked matrilines. By cooperating with the second-ranked matriline, the top-ranked matriline creates a state of dependency that fore-

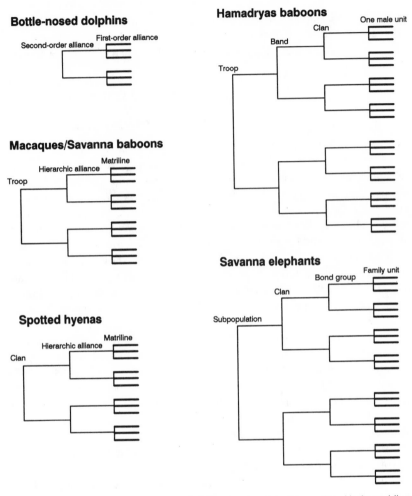

Fig. 9.4. Multitiered social systems exist in various vertebrate species. Horizontal lines indicate individuals or groups; vertical lines link cooperative partners.

stalls a "revolutionary alliance" between the second- and third-ranked matrilines. Despite these internal divisions, the entire troop or clan will cooperate during clashes with neighboring groups. In each level of complexity (summarized in fig. 9.4), individuals apparently work together according to the usual rules of genetic self interest.

In Hamadryas baboons (Kummer 1968; Sigg et al. 1982; Abegglen 1984; Stammbach 1987) and African elephants (Moss and Poole 1983; Moss 1988; Poole et al. 1988), discrete social groups show second-, third-, and fourth-level alliances (fig. 9.4). The primary social group of Hamadryas baboons is the "unit group," which consists of a single adult male, several females, and their dependent offspring. These units often compete against each other, but then coalesce into "clans" that compete against other units or clans for access to females or food. Clans merge to form "bands," which in turn form "troops." Bands and troops compete against each other for access to waterholes and sleeping cliffs. Intertroop and interband encounters can seem quite organized, with males from each clan lined up side by side in front of their females and the subadult males advancing in the lead.

In elephants, the family groups aggregate to form "bond groups," and social interactions between these individuals are just as intense as between family members (after a separation, these individuals greet each other with trumpeting, ear flapping, trunk entwining, and excited defecation; while apart, they seem to coordinate movements over long distance using low-frequency sound). Bond groups associate preferentially in a "clan," whose members interact frequently and greet each other by calmly putting their trunks in each others' mouths. Clans coalesce as a "subpopulation," whose members are intolerant of members from other subpopulations. These higher-order groups are most prominent during periods of food abundance but then disperse when food supplies diminish.

Elephant family groups (and, probably, bond groups) are matrilines (Moss and Poole 1983), and males of the Hamadryas clans are suspected to be brothers (Sigg et al. 1982; Abegglen 1984), but it is not yet known whether their higher-order alliances are also based on genetic relatedness. Even in the absence of kinship, individuals can benefit from preferential alliances because of the advantages of a clear-cut dominance relationship. As resources diminish, high-ranking groups may first align themselves with known subordinates to evict competitors of unknown fighting ability; then the dominants can oust the subordinates when the resource is only large enough for a single group.

Not all multitiered systems are obviously despotic. In some social species, higher-order alliances may essentially involve a lottery in which groups cooperate with other groups without necessarily knowing who will win. For example, male bottle-nosed dolphins form stable first-order alliances of two to three individuals, which coalesce to form second-order alliances during competition for receptive females (Connor et al. 1992). These second-order alliances will steal females from other groupings, and only one first-order alliance ultimately herds the female. Two members of the winning coalition may then mate with her simultaneously, and if there is a third male in the coalition, the identity of the "odd man out" changes from one takeover to another (Connor and Smolker 1995).

No paternity tests have yet been performed in dolphins, behavioral observations are sparse, and the kinship between coalition partners is unknown. It is possible, however, that these primary coalitions arise from mutualistic advantage. If each male in these alliances has an equal chance of fathering each offspring, then cooperation will pay as long as coalitions gain significantly greater access to females than do solitaries. A similar argument applies to the temporary coalitions of savanna baboons (Bercovitch 1988; Nöe 1990), in which pairs of males are much more successful at taking over estrous females than are lone challengers. In the dolphins, the first-order alliance is truly analogous to an individual baboon, and their extraordinary levels of cooperation apparently stem from the difficulty of sequestering receptive females in this species.

Multitiered vertebrate societies show many important similarities to human social organization, and it is striking how many of these examples involve long-lived intelligent animals. However, most of these societies are readily understood in terms of a few simple rules of despotic and nepotistic behavior. Even in cases where higher-order alliances appear to be egalitarian, they are the product of a lottery, not a social contract.

Prospecting for New Pinnacles of Complexity

The sun has set, and a herd of Cape buffalo moves away from the river, looking for a spot to bed down for the night. The bulls, cows, and calves chew their cud and occasionally call to each other. Suddenly, everyone is startled by a bright light in the sky. The herd springs into action. The bulls approach the disturbance and start thrashing the ground with their horns. The cows and calves trundle back down toward the river. An astonished biologist records the scene.

"My god, what's that object in the sky? A spaceship? Oh heavens, it's landed, and those bull buffalo—they've started digging a trench! And the cows, the cows have constructed some sort of suspension bridge. They're carrying their calves across to the other side!"

The door of the spaceship opens, and an alien voice announces, "We are Borg. You will be assimilated. Resistance is futile."

The idea that a buffalo herd could organize a complex defensive response is even more ludicrous than the notion of a pompous space alien landing in the Serengeti. (We chose the Borg because they regularly threaten the characters of "Star Trek" with a group-level cognition. Individuals function as neurons within a collective intelligence.)

True group-level adaptations/cognition in any animal as intelligent as a mammal should be elaborate, conspicuous, and unequivocal. If insects can organize fungal gardens, warfare, and bridge building, we should surely see "emerging properties" at least as impressive in the vertebrates. In our survey of some of the most complex vertebrate social behaviors, we found little evidence of such group-level adaptations. Self-interested behavior could always be seen to mold the form of society not vice versa.

With the exception of naked mole-rats, we could find no example of a division of labor that involved a long-term specialization: Every individual could alternate between foraging and acting as a sentinel, between hunting as a "center" and a "wing," and among a variety of tactics during a gang attack. The Etosha lions showed the most persistent specializations, but even here specific individuals modified their behavior in response to changes in group composition.

Are most vertebrates "generalists" because their group-level responses are too rudimentary to require specialization, or is group-level cooperation relatively undeveloped because individuals are selected to be generalists? In the absence of eusociality, all these animals are capable of independent breeding, and each individual seems well equipped to solve a variety of problems. Cooperative group sizes are typically so small that reliance on a specialist might be disadvantageous. Imagine a musical quartet in which each musician can only play a single instrument; if one person dies, the surviving trio might be unable to perform properly. In ensembles large enough to be safely redundant (like an orchestra), however, the advantages of mutualism or the force of kin selection will generally be too weak to maintain a group-level degree of cooperation.

We also could find no example of group-level complexity that justifies the sort of "group mind" envisioned by Wilson (1997b) and other neo-group selectionists. For example, Prins (1996) has suggested that Cape buffalo show "voting behavior," wherein several hundred animals assess each other's preferred destination before moving off as a single herd each day. Although group-progression patterns may indeed involve some form of collective decision making, we see no reason to invoke anything beyond a simple set of individual decision rules (e.g., chimpanzee social organization: Te Boekhorst and Hogeweg 1994; task allocation in social insects: Pacala et al. 1996).

Finally, we could find no compelling evidence that a vertebrate social system ever exceeds the sum of its parts. The coordinated evasion of dunlin flocks and fish schools is impressive only because each individual benefits from responding to the behavior of its neighbors. The most elaborate social organizations illustrated in figure 9.4 only require an ability to recognize a large number of individuals (rather than any form of group-level cognition). Though intergroup competition might be expected to provide the best possible context for group-level phenomena, group-territorial lions are hardly a paragon of cooperation, riddled as they are with "friends in need" and "fair-weather friends."

Outside of the possible exception of eusociality (with naked mole-rats providing the best-known example in vertebrates; Jarvis and Bennett 1991), the landscape of vertebrate social evolution is dynamic but ultimately leveled by the forces of self-interest.

Acknowledgments

Scott Creel, Laurence Frank, Kay Holekamp, Cynthia Moss, and Philip Stander generously provided unpublished information. We also appreciated the insightful comments of Peter Bednekoff, Laurent Keller, David Sloan Wilson, and three anonymous reviewers.

10 Conflict and Cooperation in Human Societies

John Maynard Smith

Human society is difficult to analyze because it is determined by two parallel systems of inheritance, genetic and cultural, whereby information is transmitted between generations and between individuals. One does not have to accept Marx's thesis that "Man's being determines his consciousness" to accept that what people believe, and how they behave, is culturally influenced. Yet it should be equally clear that our ability to be influenced by culture, and to construct complex societies, depends on our genetic makeup. The reason why humans live in complex societies, dependent on many learned skills, whereas chimpanzees live in rather simple ones, is ultimately that humans and chimpanzees are genetically different.

It does not follow that sociology can, or should, become a branch of biology. The relation between sociology and biology resembles that between biology and chemistry. No one expects biology to become a branch of chemistry, essentially because the presence of heredity in living systems causes the appearance of functional and adaptive properties that are absent in purely chemical or physical systems. Yet no one today could argue that biologists can get along without a knowledge of chemistry. The same ought to be true of sociology in relation to biology, although as yet it is not. We should not expect sociology to become a branch of biology, but it may well be that sociology could gain as much from genetics as genetics has already gained from chemistry.

Here I discuss human societies from three points of view, and I ask what can be learned by comparing the behavior of humans with that of other primates. The papers in Runciman et al. (1996), in particular that by R. A. Foley, treat this question in greater detail. Next, I discuss the nature of language, which is the basis of cultural, as opposed to genetic, inheritance. The books by Bickerton (1990), Jackendoff (1993), and Pinker (1994) are the sources of this section. In the next section, I present some simple, formal models of cultural cooperation. What is borrowed from biology is not any particular set of facts, but the method, originating with Darwin, of explaining complex phenomena by simple models. In the context of human social behavior, this approach was pioneered by Cavalli-Sforza and Feldman (1981) and by Boyd and Richerson (1985): the particular models described were first formulated in Maynard Smith (1983a). Finally, I discuss the difficulties that arise in modeling a society with a dual inheritance system—genetic and cultural—and with multiple levels at which selection can act.

The Comparative Behavior of Primates

Hinde (1976) emphasized that sociality in humans, and in some mammals and birds, depends on the formation of relationships between individually recognized conspecifics. In this, human societies differ from insect societies, which do not depend on individual recognition, although recognition of colony membership is present. Foley (1996) pointed out that sociality in Hinde's sense is a characteristic of anthropoid primates. Of over 175 species, only the orangutan is solitary, and even in this species, there are affinities between specific females within discrete home ranges (Galdikas 1985).

In the Old World monkeys (Cercopithecoidea), social organization depends on female kin bonding. In contrast, hominoids are socially diverse, ranging from monogamy in Gibbons, a one-male harem system in the gorilla, and a social group consisting of related males and unrelated females in the chimpanzee. Social systems in humans are so diverse that it is hard to be confident about what system(s) were present in ancestral hominids. Foley (1996) suggested that male kin bonding evolved in the human-chimpanzee clade, and hence that a patrilineal system is ancestral for hominids. Figure 10.1 illustrates his conclusions, based on a phylogenetic analysis of the behaviors of existing primates.

It is customary to emphasize the importance of increasing brain size in human evolution. However, the Australopithecines, the first hominids, show a relatively slight increase, when allowance is made for body size. The "encephalization quotient" is a little over 2.0 in extant apes, and ranges from 2.1 to 3.4 in Australopithecines. In *Homo ergaster*, at 1.6 MY, it is 3.3, after which there was a gradual increase over the next million years. The most rapid increase, however, has occurred only during the last 300,000 years. Many explanations have been proposed for this initially gradual, and more recently accelerated, increase in brain size: for example, increased group size requiring greater complexity of social interactions; skill in tool manufacture; the origin of linguistic competence. It is not necessary to choose between these and other explanations, because all may have been relevant, and they may have interacted: for example, language is obviously relevant to social interactions.

There is, however, less controversy about the physiological effects of an increase in brain size. The brain is a metabolically expensive organ, and this confronts a mother with problems in providing nutrients for her baby, before and after birth. Increased brain size has led to changes in life-history characteristics: a longer period during which the infant is dependent; longer interbirth intervals; delayed first reproduction; and greater longevity. If infant survival required male as well as female investment, this would have favored the evolution of long-term bonding between a male and one or several females.

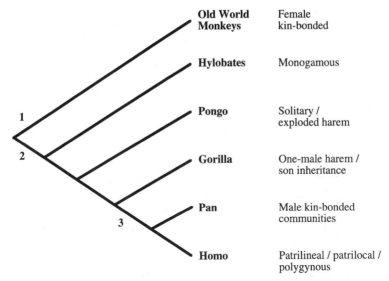

FIG. 10.1. Evolutionary relationships of Old World monkeys and apes, showing social systems. Female kin-bonding evolved only in the Old World monkey (1). Male residence appears to have been established among basal hominids (2), and male kin-bonding evolved in the chimpanzee-human clade (3). After Foley (1996; with permission of Oxford University Press).

The last 50,000 years have seen the emergence of a new phenomenon, continuous cultural change. Occasional cultural changes occur and are transmitted in other animals. What is new in humans is that such cultural innovations are cumulative, leading to behavior that no individual could learn on its own (Boyd and Richerson 1996). Cultural inheritance is not uncommon in other animals. For example, young rats acquire a preference for foods that they smell on the pelage of other rats (Galef 1988). Boyd and Richerson suggest that cumulative cultural change requires "observational learning": that is, a young individual can learn by observing the behavior of another. This contrasts with "local enhancement," which is all that is required to explain the acquisition of a new food preference in rats (Galef 1988) or the habit of opening milk bottles in great tits. Local enhancement can lead to the cultural transmission of a habit, but observational learning is required for the continued improvement of a technique.

Observational learning is not unknown in nonhuman animals. In birds, it is the basis of song dialects, but seems to be limited to that context. Boesch (1996) argued that a degree of observational learning is present in chimpanzees. He gave the following example. Some, but not all, populations of chimpanzees dip sticks into the nests of driver ants, and feed on the ants that crawl up the sticks. The chimpanzees in Gombe use a different technique from those in Tai and catch about four times as many ants per minute. Local enhancement could explain why ant-dipping is present only in some populations, but not why Tai chimps continue to use a less efficient technique, when there is nothing to stop them from using the more efficient one. But this is what we expect if young chimps acquire the technique by copying their elders. So it seems that observational learning, which is characteristic of humans, is already foreshadowed in chimpanzees.

The Nature of Language

If cumulative cultural change is the characteristic feature of human society, the capacity for language is the genetically specified property that makes it possible. Symbolic communication is common among animals. In social insects, the meanings of the symbols are genetically determined: An individual ant does not have to learn when to give a particular signal, or how to interpret one. It seems that some human signals are genetic in this sense: Facial expressions signifying pleasure, pain, anger, and so on are universal across human cultures, and their significance does not have to be learned (Ekman 1973). But the meanings of words must be learned. In captivity, apes, parrots, and dolphins have been taught to respond appropriately to a number of arbitrary signals, although of these only parrots have the motor and sensory skills to use human words efficiently. We are still disturbingly ignorant of the use of symbols in the wild by these animals. Vervet monkeys (Cheney and Seyfarth 1990) use specific sounds to indicate different classes of predators (eagles, snakes, ground predators) and the approach of other monkeys. The ability to make these signals has a genetic basis, but young monkeys must learn their detailed application. For example, adult monkeys give the "eagle" call almost exclusively to raptors, and mainly to the martial eagle, their main predator. In contrast, infants give the call to flying objects, but not exclusively to raptors; they may give it to nonraptors—for example, vultures and bee-eaters, and even to a flying leaf blowing in the wind.

It seems, then, that some apes can learn to use symbols in captivity, although their vocabulary is small compared to that of humans, and there is little evidence as yet that they use learned symbols in the wild. However, the ability to learn the meanings of a large number of symbols, although important, is not the essential feature of human language. That feature is syntax, defined by the Oxford English Dictionary as "the order of words in which they convey meaning collectively by their connection and relation." Speakers of English have no difficulty in distinguishing the meanings of the following sentences:

John gave the book to Mary.
Mary gave the book to John.
Did John give the book to Mary?

Which book did John give to Mary?
John did not give the book to Mary.

To understand these sentences, we must learn rules. Often, we can obey a grammatical rule without being consciously able to formulate it. For example, we all know that the first of these two sentences is grammatical, but the second is not:

How do you know which book John gave to Mary?
Which do you know how book John gave to Mary?

Yet few of us could formulate a rule that forbids the second sentence.

The first point to grasp is that the existence of grammatical rules makes it possible to convey an indefinitely large number of meanings with a finite number of symbols. A speaker will typically use from ten to one hundred thousand words, constructed from as few as thirty unit sounds or "phonemes." Furthermore, by arranging these words, an indefinitely large number of sentences can be constructed. The analogy with the genetic code is striking. It, too, is composed of a small set of units (in fact, only four), but by the sequence of these units, it can specify an indefinite number of proteins, and, through their interactions, an indefinite number of morphologies. The two signaling systems, linguistic and genetic, are the basis, respectively, of cultural and genetic transmission.

There is still debate about whether captive chimpanzees or other animals can learn elementary grammatical rules—for example, concerning word order—but there is no evidence for the use of such rules in the wild. There is also debate about how humans acquire grammar. It is agreed that they learn rules—for example, for forming questions, negatives, etc. There is no way in which children could learn the meaning of every possible sentence independently. But is our ability to learn grammatical rules merely an aspect of our general ability to learn by "reinforcement"—that is, by rewards and punishments—or do we have a special capacity to learn language, what Chomsky (1975) has called a "languge organ"?

There are a number of reasons for thinking that there is a competence peculiar to language.

1. Children learn language rapidly and with little reinforcement. They are rarely punished for grammatical mistakes. Thus, a child will be reproved for bad manners but not for bad grammar: for saying "fuck you," which is grammatical, but not for "I don't want no prunes," which is not.

2. There is a critical period for learning to talk. Children deprived of any linguistic input until adolescence never learn to talk grammatically, although their general ability to learn remains.

3. We learn rules that we cannot formulate consciously. (This argument is not decisive; we also learn to ride a bicycle, without knowing how we do it.)

4. Brain injury, from stroke and other causes, can lead to impairmant of lingustic ability without affecting general intelligence, or can impair intelligence while leaving linguistic ability unaffected.

5. There is recent evidence that the region of the brain responsible for learning grammatical rules is different from that responsible for storing the meanings of words. Consider, for example, the rule for learning the past tense in English. A child who has learned that the past of "walk" is "walked" will generalize and form the past of "pick" as "picked" and "watch" as "watched." She may also apply the rule, wrongly, to an irregular verb, and say "runned" instead of "ran," or "goed" instead of "went." Thus, most past tenses can be formed by following the rule "add -ed," but the past tenses of irregular verbs must be learned as dictionary items. It now appears (Pinker 1997) that brain damage to the left anterior cortex impairs the ability to form the rule, but not to learn the past tenses of irregular verbs, whereas damage to the temporal and posterior cortices impairs the knowledge of irregular verbs, but not of the rule.

6. Cases are known of genetically determined "specific language impairment" (Gopnik 1990). Individuals with an autosomal dominant mutation do not learn the rule for finding the past tense (or for other morphological word changes such as adding -s to form a plural) but can have normal general intelligence.

For these reasons, the idea that there is a specific language organ is now widely, although not universally, accepted. An evolutionary biologist is bound to ask what this organ was doing before it was used for learning language. In evolution, new organs usually arise as modifications of preexisting organs: Wings are modified legs, and teeth are modified scales. We do not know what the structure ancestral to the language organ was doing before it acquired its present function. In the long run, the answer is likely to come from genetics. If there is one gene concerned specifically with language, there are likely to be many. What do the homologues of these genes do in other mammals? It will be some time before we can answer this question.

Models of Human Cooperation

In this section, I shall discuss some simple models of human cooperation. This is a somewhat arrogant procedure. The problems have been debated from the time of Plato and Aristotle. Why should a biologist have anything new to say? If I have an excuse, it is that biologists have acquired a conviction that simple models can help to explain complex phenomena. Indeed, I would go further and argue that only simple models can help to explain complex phenomena. In the words of Boyd and Richerson (1985, p. 25), "To substitute an ill-understood model of the world for the ill-understood world is not progress." To be useful, models must be simple.

I first discuss what I call "models of rational behavior." The basic assumption is that individuals are behaving rationally, in their own best interests. The question is whether rational behavior can lead to cooperation. The particular model I discuss, the Social Contract Game, has (at least) two weaknesses. The first is that it assumes that all individuals have the same opportunities: Technically, they have the same "strategy set" available. Often, this will not be true. I therefore turn to a second game, the Class War Game. Although couched in terms of differences between economic classes, it could equally well apply to differences between other groups, for example, ethnic groups. What emerges from this analysis is the selective advantage of group loyalty. This brings into focus a second weakness of the Social Contract Game, namely the assumption that human behavior is governed by rational self-interest. In practice, myth and ritual are also important influences on behavior, particularly in conferring group loyalty. The section therefore ends with a discussion of the role of myth and ritual in human society.

RATIONAL GAMES

"The problem of organizing a state, however hard it may seem, can be solved even for a race of devils, *if only they are intelligent.* The problem is, given a multitude of rational beings requiring universal laws for their preservation, but each of whom is secretly inclined to exempt himself from them, to establish a constitution in such a way that, although their private intentions conflict, they check each other, with the result that their public conduct is the same as if they had no such intentions" (Kant; my emphasis).

What can Kant have had in mind? More precisely, how could a set of intelligent individuals be induced to cooperate to produce a result beneficial to them all? Suppose, for simplicity, that to "cooperate," C, means not to park on a yellow line, and "defect," D, means to do so. Then the payoffs to an individual might be as in table 10.1. This is a typical Prisoner's Dilemma Game. No matter what others are doing, it pays to defect: It always pays an individual to park on the yellow line. But if everyone does so, the roads are blocked: Everyone is better off if all cooperate than if all defect. But how is cooperation to be enforced?

A possible solution might be the Social Contract strategy. Every individual agrees to a contract, "I will cooperate; I will join in punishing anyone who does not." Sadly, this strategy is not stable against the "free rider." Thus, suppose that the advantage to everyone of clear roads is $+20$; the benefit from parking on double yellow lines is $+30$; the cost of being punished is -50; and of joining in punishing is -5. If all other individuals

TABLE 10.1
The Prisoner's Dilemma Game

Payoff to Individual	Policy of the Rest of the Group	
	C	D
C	20	0
D	30	10

kept to the contract, and if defectors are always detected, the payoffs to an individual are:

Cooperate and punish	$20 - 5 = 15$
Defect	$30 - 50 = -20$
Cooperate, do not punish	$= 20$

In other words, the contract is not stable against the free rider, who does not park on yellow lines, but who also does not join in punishing those who do. To be stable, the contract must be revised to read, "I will cooperate; I will join in punishing anyone who does not; I will treat as a defector anyone who does not join in punishing."

This illustrates Kant's thesis; it is possible for intelligent but selfish individuals to cooperate. Before discussing the weaknesses of this model, it is worth asking what "intelligence" means in this context. What properties must individuals have to make the Social Contract strategy possible. They must have language and a "theory of mind." Language is clearly needed, because "cooperation," which refers to a cultural trait (not parking on yellow lines), must be defined and communicated. To say that I have a theory of mind is to assert that I assume that others have a mind like mine, with similar desires and powers of reasoning. If I did not have such a theory, I would not attempt to persuade others to join me in agreeing to the contract; after all, why should they? There are real difficulties in deciding whether animals have such a theory of mind. For example, Cheney and Seyfarth (1996) argued, from studies of vervet monkeys and baboons, that there is no reason to think that a monkey, when giving a call, has in mind that another monkey may hear the call and be influenced by it. It may be that a theory of mind, like language, is unique to humans.

One weakness of the Social Contract model is that it assumes that the only reason why individuals do not do something that would be in their immediate self-interest is the fear of punishment, or, more generally, that human behavior is based solely on reason. In fact, much behavior is influenced by ritually reinforced social customs. A second weakness is that the model assumes that all participants are equal, in the sense of having the same desires and the same possible actions. This will seldom be the case. The implications of inequality are discussed in the next section. To motivate the discussion, consider the following modification to the Social Contract game. Some bright, politically minded person says, "Why should we go on punishing? It is inconvenient and costly. Let us employ a police force, and pay them to do it for us. Of course, we will have to supply them with batons." Everyone agees with this sensible proposal, and a police force is established. Then some bright, politically minded policeman says to his fellows, "Look, we have got batons, and they haven't. We will punish them if they park on yellow lines, but we can do what we like." This, of course, is a brief history of the world so far.

THE SUBDIVISION OF SOCIETY

In modern human societies, not all individuals have the same strategy set available to them. The most obvious differences are of social class: the strategies open to the owner of land or a factory are different to those open to a nonowner. This introduces a new logic, with interesting consequences. For the present, I continue to assume that individuals behave rationally. Consider the following simple Class War Game, in which there are only two groups and two strategies open to each. To make it easy to follow, imagine a game between factory owners and workers, but the logic is more general. Owners can pay high wages, PH, or low wages, PL. Individual workers can refuse to work for low wages and accept only high wages, AH, or they can accept low wages, AL. Owners who can find no workers, or workers who can find no job, get zero payoff. Owners who can find workers at low wages get 4, and at high wages get 3; workers paid high wages get 2, and those paid low wages get 1.

These payoffs are summarized in figure 10.2. We seek stable states of the system, in which most owners adopt the same strategy, most workers adopt the same strategy, and it would not pay an individual in either group to alter his strategy. There are two stable states. In one, owners pay high wages, and workers accept only high wages. In the other, owners pay low wages, and workers accept them. There are two general points about this game.

1. In either of the stable states, owners do better than workers. Stability, therefore, requires that an individual cannot choose which group to belong to. This implies laws to protect property, and a police force to enforce them.

2. Owners prefer one stable state, and workers the other.

Hence, if we want to understand the stability of such a system, we have to ask which group(s) benefit from the status quo, how they exercise power, and how group membership is controlled. To understand what leads to change, we must ask which groups benefit from change, and whether there are technical

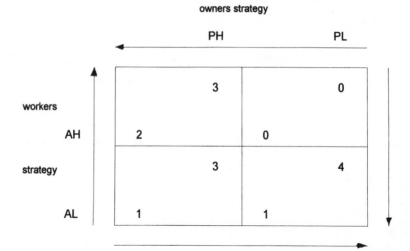

FIG. 10.2. Payoffs in the "class war" game. The strategies open to owners are PH (pay high wages) and PL (pay low wages); strategies open to workers are AH (accept only high wages) and AL (accept low wages).

developments that lead to changes in payoffs and hence destabilize a previously stable system. We also have to ask how group membership is determined.

However, the most important point to emerge from the model is that, because members of different groups benefit from different outcomes, there will be benefits to members of any group that can ensure that its members work together for the benefit of that group, even if this does not benefit all members of the society. Hence, we can expect the emergence of something analogous to the social contract discussed above binding together groups within a larger society.

We can expect the emergence of customs and rules enforcing cooperation among the members of society as a whole, and of groups within that society. Thus far, I have supposed that such customs are agreed to by rational discussion and enforced by the threat of punishment. There is, however, a second set of processes enforcing social customs, which may be much stronger: These are myth and ritual.

MYTH AND RITUAL

Gellner (1988) offered the following account of ritual.

The way in which you restrain people from doing a wide variety of things, not compatible with the social order of which they are members, is that you subject them to ritual. The process is simple: you make them dance round a totem pole until they are wild with excitement, and become jellies in the hysteria of collective frenzy, you enhance their emotional state by any device, by all locally available audio-visual aids, drugs, dance, music, and so on; and once they are really high, you stamp upon their minds the type of concept or notion to which they subsequently become enslaved.

The idea that myth and ritual are necessary for the regulation of society is not new. In the *Republic*, Plato recommended that the ruler invent a "noble lie," and then persuade people to accept it by "speech, story, and song." In *Gulliver's Travels*, Swift satirized the role of ritual by imagining a society split into two groups, characterized by whether they opened their boiled eggs at the big end or the little end. The power of myth and ritual in ensuring group loyalty is perhaps most obvious in those cases, such as in Bosnia or Northern Ireland, where there is little to bind a group together except shared symbols, rituals, and historical myths. The process in its purest form is exemplified by organizations such as the Freemasons, in which a group of individuals who initially had nothing in common are united in charitable work, and in the pursuit of a common self-interest, by nothing except a shared ritual.

Types of Inheritance, and Levels of Selection, in Human Societies

Human society is hard to understand for two reasons. First, there are two ways, cultural and genetic, whereby information is transmitted between generations. Second, selection acts at different levels, between individuals, families, groups, and societies.

For human history, it is reasonable to regard genetic predispositions as constant. Modern societies differ from that of ancient Rome for cultural, not genetic, reasons. All types of society, however, depend on the fact that humans can be indoctrinated by myth and ritual. Clearly, this depends on language, but there remains the question why humans are so easily moved by music, dance, visual symbols, and rhetoric. This susceptibility is an evolved human characteristic. Is it an accident, a mere spandrel, as Gould and Lewontin (1979) might argue? That is, is it an unselected consequence of other characteristics—for example, language—that were selected? It is hard to rule this out. My own view is that a selective explanation is more plausible. For many millions of years, our ancestors lived in small groups. Any genetically influenced characteristics enabling the members of a group to cooperate more effectively, either in surviving in a hard environment or in competition with other groups, might be favored by selection. There would certainly be selection favoring more cooperative groups, but this might be genetically ineffective if, as is typical in mammals, groups were not reproductively isolated, but exchanged members. Individual selection would be effective only if groups acquired the habit of punishing individuals who did not fit in with group customs, as suggested by the Social Contract game. This amounts to arguing that the capacity of humans to be socialized (or indoctrinated, depending on your viewpoint) evolved by individual selection, because those who lacked it were unsuccessful in a social environment.

The particular beliefs, symbols, and customs of a group would be culturally inherited, and also culturally selected, in that some beliefs would favor group survival. There is an obvious conflict, analogous to intragenomic conflict, between behavior favoring the group and that favoring the individual. Many myths have the clear effect of inducing behavior favorable to the group: A clear example is the belief that a man who dies in battle for the faith will live forever in paradise. Thus, myths have often been favored by cultural group selection. In practice, however, such myths do not wholly dominate individually motivated behavior. Modern societies, therefore, sometimes adopt a more Kantian approach, attempting to formulate laws ensuring that, if individuals do pursue their own self-interest, the result will be socially desirable: Taxes discouraging environmental pollution are an example. The matter is further complicated as societies grow larger and as subgroups appear within them. Not only is there conflict between individual self-interest and group advantage, but between groups at different levels in the hierarchy: for example, between family, ethnic, and national loyalties.

11 Laws Governing Species Interactions? Encouragement and Caution from Figs and Their Associates

Edward Allen Herre

All organisms interact with members of other species. The diversity of forms that those interactions assume is overwhelming and offers the clearest reflection of the diversity of life itself. Directly or indirectly, parasitic organisms constantly affect virtually all ecological and evolutionary processes. In counterpoint, mutualistic interactions are ubiquitous, their members often comprise ecologically dominant members of communities, and, as with parasitisms, they exert a profound influence on essentially all levels of biological organization. However, the theoretical and empirical challanges presented by the ecology and evolution of interactions among species are equal to or greater than those presented by within-species or within-genome interactions.

For example, it is widely appreciated that mutualistic relationships usually incorporate parasitic aspects, and that the converse can be equally true for parasitisms. Furthermore, the mixed nature of some of these relationships, as well as the existence of transitional intermediates, strongly suggests that outcomes of species interactions can be both ecologically and evolutionarily quite fluid (e.g., Herre 1989; Compton et al. 1991; Thompson 1994; Bronstein and Hossaert-McKey 1996; Herre et al. 1996; Nefdt and Compton 1996; Pellmyr et al. 1996; Herre and West 1997). The fundamental question is whether there are any general rules that govern the ecological and evolutionary trajectories and outcomes of interactions, or if there is simply a large collection of special cases, with no overriding principles.

After mentioning some important general properties of parasitisms and mutualisms, I present a brief overview of factors that theory suggests ought to influence evolutionary outcomes of interactions, such as patterns of ecological transmission and degree of co-speciation. I then present relevant aspects of the natural history of a series of mutualists and parasites that are associated with figs. For each group of species, I examine whether the predictions of the theory do or (in many cases) do not correspond to the degree to which an interaction is parasitic or mutualistic. Further, in keeping with one of the motivating themes of this book, I also discuss how selection at different levels of population structure influences both sex-ratio adaptations in the wasps, and the expression of virulence in the nematode parasites of those wasps (both of which influence the relationship of these organisms with the fig). Moreover, I emphasize that selection pressures resulting from different components of the fig-pollinator, parasite, nematode, and seed-disperser relationships can interact with each other in complex and often unexpected ways.

Parasitisms

In the case of parasitisms, it is the virulence (negative influence on host reproduction and survival) associated with parasites that drives their myriad influences. Therefore, central questions concerning parasitisms revolve around understanding factors that influence the expression and evolution of virulence (Levin and Pimentel 1981; Ewald 1987; Frank 1992; Herre 1993, 1995; Bull 1994; May and Nowak 1994; Nowak and May 1994; Ebert and Herre 1996). In order to frame such questions, much less answer them, it is fundamentally important to recognize that virulence is an attribute neither of the parasite nor of the host alone, but a result of the interaction between the two. It is particularly important to recognize that the outcome of that interaction usually depends on ecological context (see below). The responses to that ecological context are themselves embedded to a greater or lesser degree in evolutionary context. That is, organisms generally respond most "adaptively" to the situations most commonly encountered through their evolutionary history (Herre 1987). Nevertheless, in parasitisms, what is adaptive for the host clearly is often not adaptive for the parasite.

Furthermore, neither "host" nor "parasite" is a monolithic entity. Their populations are almost invariably composed of a variety of genotypes and strains. Because different parasite strains often have very different effects on a given host genotype, factors that influence the spatial or temporal distribution of parasite strains and of host genotypes will strongly affect the expression and evolution of virulence. Ultimately, changes in virulence observed in any particular system can result from changes selected in either the host or parasite populations, or, more likely, both. Moreover, virulence varies dramatically from system to system, and the virulence observed in any particular host-parasite system can change across space and time. All of this makes it inherently challenging to define, measure, and study virulence. Although the consequences of within-species variation have been more extensively studied in host-parasite systems (e.g., Ebert 1994; Thompson 1994), the situation with mutualisms is analogous, and all of the complexities outlined above for parasitic interactions also apply.

Mutualisms

Given that mutualisms are best viewed as reciprocal exploitations that none-theless provide net benefits to each of the involved parties, it follows that it is important to identify costs and benefits to each partner correctly and, if possible, to quantify them. Next, it is desirable to relate variation in those costs and benefits to variation in the factors that influence them (Herre 1989, 1996; Pellmyr 1989; Thompson and Pellmyr 1992; Anstett et al. 1996; Bronstein and Hossaert-McKey 1996). Finally, it is particularly important to identify the situations in which there exists a conflict of interest between the two (Herre and West 1997). Special attention should be given to the mechanisms that prevent the costs to either partner from exceeding the benefits, thereby maintaining the mutualistic nature of the interaction. For example, it appears that yuccas can curb parasitic tendencies in their mutualist moth pollinators by aborting fruits that the moths have overexploited (Pellmyr and Huth 1994).

Although there is no general theory of mutualism, several factors that can help align mutualists' interests have been tentatively identified. An important precondition for mutualisms is the potential for complementation or augmentation of functions and abilities among would-be mutualists. In such cases, the passage of symbionts from parent to offspring (vertical transmission), genotypic uniformity of symbionts associated with individual hosts, spatial structure of populations leading to repeated interactions between would-be mutualists or their descendents, and restricted options outside the relationship for one or both partners are thought to align interests and promote long-term stability. Conversely, movement of symbionts between unrelated hosts (horizontal transmission), multiple symbiont genotypes, and varied options are thought to promote the opposite effects (Trivers 1971; Axelrod and Hamilton 1981; Bull and Rice 1991; Frank 1992; Yamamura 1993; Leigh and Rowell 1995; Maynard Smith and Szathmáry 1995). It is no coincidence that many of the same situations are thought to influence the expression of virulence. Overall, this framework is logically appealing, and at least some cases appear to conform well with its predictions (e.g., Bull et al. 1991; Herre 1993, 1995; Clayton and Tompkins 1994). However, the attempt to assess the generality of this framework is necessary, and requires many carefully executed and interpreted case studies.

Fig-Associated Organisms

Here, I concentrate on the mutualistic and parasitic organisms associated with monoecious New World figs (with a few examples from Old World systems), and use them to examine the extent to which some of the proposed theories are or are not applicable. One justification for this seemingly narrow

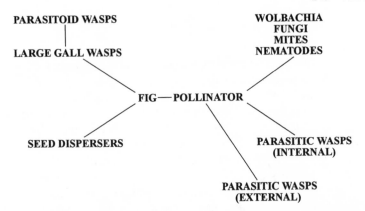

FIG. 11.1. Organisms associated with figs include: mutualist wasps, parasitic wasps (both internally and externally ovipositing) that appear to be primarily competitors with the pollinators, nematodes, fungi, mites, and *Wolbachia* (bacteria) that are phoretic or carried internally by the pollinating wasps, larger gall-forming wasps and their parasitoids, and frugivorous seed dispersers (see text).

approach is that figs and their associates provide both a wide range of types of interacting taxa (plants, insects, nematodes, fungi, bacteria, vertebrates) and a variety of types of interactions (fig. 11.1; Corner 1940; Werren et al. 1995; Kalko et al. 1996). For most of these organisms, it is possible to make fairly direct measurements of fitness or major components of fitness. Often, variation in reproductive success of one member of these interactions can be related to variation in attributes of another at various levels (e.g., species, populations, individuals) (Herre 1989, 1993; Nefdt and Compton 1996; Herre and West 1997).

These measurements can then be placed in a series of levels of ecological and evolutionary context that the above-mentioned framework suggests are important considerations. Specifically, from the basic natural histories, as well as more detailed genetic information, we know something about patterns of ecological transmission (e.g., Nason et al. 1996, 1998). Further, in most cases, we have genetic information that suggests phylogenetic relationships among many of the associated taxa, and we can therefore make inferences about longer-term associations or evolutionary tracking (Herre 1995; Herre et al. 1996; Machado et al. 1996).

Before examining the applicability of theory to these organisms, it is worth emphasizing the practical challenges to making both appropriate measurements and proper analyses that are posed by complex systems of interacting factors. For example, in the fig-wasp system many factors interact to affect the production of the basic currency of the mutualism: viable seeds and pollinator wasps (Herre 1989, 1996; see fig. 11.2). Unless the confounding effects of these factors are properly controlled for, it is very easy to

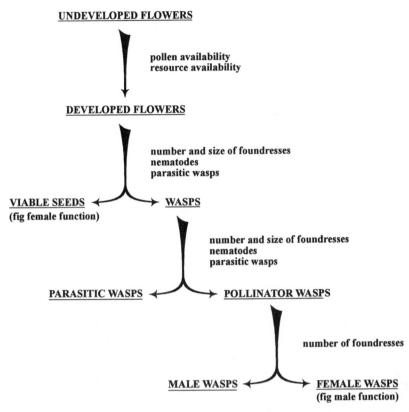

UNDEVELOPED FLOWERS

pollen availability
resource availability

DEVELOPED FLOWERS

number and size of foundresses
nematodes
parasitic wasps

VIABLE SEEDS ← → **WASPS**
(fig female function)

number and size of foundresses
nematodes
parasitic wasps

PARASITIC WASPS ← → **POLLINATOR WASPS**

number of foundresses

MALE WASPS ← → **FEMALE WASPS**
(fig male function)

FIG. 11.2. Flow chart showing stages at which different factors influence the production of viable seeds and pollinator wasps. The important point to recognize is that the production of the basic currency of the mutualism is influenced at many stages by a number of different factors, some of which are attributes of the fig, some of which are attributes of the wasp, and some of which are neither.

misinterpret the underlying relationships among variables of fundamental importance to the mutualism (West and Herre 1994; Herre and West 1997; see below).

Monoecious Fig-Pollinating Wasps

More than 750 named species of figs are found worldwide. All depend on minute wasps for pollination. The pollinator wasps, which are usually species-specific, are all members of the chalcidoid family Agaonidae, and all of them show similar life cycles (Wiebes 1979, 1995). It is important to note that the closest relatives of the wasps that pollinate figs appear to be the wasps that parasitize them (Boucek 1988; Machado et al. 1996). The monoecious figs belong to three of four recognized *Ficus* subgenera and comprise roughly half of all fig species. Both molecular and morphological evidence suggest that the life cycle of the monoecious species is ancestral to the more derived life cycles exhibited by the dioecious fig species (Ramirez 1969, 1980; Wiebes 1979, 1982; Corner 1985; Berg 1989; Herre et al. 1996; Kerdelhue and Rasplus 1996b; Anstett et al. 1997).

The reproductive cycle of the figs and the wasps begins when some number of mated, pollen-bearing foundress wasps enter a receptive fig syconium (the enclosed inflorescence that defines the genus, *Ficus*, and ultimately develops into the fig fruit); pollinate the uniovulate female flowers that line the interior; lay eggs in some of these flowers; and then die. Usually, the foundress wasps die inside the one fig fruit that they pollinate and can thus be counted (Herre 1985, 1989, 1996; Compton et al. 1991; but see Gibernau et al. 1996). The numbers of foundresses per fruit can vary greatly both within and among species, and that variation can have a number of important effects on the outcome of the wasps' interactions with the fig (e.g., Herre 1989, 1993, 1996). Unpollinated fruit are usually, but not always, aborted.

After they are pollinated, some proportion of the female flowers begins to develop. Flowers that complete development eventually produce either an intact, viable seed or an adult wasp that consumes the contents of a single would-be seed during the course of its own development. Previous studies have shown that the proportion of the flowers that develop can be strongly influenced by a combination of pollen and resource availability (fig. 11.2; Herre 1989, 1996; Anstett et al. 1996; Bronstein and Hossaert-McKey 1996; Herre and West 1997).

As final ripening of the fig fruit approaches, the wingless adult male wasps emerge from the seeds within which they matured. They crawl around the interior of the syconium, chew open seeds that contain females, and mate. The mated females enlarge the holes cut by the males, emerge from their seeds, gather pollen from male flowers, exit the fruit (through exit holes that in most species are cut by the males), and begin the cycle anew (Corner 1940; Galil and Eisikowitch 1968; Ramirez 1969; Frank 1984). After the female wasps leave, a wide range of animals eat the ripe fruit and disperse the viable seeds (Janzen 1979; Milton et al. 1982; McKey 1989; Windsor et al. 1989; Milton 1991; Kalko et al. 1996).

Ecological Patterns of Transmission and Evolutionary Patterns of Co-speciation

Recent genetic work has shown that the pollen-bearing wasps routinely disperse many kilometers, with the result that the areas covered by effective breeding populations of figs are usually a hundred or more square kilometers, an order of a magnitude larger than that documented for any other plant species (Nason et al. 1996, 1998; also see Compton 1990, 1993; Ware and Compton 1992). Therefore, the mutualistic fig-wasp system is largely characterized by an extreme horizontal transmission of the mutualist wasp. Furthermore, recent molecular data strengthens earlier morphological studies suggesting that, with a few exceptions, the wasps are species-specific (Herre et al. 1996). That specificity almost certainly arises from the species-specific chemical attractants that the figs release when receptive (van Noort et al. 1989; Ware and Compton 1992; Hossaert-McKey et al. 1994). Finally, the predominant evolutionary pattern shown is for co-speciation/co-cladogenesis among the wasp and fig lineages (Ramirez 1974; Wiebes 1979; Berg 1989; Herre et al. 1996).

Measuring Reproductive Success in the Figs and Wasps: Sorting Out the Effects of Confounding Variables

Many factors interact to affect the production of the basic currency of the fig-wasp mutualism: viable seeds and pollinator wasps (Herre 1996; see fig. 11.2). These factors include attributes of the fig (such as the number of flowers per fruit); attributes of the wasp (such as body size); attributes of the interaction between the two (such as the proportion of those flowers that develop, which is itself influenced by pollen availability [e.g., number or size of foundress pollinators] and resource availability); as well as attributes of neither (presence and densities of parasitic wasps and nematodes). It is critical to properly account for the confounding effects of these factors. Otherwise, misinterpretation of the underlying relationships among variables that are of fundamental importance to the mutualism is the likely result (West and Herre 1994; Herre and West 1997).

Specifically, the confounding of these factors often leads to the mistaken perception that there is either no relationship between viable seed and pollinator wasp production, or that the relationship exists and is positive. This would appear to be consistent with the interpretation that there is no conflict of interest between the two mutualists. Similarly, the relationship between parasitic wasp and seed or pollinator production often appears positive. This would appear consistent with the interpretation that there is no negative impact of the parasitic wasps (see below). In the absence of proper statistical control, the true underlying negative relationship between viable seed production and pollinator wasp production can be overlooked, and the underlying negative relationship between certain groups of parasitic wasps and pollinators can be missed (Herre 1989, 1996; West and Herre 1994; Herre and West 1997).

When only one foundress wasp enters a fig fruit, counting the offspring gives a direct measure of that wasp's lifetime reproductive success (Herre

1989, 1993). The variation in foundress's lifetime reproductive success can often be related to variation in a number of attributes of both the wasp and the fig fruit that it pollinated (Herre 1989; West and Herre 1994; Anstett et al. 1996; Bronstein and McKey-Hossaert 1996; Nefdt and Compton 1996; Herre and West 1997). Similarly, seed and wasp production can be measured, and their variation among fruit can be related to variation in the attributes of the individual foundress wasps (e.g., body size). Such measurements permit the documentation of how variation among individuals within both mutualist species has reciprocal effects on the reproductive success of the other.

Specifically, of the female wasps that are born in a given fruit, the larger ones appear to have a higher success rate of reaching the next fig, and, once there, the largest of these produce the greatest number of offspring (Herre 1989; Nefdt and Compton 1996; Herre and West unpubl.). Although there appears to be a slight heritable component, wasp body size is mostly explained by the dry weight of the seeds from the fruit in which they hatch. Combined, these observations suggest that wasp body size can shift the outcome of the fig-wasp interaction toward the production of wasps, but that wasp body size is most clearly affected by the fig (Herre 1989; Herre and West in prep.).

In essence, although both the wasp and the fig depend on each other for long-term survival and reproduction, their short-term interests are not necessarily aligned. An individual fig needs the foundresses to pollinate its flowers in order to produce seeds of its own. It also needs the foundresses' female offspring to disperse its own pollen. Those female offspring are useful only insofar as they produce seed with the pollen of the fig that produces them. In contrast, the foundress wasps only benefit from would-be fig seeds that are consumed by their offspring. That is, although there is a mutual, long-term interdependence between the two partners, there are clear conflicts of interest, some of which would appear to have the potential to undermine and ultimately destabilize the relationship.

Fig-Wasp Sex Ratios and Their Influence on the Fig

Because only female wasp offspring provide pollination services for the fig, the tendency of the wasps to shift sex ratio away from extreme female bias with increasing foundress numbers is not in the fig's interest. Although the details of the relationships vary among fig species, single foundresses, particularly in larger-fruited figs, often do not provide sufficient pollen to saturate the receptive flowers and maximize seed set. Generally, increased numbers of foundresses are associated with increases in both seed and wasp

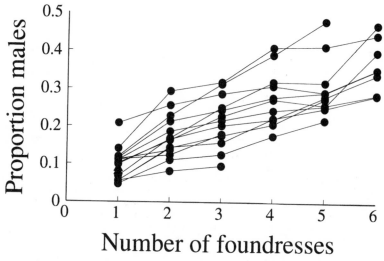

FIG. 11.3. The relationship between brood sex ratio and number of foundresses in 15 species of Panamanian fig-pollinating wasps. Within species, the brood sex ratio becomes less female-biased as the number of foundresses increases. Between species, species that have higher average foundress numbers (suggesting more outbreeding) show less female-biased sex ratios for any given number of foundresses. Because only female wasps can transport pollen, decreasing female bias runs counter to the fig's reproductive interests (see text).

production (Herre 1989). Of the wasps that are produced, however, an increasing proportion tends to be wingless males, which are of no use as fig pollen vectors (fig. 11.3; Herre 1989, 1996)

These sex-ratio shifts represent the wasps' responses to varying intensities of local mate competition (Hamilton 1967, 1979; Frank 1985; Herre 1985). Interestingly, local mate competition represents one of very few examples of a Prisoner's Dilemma-style game known from a natural system in which the payoff values can be derived from first principles. Furthermore, the key to understanding how much greater or lesser degrees of female bias are favored lies in understanding how among- and within-group selection are balanced (Colwell 1981; Wade 1985).

The payoff matrix below (table 11.1) shows the relative fitnesses of two outbred foundresses that contribute equal-sized broods to fig fruits. By "playing" a female-biased strategy (0.25 males), the foundress maximizes its absolute fitness. By "playing" an even sex ratio (0.50 males), however, the foundress insures that the relative fitness of the other wasp within the same fig fruit (deme) does not exceed her own. Cooperation (which in this case is represented by female-biased sex ratios) is only enforced by the existence of many sets of competing foundresses in multiple fruits (demes). That is, the among-deme

TABLE 11.1

Local Mate Competition Game Payoff Matrix

	Sex Ratio of Wasp B	
Sex Ratio of Wasp A	0.25	0.5
0.5	22(19)	15(15)
0.25	25(25)	19(22)

NOTE: The first value gives the payoff to wasp A for any combination of brood sex ratios employed by A and B. The value inside the parentheses gives the payoff to wasp B for the same situation. Values were calculated assuming two outbred foundresses contributing equal numbers of offspring to a common brood from which mated female offspring emigrate (Herre 1985). Notice that the female-biased ratio (0.25) "loses" to an even sex ratio (0.50) within the deme (payoff of 19 versus 22), yet female-biased sex ratios are observed because of the advantage across demes. The average payoff in demes with two even sex-ratio broods is 15; a rare genotype with a sex ratio of 0.25 (and average payoff of 19) can therefore invade, if the number of demes is high (see Colwell 1981; text).

advantage of increased productivity allows the selection for female-biased sex ratios to override the within-deme selection for even sex ratios (Taylor and Bulmer 1980; Colwell 1981; Frank 1985; Herre 1985).

Population structure also appears to play a similar role in selecting for nematode virulence, with populations characterized by higher foundress numbers permitting and even promoting increased virulence (Herre 1993, 1995). Both of these effects of increased foundress number are detrimental to the fig. These examples support the plausibility of arguments that the effects of population structure and within-deme (within-host) genetic homogeneity can be important in maintaining the beneficial effects of some types of mutualisms (see Levin and Pimentel 1981; Herre 1985; Bull and Rice 1991; Frank 1992; Leigh and Rowell 1995; Maynard Smith and Szathmáry 1995).

THE SEEDS

In monoecious figs, the flowers that develop into viable seeds (usually 40–50%) represent a large portion of the fig's investment in female function. Flowers that support the development of the pollinator wasps, in particular the females, represent a large portion of the fig's investment in "male" function (fig. 11.2; Herre 1989). Therefore, although the investment in female and male function on the part of the monoecious fig is largely reflected in seed and wasp production, respectively, the interests of the wasps are only directly aligned with the fig's investment in its male function. This raises the question of why selection on the much more numerous, and much shorter-lived, wasps to increase their own fecundity has not come at the expense of the production of any viable seeds. We might expect that selection on the pollinator wasps would, in the short term, lead to increasingly male-biased sex allocation in figs, and, in the long term, to the complete suppression of viable seed production and eventual collapse of the system. However, after at least 40 million years of possibly frantic coevolution with the wasps, monoecious figs still produce seeds (Collinson 1989; Herre and West 1997).

The mechanisms that prevent pollinators from overrunning the seeds are still not understood. In contrast with some well-studied yuccas in which overexploited fruit are generally aborted (Pellmyr and Huth 1994), there is no evidence that figs abort overexploited fruit (contra Axelrod and Hamilton 1981; also see Addicott et al. 1990; Bull and Rice 1991). In particular, some figs will retain large portions of their crops that have not been pollinated in the event of infestation by certain types of parasitic wasps (see below). Another idea had been that pollinator wasps did not possess ovipositors long enough to reach ovaries on "long-styled" flowers, and thus were only capable of producing offspring in flowers with short styles. Because measurements of pollinator ovipositors show that most ovaries are within the wasps' reach, it now seems clear that differences in style length of seed-destined and wasp-destined flowers is not a sufficient explanation for maintaining seed production (Bronstein 1992; Compton 1993; Kjellberg et al. 1994; West and Herre 1994; Nefdt and Compton 1996; see below).

Nonetheless, there currently appear to be at least two types of viable explanations for the stability that has allowed seed production in figs to persist. One possibility is that there are chemical or physical differences among flowers within a fruit that prevent a portion of them from receiving eggs or prevent them from supporting wasp development if an egg is laid on them (Verkerke 1989; West and Herre 1994; S. G. Compton pers. comm.). This explanation is based on considerations of floral anatomy and inferences from patterns of flower utilization of New World parasitic wasps (West and Herre 1994; West et al. 1996). In essence, it suggests that the figs have hit upon an as yet undescribed "unbeatable" mechanism that prevents wasp oviposition and/or development in some of the flowers, but allows it on others.

A second possibility is suggested by one of the better case studies currently available on mechanisms of stability of the fig-wasp mutualism (Nefdt and Compton 1996). This study of a series of African figs suggests that, although the foundresses have physical access to most flowers, they possess too few eggs to exploit all of them. Access was determined by comparing ovipositor lengths with style lengths of the flowers. The ovaries of most flowers were found to be within the reach of the wasps' ovipositors in most species examined, and most flowers within reach were found to be able to support a wasp's development. Furthermore, wasp egg loads were also determined in a subset of species. The norm was too few foundresses carrying too few eggs to exploit all of the available flowers, and thus a large fraction of flowers escaped oviposition.

In essence, assuming the egg counts were accurate, this explanation depends on as yet undescribed trade-offs that prevent wasps from producing enough eggs and/or from developing ovipositors that provide access to all flowers (e.g., Kathuria et al. 1995), as well as figs that can effectively limit the number of pollinators that enter any given fruit. It should be noted that slightly modified versions of this "trade-off" hypothesis and the "unbeatable seed" hypothesis are not mutually exclusive. One very intriguing possibility is that different fig wasp systems have achieved stability through different mechanisms.

Finally, figs usually expend roughly half of their seeds in feeding wasps, many of which will carry pollen. In yuccas, there is much less chance that a yucca moth developing from a yucca fruit will carry the plant's pollen. It is probably no coincidence that in yuccas only (roughly) one-fourth of seeds go to developing moths (Pellmyr and Huth 1994; Pellmyr et al. 1996). In fact, the allocation patterns (seeds and wasps) and foundress distributions appear to reflect more closely the interests of the fig rather than those of the wasps, implying that the fig is generally the dominant member in the relationship (Herre 1989, 1996; Nefdt and Compton 1996).

EFFECT OF THE SEED/WASP TRADE-OFF ON FIG REPRODUCTIVE SUCCESS

Because there is a fundamental negative trade-off between seed and pollinator wasp production (Herre and West 1997), offspring of female wasps dispersing from a given fig tree will come at the expense of seeds that would otherwise be fathered by that fig's pollen. Therefore, the fig that produces the wasps would benefit most if its wasps arrived as foundresses at another receptive fig, pollinated the flowers, induced seed production, but were themselves sterile (see fig. 11.4). One theatre in which this conflict can potentially play out is wasp body size. Increased body size appears to influence a wasp's success by increased likelihood to reach another tree, and increased reproduction (Herre 1989; Nedft and Compton 1996). Depending on the relative importance of these two effects, it appears that wasps would benefit from being as large as possible, but that the fig might benefit most from intermediate-sized wasps (Herre and West in prep.). It is worthwhile to emphasize again that more of the variation in wasp body size appears to be attributable to the environment (the fruit that they are born in) than the size of the mother (Herre 1989).

These different aspects of fig-wasp conflict are analogous to certain types of parent-offspring conflict (Godfray, chap. 6), conflict between mates (Lessells, chap. 5), as well as basic sex-allocation problems piggybacking along for the ride (Herre 1989, 1996). Some of these conflicts almost certainly

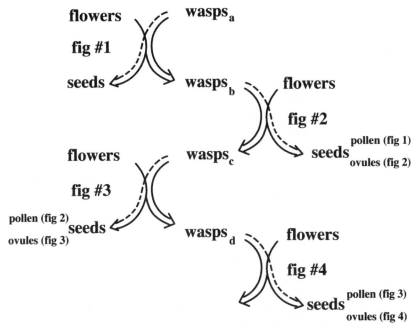

FIG. 11.4. Diagram of reproductive interests of both fig and wasp over three generations of fig pollination and wasp reproduction. Wasps (a) carry pollen to fig #1 for the production of its viable seeds, some of which are consumed to produce the wasp offspring (b). These wasps, in turn, carry pollen from fig #1 to pollinate flowers on fig #2. Some portion of those flowers develop as viable seeds (with genetic contribution from fig #1), others are eaten by the developing offspring of the wasps, and so on. The important point is that in monoecious figs, the pollinators that carry pollen from any tree reproduce directly at the cost of potential offspring of that tree. It would be in the fig's short-term interests to produce sterile wasps to carry its pollen, just as it would be in the wasp's short-term interest to be able to exploit all of the fig's seeds for the development of its offspring.

provide the selective background that has produced dioecy in figs on at least two occasions over the course of evolution. The interested reader should consult the increasing literature concerning the ecology and evolution of dioecious figs (Ramirez 1969, 1980; Wiebes 1979, 1982; Kjellberg et al. 1987; Berg 1989; Herre 1989; Kjellberg and Maurice 1989; Gibernau et al. 1996; Kerdelhue and Rasplus 1996b; Patel 1996; Spencer et al. 1996; Anstett et al. 1997).

THE REASON FOR WASP POLLINATION

Across the different species of fig-pollinating wasps, both active and passive pollination are found. In wasp species that actively pollinate, there are usually specialized morphological structures in which pollen is stored, and there are deliberate behaviors for collecting pollen in the ripe, natal fruit and then depositing it on the stigmatic surfaces in the receptive syconia (Ramirez 1969; Frank 1984). In species with passive pollination, specialized structures and behaviors appear to be absent. Recent work has discovered structures and behaviors that were previously missed, however, so that specialized structures and active pollination are perhaps more common than has been thought (Ramirez and Malavasi 1997).

Superficially, there appear to be several reasons why the pollinators perform these services for the figs. First, the fig trees usually abort unpollinated fruit. Given that, in most cases, the foundresses do not leave the fruit they enter, this binds the fate of the foundress wasp's progeny to that of the fruit it enters. Second, the production of wasps is generally linked to the proportion of flowers that develop. Therefore, to the extent to which flower development is linked to pollination, the reproductive success of the wasps is linked to their own capacity to pollinate (Herre 1989; Herre and West 1997). Third, pollinated flowers may provide a better resource for developing wasps (Verkerke 1989). Furthermore, this relationship between pollination success and wasp success may be linked to the species-specificity that is generally observed in these systems; usually the "right" pollinator species does the best job both in terms of seed and wasp production (Compton 1990; Ware and Compton 1992; see Kerdelhue and Rasplus 1997).

Nonetheless, there are some problems with this straightforward view of factors that align the fig-wasp interests and thus potentially stabilize their relationship. At least three fig species are known to have more than one species of associated pollinator (Galil and Eisikowitch 1968, 1969; Wiebes 1979; Michaloud et al. 1985; Kerdelhue and Rasplus 1997). In fact, of the two "pollinators" associated with *Ficus sycomorus*, *Ceratosolen arabicus* and *C. galili*, the females of the latter species do not pollinate. *Ceratosolen galili* has pollen pockets, yet does not use them and is not normally associated with any viable seed production. The clearest interpretation is that *C. galili* has lost the behavioral component needed for pollination. It is the sole member of the genus, indeed among all of the known pollinator species, that has effectively become a parasite on the system (Galil and Eisikowitch 1968, 1971; Compton et al. 1991). *Ceratosolen galili* is not the sister taxon to the actual pollinator (Herre et al. 1996). Interestingly, this is also the case with nonmutualistic yucca moths (Pellmyr et al. 1996). In contrast, a reversion to nonpollinating has happened more than once among the group of moths that pollinate yuccas.

The existence of a pollinator-turned-parasite raises several questions. One concerns how this wasp has bypassed mechanisms that seem to keep other systems in check. The parasitic *C. galili* does not depend on the pollinator, *C. arabicus*, to pollinate the fruit. Seedless fruit will not abort in the absence

TABLE 11.2

Attributes of Pollinator Wasp-Fig Systems

Wasps that pollinate figs form a monophyletic group

Within the pollinator clade, only one case is known in which the species no longer pollinates

Closest sister taxa are all parasites of mutualism of fig and fig pollinator wasp

With few exceptions, pollinator wasps are species-specific

Preliminary analyses suggest a high degree of co-speciation with host figs

Pollinator wasps are horizontally transmitted from host to host

All pollinate and oviposit from inside the fig fruit

Many factors influence the outcome of the fig-wasp interactions

Complex system of alignments and conflicts of interest between partners

Partial alignment of interests with figs occurs in male function of fig

Female (seed production) function of figs at odds with wasp short-term interests

---

of the pollinator if *C. galili* wasps are present (Compton et al. 1991). Is this due to some special trick of *C. galili* or could a nonpollinating *C. arabicus* wasp accomplish the same result? In fact, in the most carefully studied sites, roughly 66% of all fruits were occupied by *C. galili*, whereas only 40% of the fruit were occupied by the legitimate pollinator, *C. arabicus*. Although there are possible disadvantages in fecundity and survival that comes with the smaller size of *C. galili*, it is not clear that these are sufficient to prevent them from completely displacing the pollinator, at least locally. Finally, if this pollinator-turned-parasite wasp cannot only succeed in parasitizing but even come to dominate, why does this not occur more often? In *F. sycomorus*, *C. arabicus*, and *C. galili*, are we witnessing an ongoing collapse that could occur in any of the other fig-wasp mutualisms?

The ability of the "correct" pollinator species to consistently outcompete the odd crossover may be an important component in preventing this (e.g., Compton 1990; Ware and Compton 1992; Kerdelhue and Rasplus 1997). It may also be important that *F. sycomorus* is a relatively large-fruited fig. Generally, large-fruited figs receive several foundresses (Herre 1989; Compton et al. 1991), and the consequences of what any one foundress does in any particular fruit (e.g., not pollinate) are, on average, covered to a greater extent than in systems in which there are normally fewer foundresses. Such a situation may facilitate a pollinator's evolutionary experimentation with increasing degrees of parasitism (also see Thompson and Pellmyr 1992). (See table 11.2.)

FIG. 11.5. Photograph of the pollinator (*Pegoscapus hoffmeyeri*), externally ovipositing parasitic wasp (*Idarnes* sp.), large gall-forming wasp (*Apocerus* sp.), and parasitoid of the gall-former (*Physothorax* sp.) associated with *Ficus obtusifolia*.

Fig-Parasitizing Wasps

Figs also support a diverse community of parasitic nonpollinating wasps (Gordh 1975; Hamilton 1979; Janzen 1979; Ulenberg 1985; Boucek 1988, 1993; Bronstein 1991; Compton et al. 1991; Compton and Hawkins 1992; Compton and van Noort 1992; Hawkins and Compton 1992; Compton 1993; West and Herre 1994; Cook and Power 1996; Kerdelhue and Rasplus 1996a; van Noort and Compton 1996; West et al. 1996). All chalcidoid wasps that depend on the tissues of the syconium for completing their life cycles (both the pollinators and the majority of nonpollinators) have been grouped in the family Agaonidae. According to this classification, all pollinating fig wasp genera belong to the subfamily Agaoninae, whereas nonpollinators are grouped into five different subfamilies (Epichrysomallinae, Otitesellinae, Sycorictinae, Sycophaginae and Sycoecinae) (Boucek 1988). This scheme is largely supported by molecular data (Machado et al. 1996; Machado and Herre unpubl.). Thus, given the phylogenetic relationships, an understanding of the biologies of these parasitic groups is very likely to contribute to understanding the evolutionary origins and mechanisms underlying the stability of the mutualism between the figs and the pollinators (West and Herre 1994; Kerdelhue and Rasplus 1996a).

In the New World, all nonpollinating wasps oviposit from the exterior of the syconium. These externally ovipositing wasps seem to comprise three very ecologically distinct groups (see fig. 11.5): (1) a group that is similar in body size to the pollinators and that appears to compete with them for the same resources for larval development (e.g., *Critogaster* and some groups of *Idarnes*); (2) a group of relatively larger wasps that lay their eggs in the flowers or fruit walls, inducing the formation of large galls in which the larvae develop. The presence of these galls appears to prevent unpollinated fruit from being aborted (e.g., *Aepocerus* and *Idarnes* (incerta)); (3) a group of true parasitoids of the larger gall-formers (e.g., *Physothorax*) (West and Herre 1994; West et al. 1996). In addition to wasps of these types, some Old World nonpollinators enter the syconia to oviposit. These wasps include all members of the subfamily Sycoecinae, a few members of the subfamily Otitesellinae, and the genus *Sycophaga* from the subfamily Sycophaginae (Abdurahiman and Joseph 1967; Boucek 1988).

EXTERNALLY OVIPOSITING COMPETITORS OF THE POLLINATORS

Idarnes (subfamily Sycophaginae) is a genus of nonpollinating wasps associated with figs in the New World subgenus Urostigma (Americana), and *Critogaster* (subfamily Sycorictinae) is a genus of nonpollinating wasps associated with figs of the New World subgenus Pharmacosycea. Unlike pollinator wasps, *Idarnes* and *Critogaster* females do not enter the fig. Instead, they penetrate the fig wall from outside with their characteristically long ovipositors and lay eggs in the interior seed layers of the fruit. Like the pollinators, individual larvae of these parasites develop at the expense of one flower within the fig fruit. *Idarnes* and *Critogaster* females emerge and leave without collecting pollen. Although their ecologies and effects on host wasps and figs appear to be quite similar, *Idarnes* and *Critogaster* are only very distantly related (Boucek 1988; Machado et al. 1996; West et al. 1996). In fact, in both cases, these wasps are more closely related to other genera of nonpollinator wasps that have very distinct ecologies (e.g., large gallers or internally ovipositing parasitic wasps; see below) than they are to each other (Herre 1996; Machado et al. 1996; West et al. 1996).

Idarnes (and *Critogaster*) parasites have a detrimental effect on the reproductive success of their hosts. Specifically, after statistically controlling for confounding variables (e.g., foundress number, etc.), there is a clear negative correlation between the number of *Idarnes* (and *Critogaster*) wasps emerging from a fruit and pollinator wasp production (West and Herre 1994; West et al. 1996). In contrast, there is no significant correlation between the number of *Idarnes* wasps emerging and viable seed production. *Idarnes* also are found to develop in unpollinated fruit that are retained, apparently because of the influence of large galling wasps (Bronstein 1991; West and Herre 1994). Therefore, *Idarnes* are not obligate parasitoids of the pollinators, nor do they depend on them to pollinate the flowers.

IMPLICATIONS FOR FIG-POLLINATOR STABILITY

Both the pollinators and the externally ovipositing parasites in the New World figs tend to develop in seeds closer to the interior of the fruit, which are predominantly derived from flowers with short styles (Herre 1989; West and Herre 1994). It is odd that the parasites that are oviposit from the outside of the fig do not preferentially develop in seeds derived from longer-styled flowers that are closer to outside of the fruit, closer to where these wasps oviposit. This suggests that the longer-styled flowers are not available for use or that they provide a considerably inferior resource for developing wasps than do short-styled flowers. Therefore, considering the observed negative relationship between these parasites and pollinator production, and the lack of any relationship between these parasites and good seed production in New World figs, it is difficult to embrace the idea that all flowers are equally available for oviposition or equally conducive to wasp development.

At least in the New World systems, these externally ovipositing parasitic wasps appear to be competing with the pollinators for a subset of the flowers (predominantly short-styled) that either have the potential to develop as seeds or to support the development of wasps. These observations suggest that these wasps are exploiting the opportunities that the fig presents to the pollinators and form the basis of the inference that figs have some unexploitable flowers that cannot be used by either of these groups of wasps, and that the mechanism involved is not simply spatial position within the syconium (West and Herre 1994; West et al. 1996). A modified form of this idea consistent with the Nefdt and Compton (1996) observations on the African systems is that some subset of flowers is only marginally exploitable, or provides a potential, but inferior, resource for developing wasps.

Published reports suggest that the Old World genera *Sycoscapter* and *Philotrypesis* have similar ecologies and similar effects (Compton et al. 1991, 1994; Compton and van Noort 1992; Kerdelhue and Rasplus 1996a,b; F. Kjellberg pers. comm.). However, directly comparable studies need to be conducted in both New and Old World fig systems in order to determine clearly whether these different wasps are in fact exploiting their host figs in a similar manner, and, more interestingly, whether the host figs rely on different mechanisms to maintain the stability with their associated wasps (Herre 1989; Nefdt and Compton 1996; West et al. 1996; Herre and West 1997).

EVOLUTIONARY PATTERNS OF CO-SPECIATION

Both morphological and molecular data suggest that these and other parasitic wasps have co-speciated with their hosts (Gordh 1975; Ulenberg 1985; Machado et al. 1996). Both the *Idarnes* and *Critogaster* parasites also show similar degrees of co-cladogenesis (suggesting co-speciation) with the figs when compared to the pollinators (Herre et al. 1996; Machado et al. 1996). As is the case with the pollinators, these parasites are also horizontally transmitted. Thus, neither evolutionary nor ecological patterns of transmission tell us anything with respect to the parasitic/mutualistic nature of these wasps.

The major difference between these wasps and the pollinators appears to result from the fact that the parasitic wasps lay eggs from the outside of the fruit, and further that these wasps tend to lay eggs in several different fruits. Beyond making it difficult for these wasps to perform any pollination ser-

TABLE 11.3
Externally Ovipositing Parasitic Wasps, "Competitors"

External oviposition

Members of different families of nonpollinators

Often nearest relatives show very different ecologies (e.g., large gall-formers)

Predominant negative effects on pollinators, not seed production

Pollination and pollinators are not needed for these wasps to exploit the fig

Do not appear to be parasitoids of pollinators

Appear to exploit a similar subset of the fig's flowers as the pollinator

Suggests that fig controls access to flowers, and not by spatial arrangement alone

Co-speciation with hosts (as with pollinators)

Horizontal transmission (as with pollinators)

vices, perhaps this decouples the parasites' interests from those of any particular fig fruit that it lays its eggs in, and thereby helps explain the relative mutualistic and parasitic tendencies of these different wasps. Unfortunately for this attractive idea, other groups of parasitic wasps have life histories that caste doubt on this notion (see below). (See table 11.3.)

INTERNALLY OVIPOSITING PARASITIC WASPS

All members of the subfamily Sycoecinae, a few members of the subfamily Otitesellinae, and the genus *Sycophaga* of the subfamily Sycophaginae, are reported to oviposit from the inside of the fruit, much as the legitimate pollinators do. Furthermore, these species generally appear to be using the flowers that the pollinators would otherwise use (much as external competitors) (Galil and Eisikowitch 1968, 1969, 1971; Compton et al. 1991; Kerdelhue and Rasplus 1996a,b). The females of these species emerge and leave without collecting pollen, and thus, with a few possible exceptions in which pollen may be passively transferred (see Newton and Lomo 1979), they are parasitic on the system, as are the externally ovipositing species.

Interestingly, parasites of this type are found to be associated only with Old World fig species (Abdurahiman and Joseph 1967; Boucek 1988; Compton et al. 1991; van Noort and Compton 1996). Either these types of wasps did not arrive or had not yet evolved when the New World was colonized by the *Ficus* groups now present, or they subsequently became extinct. Combined with the observations and inferences mentioned above, the interesting possibility that New and Old World figs differ in fundamental ways with respect to their interactions with their associated wasps deserves consideration (West and Herre 1994; Kerdelhue and Rasplus 1996a; Nefdt and Compton 1996; West et al. 1996; Herre and West 1997).

Like the pollinators and the externally ovipositing parasites, internally ovipositing Sycoecines associated with a diverse group of African figs are generally species-specific. Furthermore, both pollinators and parasites associated with any given fig show remarkably similar head morphology. Although an independent test (e.g., molecular-based phylogenies) is desirable, this pattern suggests co-speciation and co-evolution with their hosts (see van Noort and Compton 1996). Thus, as with the other wasps, both mutualistic and parasitic, it appears that these wasps are horizontally transmitted and show a high degree of co-speciation with their host figs.

That these *parasites* enter the fig fruit in order to oviposit undermines the idea that *pollinators* entering the fruit to oviposit helps to explain the maintenance of the mutualism. In a certain sense, it would seem obvious why the external ovipositing species are parasites, but it is unclear why the internally ovipositing species (*Sycophaga*, etc. or the *C. galili*, for that matter), which are often the most abundant insects found in the fruit, do not establish mutualistic relations with the fig (but see Newton and Lomo 1979).

It is also unclear why these internal parasites do not take over the systems and drive the pollinators (and ultimately the figs) to extinction. One testable explanation combines the observation that, within species, large body size gives advantages in wasp survival and reproduction (Herre 1989; Nefdt and Compton 1996), with the suggestion that pollinated flowers provide better nourishment for developing wasp offspring (Verkerke 1989). Clearly, many, if not all, species of both internal and external parasitic wasps can develop in the absence of pollination or pollinators. These wasps may develop to larger size on flowers that have been pollinated. This could provide a mechanism for maintaining the pollinators in these systems, depending on the strength of the effects. Thus, if short-styled flowers also provided superior resources for developing wasps, then the combination of the two effects would have the potential to maintain both seed and pollinator wasp production in figs. (See table 11.4.)

LARGE GALL-FORMING WASPS

A diverse group of relatively larger wasps is also associated with figs of the New World section Urostigma (e.g., *Aepocerus*, *Heterandrium* (Otitesellinae), and *Idarnes* (incerta) (Sycophaginae); see fig. 11.5). These wasps can be more than 10 times the size of the pollinators. They lay their eggs in the flowers or fruit walls, inducing the formation of large galls in which their larvae develop. Ecologically, they appear to drain resources and decrease both pollinator (fig male) and seed (fig female) production in the fruits in

TABLE 11.4
Internally Ovipositing Parasitic Wasps

Internal oviposition

Members of different families of nonpollinators

Old World taxa

Often nearest relatives show very different ecologies (e.g., large gall-formers, or external ovipositers)

Predominantly negative effects on pollinators, not seed production

Pollination and pollinators are not needed for these wasps to exploit the fig

Appear to exploit a similar subset of the fig's flowers as the pollinator

Apparent co-speciation with hosts (as with pollinators)

Horizontal transmission (as with pollinators)

which they are present (West et al. 1996; see also Cook and Power 1996; Kerdelhue and Rasplus 1996a). As with the pollinators and other parasites, the large-gall formers also generally appear to be species-specific, and generally show co-cladogenesis with the host (Machado and Herre unpubl.). Like the other species in these systems, these wasps are also horizontally transmitted.

Interestingly, molecular work supports the proposition that the different species that share this type of ecology are also only distantly related to each other. Wasps that form large galls have therefore arisen independently several times during the radiation of the five families that comprise the fig parasites (as have the "competitors"). That work further suggests that the closest relatives to some of these wasps (e.g., *Idarnes* (incerta)) are the "competitor-type" parasitic wasps that often parasitize the same groups of figs (the *Idarnes* already discussed). Both morphological and molecular studies of these genera indicate that these "competitor" and "large-gall-forming" members of *Idarnes* are more closely related to the Old World genus *Sycophaga*, whose ecology entails entering the fig syconium to lay eggs, than they are to other externally ovipositing wasps that show similar, if not identical ecologies (Boucek 1988, 1993; Herre 1996; Herre et al. 1996; Machado et al. 1996; West et al. 1996). Thus, it appears that the various basic ecologies associated with parasitizing figs have evolved separately on several occasions in both the New and Old World. Given this pattern in the parasites, it is conspicuous that the pollinators form a monophyletic group and that pollination has apparently evolved only once (Ramirez 1974; Wiebes 1979, 1982; Boucek 1988, 1993; Machado et al. 1996; Herre et al. 1996; West et al. 1996; but see Newton and Lomo 1979).

The presence of these gall-forming wasps seems to prevent unpollinated fruit from being aborted, and thus these wasps do not seem to require the

TABLE 11.5
Large External "Galler" Parasitic Wasps

Induce the formation of large galls from fruit wall or flower tissue
Drain resources away from both seed and wasp production
Can prevent the abortion of unpollinated fruit
Taxonomically diverse set of species
Closest relatives often exhibit very distinct ecologies
Horizontally transmitted
Generally high degree of species-specificity
Predominance of co-speciation with host

presence of the pollinators in order to utilize the syconium. Although it is not clear how the large gallers prevent the abortion of unpollinated fruit, it is likely that these wasps can produce substances that mimic (or interfere with) hormonal signals in the fig. If so, cracking the hormonal code of the fig must have been a great (and possibly ongoing) evolutionary achievement for these wasps (as well as a great threat to the figs). Given the multiple origins of parasitic wasps with such different ecologies, it appears that the hormonal manipulation developed by the different types of wasps must allow a great deal of evolutionary flexibility associated with the multiple radiations into what appear to be a multiply repeated suite of fig-exploiting ecologies (Herre 1996; Machado et al. 1996). Deciphering both the fig and wasp hormonal codes and then placing them in a phylogenetic context would be a fascinating research challenge. (See table 11.5.)

PARASITOIDS OF GALL-FORMING WASPS

The wasps that are parasitoids of the large-gall makers (e.g., *Physothorax*) are almost certainly mutualists for the fig and the pollinators to the extent that they control the abundances of the parasites. Theoretical analysis of aggregation patterns of the parasitoids on gall-forming hosts suggests that they have this capability (West et al. 1996). Thus, it appears that ecological control of one of the most obvious threats to the New World fig-pollinator mutualisms can be acounted for.

As with the other taxa discussed, preliminary molecular data suggest that these also appear to be species-specific and also appear to be co-speciating with their hosts (Machado and Herre unpubl.). As with the others, these wasps are horizontally transmitted. Furthermore, as is the case in the majority of the nonpollinating wasps that are parasitic on figs, these beneficial (to the fig) parasitoids oviposit from the outside the fig fruit. Therefore, if we

TABLE 11.6
Parasitoids of Gall-Forming Wasps

Parasitoids of large-gall-forming wasps
Appear to be capable of controlling gall-former populations
Horizontally transmitted
Predominantly species-specific
Appear to co-speciate with hosts

consider the wasps associated with the figs collectively, none of the factors that theory suggests might be important in influencing relative benevolence of mutualists or parasites gives any consistent pattern. Mutualist or parasite, all are horizontally transmitted, all are for the most part species-specific, and all of their radiations show a strong tendency to co-speciate with the figs (Herre et al. 1996; Machado et al. 1996; Machado and Herre, unpubl.). Have these ideas just declared bankruptcy? (See table 11.6.)

Nematode Parasites of the Pollinators

Just as distinct species of fig-pollinating wasps are generally associated with distinct species of host figs, morphological and molecular work on the Panamanian species has established that distinct species of nematodes of the genus *Parasitodiplogaster* are associated with distinct species of host fig wasps (Poinar and Herre 1991; Herre 1995, unpubl.). In nematode-infested fig fruits, immature, dispersal-phase nematodes crawl onto newly emerged female fig wasps and are thereby carried to the next fig. Before the host wasp leaves its natal fruit, the nematodes enter the body cavity of the wasp and, at some point, begin to consume it and grow. Later, up to 20 or more adult nematodes (with a median of roughly 6 or 7) emerge from the dead wasp's body (Herre 1993, 1995), mate, and lay eggs within the fig fruit in which the host wasp has laid her eggs. The nematodes' eggs hatch before the emergence of the next generation of fig wasps, and the nematodes begin *their* cycle anew (Poinar 1979; Poinar and Herre 1991; Herre 1993, 1995; Giblin-Davis et al. 1995).

The natural histories of fig-pollinating wasps and the nematodes that parasitize them (Poinar 1979; Poinar and Herre 1991; Giblin-Davis et al. 1995) permit the direct measurement of several parameters that theory identifies as important to the evolution of virulence (Herre 1993, 1995). Specifically, in nearly ripe fig fruits that have been pollinated by only one foundress wasp, the presence of immature nematodes can be used to determine whether that individual wasp was infected. Therefore, within species of fig wasps, the

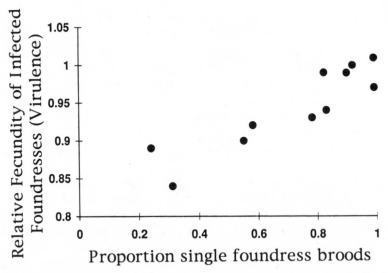

FIG. 11.6. The relationship between virulence (calculated as the proportion of offspring that nematode-infected foundresses produce relative to uninfected foundresses) and average proportion of single foundress broods for 11 species of fig-pollinating wasps. As foundress number increases, opportunities for horizontal transmission increase. Species in which the opportunities for transmission are increasingly horizontal harbor more virulent species of nematodes.

number of wasp offspring associated with nematode-infected single foundress wasps can be compared to the number of offspring associated with uninfected single foundresses in order to estimate the nematodes' effects on the fecundity (one major component of lifetime reproductive success) of their host fig wasps (i.e., virulence). Across species of fig wasps, virulence estimates can be compared with differences among them in the opportunities for nematode transmission.

Long-term studies indicate that fig wasp species vary characteristically in population structure (number of foundresses). Studies initiated over 15 years ago (roughly 180–200 fig wasp and nematode generations) in the vicinity of the Panama Canal show that the different host wasp species present a continuum of population structures (distributions of numbers of foundresses per fig fruit) (Herre 1985, 1987, 1989, 1993, unpubl.). Species in which the average number of foundresses is consistently higher present their parasitic nematode populations with relatively increased opportunities for horizontal transmission. Additionally, increased opportunities for horizontal transmission are linked with the increased mixing of unrelated nematodes within individual hosts. As previously reported, these are the situations in which the nematode species with the greatest estimated virulences are found (fig. 11.6; Herre

TABLE 11.7
Nematode Parasites of the Pollinators

High degree of species-specificity with host wasps
High degree of concordance (implying a predominance of co-cladogenisis) with host wasps
Ancient association with the wasps
Range of opportunities for transmission
Range of negative effects (virulence) on host fitness
Most virulent species are associated with highest opportunities for transmission

1993, 1995). Finally, theory can enjoy a modest success. But what is really going on here? (See table 11.7.)

Figs and Seed Dispersers

In the vast majority of cases, pollinator wasps are required in order for figs to develop fully ripe fruit. Without the wasps to catalyze the usually bountiful fruit production, a wide array of frugivorous animals would be that much closer to starvation. In turn, the various frugivores disperse the fig seeds (Janzen 1979; Milton 1991; Kalko et al. 1996). Although fig fruit have been characterized as low-quality resources for frugivores, more recent and careful analyses show that fig fruits can be quite high in certain nutrients (e.g., calcium, protein, as well as sugars) (Kalko et al. 1996; Herre 1996; O'Brien et al. 1998; Wendln et al. unpubl.). Figs invest a great deal of energy and resources into their fruit. Why?

Although the fruits of many different species of figs are eaten by a wide range of frugivorous animals, closer inspection in New World systems shows at least two types of specificities. First, different sets of frugivorous species are associated with the fig species possessing relatively small fruits that ripen to a red color, and the species with green-ripening fruits that are much more variable in size. Mainly birds take the former during the day, and bats predominantly take the latter during the night (Kalko et al. 1996; Korine, Kalko, and Herre in prep.). Furthermore, within the species with green ripening fruit, there is a very clear association between size of the fruit of a given species and the body size of the bat species that prefer them. In general, the bat species that constitute the principal frugivores/seed dispersers of larger species of figs are larger, travel longer distances with the fruit, and have larger home ranges. The implication is that, on average, they disperse seeds over larger areas (Herre 1989, 1996; Kalko et al. 1996). Assuming that increased seed dispersal is advantageous, selection for dispersal may favor larger fruit.

Fruit size is related to a wide variety of what would otherwise appear to be unrelated traits. Both across and within species, fruit size (specifically, the number of flowers per syconium) affects the wasp reproductive success; with all else equal, more flowers generally lead to greater wasp reproductive success (Herre 1989; Herre and West 1997). In most large-fruited species, however, foundress number is relatively high. High foundress numbers are associated with wasp sex ratios that are less female-biased, which means less efficient pollen dispersal for the fig (Herre 1985, 1987, 1989). Because wasp offspring generally do not increase proportionately to foundress number, higher foundress number also means lowered reproductive success of the average foundress (Herre 1989). In addition, higher foundress numbers are related to increased opportunities for nematode transmission, increased genetic heterogeneity of nematodes that infect single foundresses, and consequently higher nematode virulence (Herre 1993, 1995).

Larger fruit also show an increased absolute and relative dependence on evaporation in order to produce fruit temperatures cool enough to maintain viable wasps (Patiño et al. 1994). The result is that producing the wasps that carry the pollen is also more expensive in terms of water loss in large-fruited species. Moreover, in both the Old World (Compton and Hawkins 1992) and New World (West and Herre unpubl.) figs, larger-fruited species generally harbor more parasitic wasps, both in number of species and often in number of individuals. Finally, as mentioned before, large, multifoundress fruit may present an open door to internal parasites such as *Sycophaga*, or to pollinators with overt parasitic tendencies, such as *Ceratosolen galili*. Given what appear to be a number of disadvantages associated with larger fruit size, why do large-fruited species exist? Perhaps the advantages from increased seed dispersal balance the multiple disadvantages mentioned above. All of this is consistent with the interpretation that selection generated from seed dispersers produces changes (fruit size) that impinge on other aspects of the fig biology. That is, one set of mutualists associated with figs appears to impose selection that can oppose the interests of others.

Conclusions

In the introduction I noted the great importance and diversity of parasitic and mutualistic interactions and asked if there were some simple rules that govern them. After stating some of the candidates for simple rules, I discussed a great deal of very detailed natural history concerning the different mutualists and parasites associated with figs. The presentation of this detail is crucial for several reasons.

First, in order to test the applicability of theory, it is essential to understand the mechanics that underlie these interactions. In order to do that, it is critical to understand that these are complex systems and that multiple factors influence outcomes. For example, taking the simple (usually positive) correlations between pollinators and seeds, or those between parasites and pollinators, at face value, it is difficult to understand that the relationships are in fact either partially or completely dominated by antagonistic elements. Failing that, with respect to understanding the relationships or using those relationships to test theory, you are lost.

Furthermore, it is frequently important to understand the interactions among several different types of species in order to provide the context to properly pose and test evolutionary hypotheses about any of them. For example, understanding the biology of the different types of parasites is almost certainly important in understanding either the stability of the fig-pollinator interactions, or the evolutionary shifts of figs from monoecy to dioecy. It is not an accident that different facets of natural history relevant to the discussion of fig-pollinator stability were discussed over several sections concerning several different wasp taxa associated with the figs.

Moreover, the degree of interrelationships among what would seem to be unconnected characteristics of the figs and their associates is both extraordinary and unexpected. In essence, the central message is the importance of viewing any one attribute of this system (and by implication any system) within the context of the other aspects, and of clearly understanding the functional interrelationships among them.

It should be noted that the examples presented as tests of theory have involved analyses at very different scales of biological organization. Specifically, I have considered how differences (and similarities) among broadly defined taxa (pollinators, competitors, etc.) in characteristics such as ecological dispersal or evolutionary tracking are associated with parasitism or mutualism. I have considered how differences among species within a taxon may be associated with differences that increase or decrease detrimental effects on an interacting species (virulence in the nematodes, sex ratios in the pollinator wasps). Finally, I have considered how differences among individuals within one species may affect the other (e.g., body size in the pollinators). With respect to the latter, it appears that analyses of the consequences of variation among individuals may provide the key to understanding the mechanisms underlying fig-wasp stability.

At the level of the stereotyped entities such as "pollinator wasps," "externally ovipositing parasitic wasps," "internally ovipositing parasitic wasps," "large gall forming wasps," "parasitoid wasps," the "rules" relating ecological transmission patterns or evolutionary tracking to greater or lesser beneficial effects simply fail. In contemplating this failure, other, more general, examples should also be considered. For example, it is worth bearing in mind that gonorrhea, syphilis, herpes, and HIV all have similar transmission patterns (mostly horizontal) but very different effects on their hosts. Similarly, mycorrhizal fungi, zooxanthellae (symbiotic algae associated with corals), and nitrogen-fixing bacteria are also mostly horizontally transmitted. Yet, the effects they have on their hosts range from commensal to beneficial to indispensable. The nematode example provides the most clear-cut case for making a statement about a connection between mode of transmission and virulence. Why do "the rules" work in the one case and apparently not in the others?

In the specific case of the fig-pollinating wasps and their nematode parasites, increased opportunities for horizontal transmission decouple the parasite's reproductive interests from those of its host. Therefore, in this system, opportunities for horizontal transmission per se only serve to release a constraint against virulence. It is within-host competition of different parasite genotypes that is identified as the driving selective force toward virulence, much as within-deme selection favors less female-biased sex ratios in the pollinator wasps (Frank 1985, 1992; Herre 1985, 1993, 1995).

More generally, the much maligned rules are much more helpful in understanding differences among species within a group of very similar organisms (e.g., differences in virulence among nematodes, or differences in sex ratio among pollinator wasps), in which a few variables that theory identifies as key shift (respectively, transmission, local mate competition), than they are with respect to explaining differences among much more dissimilar entities (Read and Harvey 1993). As is the case with making single-factor analyses within complex systems, among-taxa comparisons confound many factors.

These examples suggest that "rules" (e.g., links between transmission and virulence, links between long-term association and niceness, etc.) are subject to contravening influences. For many reasons, it is unlikely that rules governing among-species interactions across a wide diversity of types of interactions will be found that will even approach the power of Hamilton's Rule in explaining within-species interactions. Instead, the task appears to be to carefully document many systems, especially very similar systems in which only a few key variables change, and then to attempt to determine if there is a hierarchy of rules. Ultimately, a great deal of attention will need to be paid to differences in details of the natural history of the different systems.

This piece started out as an exploration of applicability of rules. It has evolved by necessity into a discussion of details of the natural history of a series of unexpectedly interrelated examples of interactions involving figs. The central message should not be the lack of rules with any explanatory power concerning the outcomes of species interactions. Instead, understanding why the rules work in the cases that they do is crucial, as is the appreciation that context and scale determine the applicability of those rules we presently recognize.

Addenda

1. Based on the average number of offspring that one foundress is capable of producing, Panamanian fig fruits pollinated by multiple foundresses frequently receive several times more eggs than the number of flowers per syconium (Herre 1989). Yet they still produce intact seeds. Therefore, the hypothesis of egg limitation (e.g., Nefdt and Compton 1996) cannot account for the maintenance of seed production in these species.

2. Recent molecular studies of Rasplus et al. (1998) show that the pollinating and nonpollinating wasps that had been classified in the family Agaonidae (sensu Boucek 1988) are not monophyletic. These results are consistent with results reported earlier by Machado et al. (1996). Although it is presently unclear which taxa show the closest affinities with the pollinators, the hypothesis that the pollinators' closest relatives are parasites of the fig-wasp mutualism cannot be rejected, and it is clear that there have been multiple instances of both convergent and divergent evolution among wasp taxa associated with figs.

Acknowledgments

For extensive discussion and useful coments during the writing of this paper, I thank Betsy Arnold, Elisabeth Kalko, Egbert Leigh, and Stuart West. The empirical and conceptual foundations contributing to this work were laid during the course of a series of very enjoyable collaborations with Elisabeth Kalko, Carlos Machado, John Nason, Sandra Patiño, Mel Tyree, Don Windsor, and Stuart West. For additional useful discussions and comments, I thank Steve Compton, Cees Berg, Koos Wiebes, William Hamilton, Drude Molbo, Dieter Ebert, John Thompson, Finn Kjellberg, Judie Bronstein, and Rhett Harrison. Laurent Keller has circumnavigated the Sea of Patience on my behalf.

12 Lineage Selection: Natural Selection for Long-Term Benefit

Leonard Nunney

Natural selection can occur at any level of biological organization and at any time scale. As a result, it is not unusual for conflicts to arise (Leigh 1971). For example, a character that is favored at the individual level over a few generations may not be advantageous when viewed from the perspective of population persistence over several thousand generations. Traditional population genetic theory gives us the tools to model such conflicts and shows us that, in general, they are resolved in favor of the shorter time scale. For example, consider the case in which an episodic event, such as an epidemic, favors one resistant genotype, but the nonepidemic conditions favor an alternate form. Between epidemics, selection reduces the frequency of the resistant genotype, and if the period between epidemics is long enough, the nonresistant genotype will spread to fixation. This occurs even if the loss of resistant individuals results in the next epidemic being uniformly lethal. Only when the time scales of the opposing types of selection are not so disparate can a polymorphism be maintained (at least under certain conditions; see Haldane and Jayakar 1963). Thus, our models reinforce the truism that selection is blind to the future: Strong selection acting on individuals on a short time scale will drive a trait to fixation, even if that trait, once fixed, results in the future extinction of the population.

Does this result mean that evolution is always controlled by the shortest time scale operating? Clearly it is not. For example, cells make up individuals just as individuals make up populations, and events acting on the time scale of the cellular level (such as somatic cell growth) are not the primary processes driving the evolution of individuals. We can resolve this apparent paradox by invoking kin selection: The 100% relatedness of the cells of an individual provides the needed link between the short-term interests of the cell and the longer-term interests of the individual (see, e.g., Maynard Smith and Szathmáry 1995). This explanation is sufficient to account for the persistence of multicellularity. It is not the whole story, however. In modeling kin selection (Hamilton 1964a,b), we implicitly (and appropriately) view selection on the time scale of the individual; however, the conflict between the short- and long-time scale persists at the cellular level. This conflict is apparent in the continued occurrence of cancer. To an evolutionary biologist, cancer is the result of proliferating cells (successful on their time scale) jeopardizing the survival of the population of cells (the individual).

On the time scale of the cells, each multicellular individual is a single lineage that derives from a single zygote. The cells of a lineage share certain properties because of their common genetic origins. Similarly, although on a different time scale, the members of a species all belong to a single lineage with common ancestry and share some characteristics as a result. (In fact, in some cases, it may be appropriate to consider a species as several isolated lineages). The similarity of the members of a lineage creates the potential for lineage selection, which is defined as selection favoring lineages that suppress strategies successful only on a short time scale (Nunney 1999). Lineage selection operates whenever there is a conflict between selection acting on two different time scales (which we generally view as two different levels) and acts to preserve the longer-term strategy.

Lineage selection has its conceptual roots in the 1960s and 1970s, in the debate over the conflict between individual-level (short-term) and group-level (long-term) selective pressures (see Williams 1971). The initial discussion of this conflict revolved mainly around two topics: the maintenance of sexual reproduction and the evolution of population self-regulation. In particular, Maynard Smith (1964) and Williams (1966a) emphasized that, in general, a trait could not be favored because of its long-term benefit to the group if the trait had a short-term fitness disadvantage, that is, group selection could not usually overcome the force of individual selection. Theoretical models provided broad support for this conclusion (reviewed by Maynard Smith 1976). Subsequently, discussion of group selection became complicated by different researchers defining it in different ways (see Nunney 1985); however, the debate had some very important effects. It led to (1) the search for short-term advantages for sex; (2) the general rejection of the hypothesis of population self-regulation proposed by Wynne-Edwards (1962); and (3) a general awareness that the evolution of a trait should not be justified in terms of its benefit to the group, unless rather special conditions prevail (e.g., close kinship).

Our ability to recognize the erroneous use of "group-selection arguments" remains a valuable legacy of the early debate. But why do we suspect that such arguments are erroneous? In his debate with Wynne-Edwards in the pages of *Nature*, Maynard Smith (1964) raised the problem of the "cheat," that is, a mutant genotype that displayed a selfish, antisocial phenotype. He pointed out that a cheat of this type would spread in the population regardless of the negative consequences for the group, unless the total population was subdivided unrealistically into many highly isolated groups. Wynne-Edwards (1964) responded with the argument that the social behavior driving his hypothesis was an ancient component of animal biology and, because of this ancient origin, mutant cheats simply could not occur. Although this argument is untenable, it remains true that if a "cheating" mutant does not arise, the group-selected trait would be maintained.

The conflict between short- and long-term strategies that results in lineage selection can be viewed in this larger framework of selfish cheats versus nonselfish altruists. In this chapter, I first discuss this framework and in doing so introduce some possible cases of lineage selection. I then illustrate the potential of lineage selection using two examples involving conflicts at very different levels of organization: the maintenance of sex (species vs. individual levels) and the avoidance of cancer (single cell vs. individual levels). In each case, lineage selection does not alter the fitness relationships of the traits; instead, it exploits variation in the genetic architecture of lineages to minimize the occurrence of the trait that is advantageous in the short term but disadvantageous in the long term.

Defense against Cheats

The role of cheats and their suppression has been recognized as central to the likelihood of successful group selection since the early 1960s. In an altruistic community, selfish cheats can spread if they can avoid the prevailing rules of association and exploit altruistic neighbors. This importance of association is formalized in the theory of kin selection (Hamilton 1964a,b): Selection for altruism is successful only if relatives preferentially help each other. More generally, the positive association of similar genotypes is always required to drive the spread of altruistic traits by group selection (Nunney 1985). Although it can be argued that the conditions permitting the evolution of altruism also preclude the invasion of cheats (Wade and Breden 1980), this only applies if the cheats follow the same rules of association (Nunney 1985). Because cheats, by definition, are not following rules, we expect group-selected systems to evolve policing mechanisms (Frank 1995).

Cheats can arise from different sources; in particular, they may invade from outside the group, or they may originate within the group. Those coming from outside can be excluded from an altruistic community provided there is group (generally, kin) recognition. Lacy and Sherman (1983) developed the conceptual framework for studying kin recognition, and it is apparent that kin (or colony) recognition is indeed widespread among social groups and is used as the basis for colony defense (e.g., in the social insects; see Crozier and Pamilo 1996). Grosberg and Quinn (1989) made an important generalization when they noted that such colony defense is often likely to be driven by aggression toward nonmatching phenotypes, rather than by the acceptance of similar phenotypes. This pattern is particularly relevant to colony formation among related and unrelated marine invertebrates. McKean and Zuk (1995) have suggested that there are important similarities between such recognition systems and the immune system. Thus, for example, cancer cells can be viewed as selfish cheats that exploit the resources of a group (the individual) and that are destroyed by the immune system when they are recognizably different. A second example of the suppression of internal cheating is provided by "worker policing" (Ratnieks 1988). In social Hymenoptera, haplodiploidy means that the workers, if they have functioning ovaries, can produce sons from unfertilized eggs. In honeybee colonies, workers indeed lay eggs, but these are generally detected and destroyed by other workers (Ratnieks and Visscher 1989). These last two examples, immune policing (for cancer cells) and worker policing (for worker eggs) are almost certainly, in part, the products of lineage selection, that is, the evolution of these traits was probably influenced by antagonism between the short-term and long-term fitness effects of the underlying traits (cancer and worker reproduction, respectively). It is likely, however, that other factors were very important in the evolution of both of these functions. For example, the primary value of the vertebrate immune system is in recognizing interspecific (pathogen or parasite) invaders. It is therefore an open question how much of cancer detection by the immune system is simply a by-product of disease prevention, and how much is specifically tailored by lineage selection. Similarly, worker policing may have originated to detect and destroy colony parasites and was only later recruited by lineage selection to detect eggs laid by workers.

In both of the policing examples just described a specialized fraction of the group seeks out and destroys the product of cheating. An alternative evolutionary path acts to make the occurrence of cheats within the group less and less likely. Wynne-Edwards (1964) recognized this possibility when he argued that certain behavioral cheats could not exist because of evolutionary constraints. His appeal to the "impossibility" of a cheat was unacceptable because there was no clear justification of how this situation could arise. However, under the right conditions, precisely such a situation can arise through lineage selection. The sequestering of the germ line is perhaps the end result of this type of lineage selection. Buss (1987, p. 181) considered this "the triumph of the level of the individual over the level of the cell" because it prevents somatic variants from entering the gametes, a view supported by the theoretical work of Michod (1996; see also chap. 3). The germ line limits somatic cheating, because growth beyond that required to maximize the fitness of the germ line cannot be favored on a time scale greater than one generation.

Lineage selection reduces the likelihood of cheating over the long term provided lineages that give rise to cheats at a low frequency are more successful over the long term than are lineages that give rise to cheats at a high frequency. Thus, the action of this form of lineage selection relies on variation among lineages in the frequency with which they give rise to cheats. To understand the process more clearly and to be able to use it in a predictive way, we need to be able to model it. To this end, I consider two examples of lineage selection in which different aspects of the process can be examined quantitatively.

The Maintenance of Sexual Reproduction

Early in this century, East (1918), and later Fisher (1930) and Muller (1932), proposed that sexual species are more successful than parthenogens because of the ability of sexuals to evolve more rapidly in response to natural selection. This hypothesis was generally accepted, and subsequent theoretical work validated the basic argument (see Kondrashov 1993). In general, the difference in the rate of evolution of sexual versus asexual populations increases as the number of loci simultaneously involved in allelic change increases, and only in small populations subject to strong selection does this advantage disappear.

Even though the underlying logic was sound, the evolutionary rate hypothesis was undermined when Maynard Smith (1971) recognized that sexual individuals often experience a halving of their reproductive potential when compared to asexuals. This result was based on the realization that, in general, the fecundity of females is fixed, irrespective of their sexual strategy. This is true whenever a male's investment in his offspring is limited to his sperm, that is, when he neither provisions his mate nor his offspring. Under such conditions (and given a 1:1 sex ratio), the productivity of an asexual female is twice that of a sexual female because she produces twice as many daughters. A parallel argument has been made by Williams (1988). He noted that the cost of sex can be viewed from a more genetic perspective, because, under asexuality, two copies of genes are passed on to offspring rather than one. Regardless of the argument one favors, the twofold advantage requires that any hypothesis accounting for the maintenance of sex must invoke a process capable of overcoming this fitness loss, and it seems unlikely that evolutionary flexibility can achieve this.

Since 1971, a variety of additional hypotheses have been proposed and older ones reevaluated (reviewed by Kondrashov 1993). They can be divided into two broad categories depending on whether they invoke short- or long-term fitness benefits. Each of the short-term models proposes a different mechanism by which short-term selective forces overcome the twofold disadvantage. Their validity rests on empirical support for the strong short-term selection that they invoke. The long-term models, however, raise the complexity of selection acting at different time-scales, and here I focus only on this issue.

As noted above, the original long-term model proposed that the enhanced ability to evolve in response to changing environmental conditions increased the probability that a species would persist. The remaining models, which also predict lower extinction rates for sexual species, focus on mutation accumulation. Sexual reproduction prevents the action of "Muller's ratchet" (Muller 1964; see also Haigh 1978; Lynch et al. 1993) and a variant of this idea, "Muller's hatchet" (Kondrashov 1982). Both of these hypotheses emphasize that an asexual clone is much more vulnerable to deleterious mutations than a sexual group. Muller's ratchet is the continuous stochastic accumulation of deleterious mutations in small or highly mutable asexual populations. In particular, if all lineages in an asexual population have at least n deleterious mutations, then, given that the chance of a back mutation is negligible, no lineage will ever have less than n mutations. In a small population, these lineages may continue to accumulate mutations in spite of selection, and Muller's ratchet will turn when the last remaining lineage with only n mutations gains another. Sexual populations avoid this effect because of sexual recombination: If two individuals with n deleterious mutations (at different loci) mate, then some of their offspring will carry fewer than n mutations. Muller's hatchet reflects the fact that, at equilibrium, asexual populations can have a much higher genetic load than a sexual population if the number of deleterious mutations carried by an individual reduces fitness in an accelerating fashion. This difference arises because sex, by creating genetic variance, is more efficient at eliminating deleterious mutations. Thus, both of these mechanisms favor sexual reproduction.

There is good evidence that a long-term advantage to sexual reproduction does exist. Based on the taxonomic distribution of parthenogenetic species, it appears that parthenogenesis must have evolved many hundreds of times, yet in almost all taxonomic groups, the closest relatives of a parthenogenetic species are extant and sexual (see Bell 1982). These findings suggest that the asexual lineages present today are usually of recent origin, which can be explained if asexuality results in a relatively high extinction rate (Stanley 1975). Indeed, criticism of long-term hypotheses does not revolve around researchers' doubting the reality of a long-term benefit to sex; instead, it concerns the extreme imbalance between the immediacy of the twofold per-generation cost of sex versus the delayed benefits. The selective advantage of an enhanced evolutionary rate (or the avoidance of mutation accumulation) is small when considered over a single generation; hence, as Maynard Smith (1978) emphasized, supporters of a long-term hypothesis must explain why asexual "cheats" are not constantly arising and displacing sexuals as a result of their superior individual fitness.

Maynard Smith (1971) recognized this as a problem of group selection: How can a trait beneficial to the species (group) over the long term be retained in opposition to short-term individual selection? Van Valen (1975) quantified the problem and showed that the equilibrium frequency of sexual lineages, when the twofold advantage of asexuals causes them to displace their sexual ancestors, is $(1 - U/E)$, where U is the rate of production of fully viable asexuals, and E reflects the excess extinction experienced by asexuals (i.e., $E_{asex} - E_{sex}$). Thus, a long-term advantage can account for the maintenance of sex only if $E > U$. Maynard Smith's (1978) point was that there seemed to be no a priori reason why this inequality should true. In fact, the precise opposite seems likely, because E, the per-generation advantage of sex resulting from extinction (or speciation), is very small. However, U may be relatively large because, although the chance of an individual's being an asexual mutant is small, U is the product of this and the population size of the species. The problem of why the inequality $E > U$ should, in general, be satisfied can be resolved by recognizing the action of lineage selection. Recall that lineage selection acts to resolve conflicts between long- and short-term effects of selection in favor of the long term. In this case, the conflict is between long-term species-level success and short-term individual-level fitness. In its action, lineage selection exploits differences among lineages. These fixed differences are assumed to be due to the accumulated effects of short-term processes (either selection or drift) that may have nothing to do with the conflict between sexual and asexual reproduction. For example, if some lineages evolve genetic imprinting (discussed below), then these lineages may be much less likely to give rise to asexual mutants. It is possible that in plants, the evolution of self-incompatibility could create a similar effect.

To examine this conflict between the long- and short-term, I developed a model that shows how lineage selection can drive U to a very low value, thus allowing sexual reproduction to prevail (Nunney 1989). Note that when the inequality holds, sex is retained even though asexual reproduction may be the most advantageous strategy most of the time. My model of lineage selection has the following features. First, the "group selection" conflict between strong short-term selection and weak long-term selection is included in the following way: (1) Over the short term, asexuals very rapidly displace sexuals because of their twofold advantage; and (2) over the long term, asexual species have a higher extinction rate than sexual species. Note that the *cause* of the extinction rate difference is not specified; it only matters that a difference exists. As a result, this approach applies to all long-term hypotheses. Second, the number of species is held constant. Following an extinction, another extant lineage is chosen at random to "speciate" and thus replaces the lost line. It is very unlikely that asexual lines are able to "speciate" at the rate of sexuals, as assumed in the model; however, this inequality would simply serve to accelerate lineage selection. Third, the opportunity for lineage selection is created. Lineage selection, like any other form of natural selection, operates on genetically based variation. In this

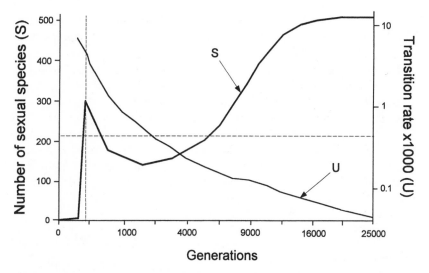

FIG. 12.1. The effect of lineage selection on the maintenance of sex. Initially, a high overall extinction rate favored the spread of sex (solid curve, S); however, after 200 generations (vertical dashed line), the extinction rate was lowered to favor asexuality. Lineage selection continuously lowered the transition rate (solid curve, U) until it was below the critical level for the spread of sex (horizontal dashed line), at which point sex spreads to near fixation. For further details, see text. Redrawn from Nunney (1989; with permission of *Evolution*).

case, the variation is among species, and it takes the form of differences in the rate at which species give rise to fully viable asexual forms.

Simulations of the model showed that the effect of lineage selection is to favor sexual species that generate asexuals at a very low rate and, as a result, to favor the maintenance of sex. The mechanism is very simple. (1) Sexual lineages that are prone to give rise to asexuals, soon do so, and strong short-term selection causes the rapid transition of these lineages to asexuality. (2) The resulting parthenogenetic species have a relatively high extinction rate, so these lineages tend to disappear over the long term. (3) The remaining species are primarily those that retained sex, and this retention is a direct consequence of their low rate of forming asexuals. Thus, sex is retained, not because the twofold advantage of asexuals is overcome, but because the species that persevere lack the opportunity to become parthenogenetic.

The simple, but inexorable, process of lineage selection is illustrated by the simulation shown in figure 12.1. The main points can be summarized by recognizing that the simulation has three phases. The first phase illustrates that a high extinction rate is one of the mechanisms that could have originally favored the spread of sex. Note, however, that, in reality, the mechanism originally favoring the spread of sex is quite likely to be unrelated to the mechanism that now maintains it. The other two phases show how lineage selection acts to maintain sex once it has become established. Phase 1: Sexual reproduction spreads (generations 0–200). All 500 species were initially asexual, but they were in an environment with a very high extinction rate. I set this rate high enough to favor sex, given the initial rate of asexual production, i.e., $E > U$. The percentage of sexual species (S) increased to 60%. Phase 2: Asexual reproduction spreads as the result of an environmental change (gens. 200–2,000). To initiate this phase, I assumed that the environment became more benign and lowered the extinction rates by a factor of 100 (such that $E < U$). Over the next 1,800 generations, the proportion of sexual species dropped to 28%. Phase 3: Sexual reproduction is favored because of the accumulated effects of lineage selection (gen. 2,000 on). By generation 2,000, the average rate at which asexuals arose in lineages had dropped low enough through selection to favor sexual reproduction. The decline in U continued through the end of the simulation, by which time, sex completely predominated.

As I noted above, the driving force of lineage selection is the variation among lineages in U, the chance of a viable asexual form occurring once (or very few times) within the species per generation. This rate is proportional to population size and could, in some circumstances, be quite large. For this simulation (fig. 12.1), all lines were initially set at $U = 0.01$, and it was assumed that this rate translates directly into the probability that sexual reproduction is lost, because very strong selection favors the spread of asexual clones once they arise. At each speciation event (following each extinction), the daughter species had a 20% chance of having a different value of U (30% higher or lower), with the constraint that $10^{-2} \geq U_s \geq 10^{-5}$. These mutational changes alone would drive U lower (to roughly 10^{-3}), but not low enough to favor sexual reproduction, that is, phase 3 would never be reached.

Although the circumstances required for lineage selection appear to be in place, it remains difficult to establish the importance of lineage selection in maintaining sexual reproduction. The time scale of its action is very long, and alternative explanations are difficult to rule out; however, there is a strong circumstantial case. The expected result of lineage selection is that the clades least likely to give rise to an asexual are the most successful. The very rare occurrence of parthenogenesis in vertebrates, with its usual (and possibly complete) association with hybridization, is certainly consistent with this expectation. One process that makes abandoning sex particularly difficult is genetic imprinting, because two epigenetically identical gene copies (both imprinted, or both not imprinted) can lead to defective development. It has been suggested that imprinting is individually advantageous to female mammals because it protects maternal control over the allocation of resources to offspring (Haig 1993b); however, once evolved, it is very effective at preventing parthenogenetic reproduction.

Another source of indirect evidence appears at first sight to run counter to the expectations of lineage selection. This concerns the widespread occurrence of asexual forms in some successful groups, such as the branchiopod crustaceans and the parasitoid Hymenoptera (see Bell 1982). Here the mechanism inducing asexuality becomes important. Lineage selection predicts that lineages with a genetic predisposition to give rise to asexual individuals will tend to decline over the long term. So why does parthenogenesis remain relatively common? It is becoming increasingly likely that parthenogenesis in many arthropods has little to do with the genome of the organism and a great deal to do with an intracellular parasite. It appears that these parasites (e.g., *Wolbachia*) may bypass the usual genetic constraints of the host. These parasites are under strong selection to induce parthenogenesis in their host, because they are only passed on through the female cytoplasm. For example, Stouthamer et al. (1990) showed that an all-female species of parasitoid wasp could be "cured" of asexuality through the use of antibiotics.

Vegetative reproduction is another strategy for "avoiding" sex. This is a difficult option for many animals because of the sequestering of the germ line, with the result that somatic growth beyond that required to favor the germ line has ceased to be advantageous. On the other hand, plants do not have this restriction (see Williams 1975). It would be interesting to know how difficult it is for higher plants to give up sexual investment. If lineage selection acts at this level, then fully viable mutants suppressing all sexual function should be very infrequent. We know that producing sexual sterility is relatively easy, as can be seen from artificial selection for doubled flowers and seedless fruit; however, these mutants still invest in the process of sexual reproduction, even though they produce no offspring.

This form of lineage selection, favoring species lineages based on variation in their genetic architecture, is a form of species selection (Stanley 1979). This in turn can be subsumed under the more general category of clade selection (Stearns 1986), which can be applied at any phylogenetic level. I have avoided these terms because there is no general agreement over whether these hierarchical terms should be restricted to cases in which there is some emergent property at the reference level (Vrba 1989) or whether they should be used more broadly as a description of processes occurring at any chosen level (Williams 1992). My own preference is that the purely descriptive usage should be avoided, because it fails to provide insight into evolutionary mechanisms (Nunney 1993). The term *lineage selection* describes a particular evolutionary mechanism that results in an emergent property at the level with the longest time scale. As with clade selection, different examples may apply at different hierarchical levels. To illustrate this flexibility, I now consider a second example of lineage selection that acts not at the species level, but at the individual level.

Protection from Cancer

Cancer cells reap a short-term reproductive benefit at the expense of the long-term success of the lineage (the individual) from which they arise. This suggests an analogy between the effects of cancer on an individual and the effects of asexuality on a species as described in the previous section. For the case of sex, it was shown that lineage selection acts to reduce the likelihood of asexual mutants. Similarly, for the case of cancer, lineage selection acts to reduce the likelihood of cancerous mutants. In particular, I have argued (Nunney 1999) that lineage selection will act to decrease the likelihood of such mutants by increasing the number of genes recruited to control cell proliferation. Of course, this increase is finite; as the complexity of the control increases, the effectiveness of selection decreases.

I expect lineage selection to act somewhat independently in different tissues. This expectation stems from tissue-specific differences in the number of divisions that cells undergo during the reproductive life of an individual. These differences are important because they influence the likelihood of somatic mutation. In addition, different cell types are under the control of different genes, which may require different growth controls. Both of these features also vary across organisms. Thus, the evolutionarily stable degree of complexity that controls a given cancer depends on both the tissue and the organism.

Tumors go through complex evolutionary changes within individuals as they become progressively more adapted for growth (see Shackney and Shankey 1997), including the loss of some chromosomal regions and the increased ploidy of others. However, of particular interest to the evolution of cancer avoidance are the very early stages of tumerigenesis, when the switch is made from a normal "altruistic" cell to a "selfish" precancerous one. This switch results from mutation, but the number of mutations required can vary, depending on the type of cancer. Here I wish to show that this variation is both expected and predictable, based on the action of lineage selection.

Retinoblastoma is a tumor of the retina that can be inherited. When inherited, it acts like a simple dominant Mendelian trait and usually results in tumors in both eyes, whereas the very rare nonfamilial cases affect only one eye. Knudson (1971) realized that what is inherited is not a dominant allele but a recessive mutation in a tumor-suppressor gene (RB), and proposed his "two-hit" model for retinoblastoma. Given this model, the onset of familial retinoblastoma results from the somatic mutation of the second, previously functional, gene copy in heterozygotes. In nonfamilial cases, both copies of the *RB* gene must mutate, which occurs at a frequency of about 1/30,000. Hethcote and Knudson (1978) modeled this process and estimated a somatic mutation rate in *RB* of about 4×10^{-7}/daughter cell/division ($= u$) in a pair of retinoblasts that peak at a combined size of approximately 4×10^6 cells ($= C$). Why a "two-hit" model? Wouldn't the requirement for three or even four mutational "hits" better protect us from this cancer? We can use the concept of lineage selection to evaluate this issue.

Cell growth can be controlled either positively or negatively. Proto-oncogenes positively transmit growth signals. As a result, mutations that cause these genes to transmit a false signal are dominant. In contrast, tumor-suppressor genes (such as *RB*) inhibit growth, and loss-of-function mutations in such genes are recessive. We can model the effect of somatic mutation on the likelihood of cancer, given these two basic forms of control (Nunney 1999). Thus, a "single-hit" model would be exemplified by a tissue in which cell growth is stimulated by a single proto-oncogene. Given simple geometric growth of the tissue, the probability of at least one cancer-producing mutation occurring during growth is approximately $1 - \exp(-4Cu)$. Thus, if u is about 10^{-7}, then cancer is likely to occur more than 10% of the time if $C > 250,000$ cells. Recalling that the tiny stem-cell population of the retinoblasts is about 4×10^6, it becomes apparent that there would be strong selection to recruit additional safeguards into such a simple regulatory system.

This discussion raises the question of how strong selection must be in order to drive the recruitment of additional safeguards. If selection is weak enough, then genetic drift will prevent directional selection from being effective. Wright (1931) formalized the relationship between selection and drift and showed that we can expect selection to overcome genetic drift provided $N_e s > 1$, where N_e is the effective population size, and s is the selection coefficient. In the present context, this result shows that selection depends not only on the number of cells in an individual's tissue, but also on the number of individuals present. N_e is the effective number of individuals (lineages) in the population and is a measure of the size of the population passing genes to the next generation (reviewed by Caballero 1994; Nunney and Elam 1994). The second parameter, s, is the average cancer-induced loss of fitness of individuals with the prevailing level of anticancer safeguards. Specifically, s equals the product of the probability of an individual getting cancer and the fitness loss to the individual when cancer strikes. Thus, for a cancer that is lethal before reproduction, s equals the chance of an individual's getting cancer and, based on the inequality $N_e s > 1$, selection will be effective if this chance is greater than about $1/N_e$. If we assume that the prevailing level of safeguard is a single hit and $N_e = 10^4$, then selection will favor increasing the safeguards over unregulated cell growth for any tissue with more than 250 cells!

The next simplest regulation is provided by a single tumor-suppressor gene such as *RB* (an alternative model would be regulation mediated through two proto-oncogenes). Because these gene products act to suppress growth, two mutational hits are needed for cancer to develop. Again, assuming geometric tissue growth, and given that u is small, the probability that the two necessary mutations accumulate in a single cell is approximately $4C(k-1)u^2$ (Nunney 1999), where k is the number of rounds of cell division needed under geometric growth. Under this two-hit system, and given the same parameters used in the one-hit example, cancer will occur with a frequency of 10^{-4} or more only if the tissue is about 10^8 cells or more. This corresponds to about 0.1g of tissue, which suggests that the two-hit system is evolutionarily stable only for very small tissues.

Combining the last two models gives a three-hit scenario, in which cell growth is regulated by a single proto-oncogene and a single tumor-suppressor gene. For large tissues ($k \gg 1$), the probability that a final mutation occurs in a lineage already containing the other two mutations is approximately $12C(k-1)^2u^3$ (Nunney 1999). As before, assuming that $u = 10^{-7}$ and $N_e = 10^4$, selection for enhanced protection through lineage selection occurs only in tissues greater than about 10^{13}, which is about the number of cells in a medium-sized dog.

From these three models, it appears that: (1) the one-hit model is unlikely ever to apply; (2) the two-hit model is expected only when the mutable tissue is small; and (3) the three-hit scenario appears to be an evolutionarily stable strategy for moderately sized mammals, unless their effective population size is large. An important assumption of these models, however, is that tissue grows to a fixed size and then stops proliferating. By contrast, most tissues continue proliferating throughout life, and the effective number of cell divisions can thus be much greater than 10^{13}. For example, the epidermis, the epithelial cells of the gut, various reproductive system cell populations, the hemopoietic cells of the bone marrow, and the lymphatic cells of the thymus, spleen, and lymph node all have rather high turnover rates, generally in the range of 2–10 days (see Cameron 1971). Thus, a high risk of accumulated somatic mutation continues throughout life, and it is probable that additional cell-growth regulation would be recruited through lineage selection (Nunney 1999).

The lineage selection argument suggests that the two-hit, three-hit, or multi-hit scenarios for the genesis of cancer reflect evolved systems that regulate uncontrolled cell proliferation. Lineage selection favors individuals that can suppress a fitness loss from cancer by recruiting additional growth regulation. The degree of complexity of the regulation (and hence the number of somatic mutations needed to destroy it) is expected to increase with the number of cell divisions occurring in a tissue between conception and the cessation of reproduction. Furthermore, the recruitment of regulatory mechanisms is expected to be opportunistic; thus, the specific genes involved may vary among tissues and among taxonomic groups.

I have suggested (Nunney 1999) that the model of lineage selection makes two strong predictions. The first concerns the effect of genetic diseases that increase somatic mutation rates. Tissues experiencing the largest number of cell divisions should exhibit increased rates of cancer. This prediction derives from the expectation that these tissues are those protected by a "multi-hit" system. If the number of hits required is n, then an increase by x-fold in the somatic mutation rate will enhance the probability of n hits by x^n. It is notable that the examples of inherited DNA-repair abnormalities listed by Hall et al. (1995) all increase the incidence of one or more such cancers: lymphoma, leukemia, skin cancer, and colon cancer. The second prediction is that the selection for larger size and for a longer reproductive life will be accompanied by selection to recruit new levels of cell-growth regulation. Different systems will be influenced differentially, because the change in the number of cell divisions that induces effective lineage selection will vary from tissue to tissue. In general, however, larger, longer-lived animals will be those with the most complex growth controls.

Conclusion

Leigh (chap. 2) emphasized the potential conflicts between levels of organization and concluded that features must evolve either to eliminate or limit such conflict. Elimination can occur when, for example, the twofold cost of sex is avoided through the evolution of a sexual strategy lacking this cost, such as monogamy with paternal care (Leigh 1977). In contrast, lineage selection is a mechanism that minimizes this conflict, not by elimination, but by limiting its occurrence. The potential conflict remains, but it is avoided through the suppression of the "antisocial" or "selfish" trait. To be effective, lineage selection requires the modular organization emphasized by Leigh, that is, the lineages must remain distinct long enough for the effects of both long- and short-term selection to influence their success.

The important feature of lineage selection is that it acts on the genetic architecture of lineages. This is very different from the usual effect of natural selection. Usually, if a trait is subject to natural selection, then any response involves the genes that control the trait. By contrast, lineage selection does not change the trait itself. In the two examples that I considered in detail, the traits sexual versus asexual and the traits normal cell versus cancerous cell are not altered. Instead, lineage selection works on the serendipitous recruitment of genes that can act to suppress the "selfish" trait, but these genes are not involved in determining the trait. In these examples, the suppression involved minimizing their occurrence. Another possible outcome of lineage selection is the suppression of the trait through some form of policing (see Frank 1995). Two possible examples were discussed: the immune system policing of cancer cells in vertebrates, and worker policing of worker-produced eggs in eusocial Hymenoptera.

Recognition of lineage selection allows us to have a different perspective on hierarchical selection. In the case of the maintenance of sex, the view that sex was maintained because of its beneficial effect on increasing the rate of evolution and/or reducing the rate of extinction was largely discarded following Maynard Smith's (1971) recognition of the twofold short-term disadvantage of sex. Lineage selection provides a mechanism that can make the short-term disadvantage largely irrelevant because fully viable asexual competitors rarely occur. As a result, sexual reproduction can persevere despite a large short-term disadvantage, provided it is superior to asexuality over the long term (Nunney 1989).

Lineage selection also provides some novel insights into the genesis of cancer by providing a framework to view cancer with an evolutionary perspective (Nunney 1999). In particular, it emphasizes that the cells of different tissue types within the same organism, and of the same tissue types in different organisms, may often evolve different mechanisms for the prevention of cancer. However, although the specific genes involved in the prevention of cancerous cell proliferation may not always be predictable, the theory of lineage selection does predict its complexity, based primarily on the number of divisions that a cell undergoes.

In summary, lineage selection shows how evolutionary conflicts can be resolved. When selection favors opposing traits on two different time scales, lineage selection acts to favor the longer time scale. Thus, the fact that sexual reproduction has been advantageous in the past may have driven lineage selection that makes it difficult to lose sex in the future. The future thus may be more important in evolution than we thought, provided, of course, that it is the same as the past!

Literature Cited

Abdurahiman, U. C., and K. J. Joseph. 1967. Contributions to our knowlege of fig insects (Chalcidoidea: Hymenoptera) from India. II. New fig insects from Delhi and correction of a mistaken identification. *Bull. Entomol.* 8:48–57.

Abegglen, J. J. 1984. *On socialization in Hamadryas baboons.* Cranbury, N.J.: Associated University Presses.

Addicott, J. F., J. L. Bronstein, and F. Kjellberg. 1990. Evolution of mutualistic life cycles: yucca moths and fig wasps. In *Insect life cycles*, ed. F. Gilbert, 143–161. Berlin: Springer-Verlag.

Aguadé, M., N. Miyashita, and C. H. Langley. 1992. Polymorphism and divergence in the *Mst26A* male accessory gland gene region in *Drosophila*. *Genetics* 132:755–770.

Alatalo, R. V., and A. Lundberg. 1990. Polyterritorial polygyny in the pied flycatcher. *Adv. Study Behav.* 19:1–27.

Alatalo, R. V., A. Carlson, A. Lundberg, and S. Ulfstrand. 1981. The conflict between male polygamy and female monogamy: The case of the pied flycatcher *Ficedula hypoleuca*. *Am. Nat.* 117:738–753.

Alcock, J. 1998. *Animal behavior: An evolutionary approach*, 6th ed. Sunderland, Mass.: Sinauer.

Alexander, R. D. 1974. The evolution of social behavior. *Annu. Rev. Ecol. Syst.* 5: 325–383.

Alexander, R. D., K. M. Noonan, and B. J. Crespi. 1991. The evolution of eusociality. In *The biology of the naked mole-rat*, ed. P. W. Sherman, J. U. M. Jarvis, and R. D. Alexander, 3–44. Princeton, N.J.: Princeton University Press.

Allen, M. F. 1991. *The ecology of Mycorrhizae.* Cambridge: Cambridge University Press.

Ameisen, J. C. 1996. The origin of programmed cell death. *Science (Wash.)* 272:1278–1279.

American Genetics Society Symposium for the Evolution of Sex. 1993. *J. Hered.* 84.

Anderson, D. J. 1990a. Evolution of obligate siblicide in boobies. II. Food limitation and parent-offspring conflict. *Evolution* 44:2069–2082.

———. 1990b. Evolution of obligate siblicide in boobies. I. A test of the insurance-egg hypothesis. *Am. Nat.* 135:334–350.

Anderson, P. 1997. Kinase cascades regulating entry into apoptosis. *Microbiol. Rev.* 61:33–46.

Anderson, R. M., and R. M. May. 1982. Coevolution of hosts and parasites. *Parasitology* 85:411–426.

Andersson, M. 1994. *Sexual selection.* Princeton, N.J.: Princeton University Press.

Anstett, M. C., J. L. Bronstein, and M. Hossaert-McKey. 1996. Resource allocation: a conflict in the fig/wasp mutualism? *J. Evol. Biol.* 9:417–428.

Anstett, M. C., M. Hossaert-McKey, and F. Kjellberg. 1997. Figs and fig pollinators: Evolutionary conflicts in a coevolved mutualism. *Trends Ecol. Evol.* 12:94–99.

Arak, A., and M. Enquist. 1993. Hidden preferences and the evolution of signals. *Philos. Trans. R. Soc. Lond. B* 340:207–213.

Arnold, G., B. Quenet, J. M. Cornuet, C. Masson, B. Deschepper, A. Estoup, and P. Gasqui. 1996. Kin recognition in honeybees. *Nature (Lond.)* 379:498.

Arnqvist, G., and L. Rowe. 1995. Sexual conflict and arms races between the sexes: A morphological adaptation for control of mating in a female insect. *Proc. R. Soc. Lond. B* 261:123–127.

Aron, S., L. Passera, and L. Keller. 1994. Queen-worker conflict over sex ratio: A comparison primary and secondary sex ratios in the Argentine ant, *Iridomyrmex humilis. J. Evol. Biol.* 7:403–418.

Aron, S., E. L. Vargo, and L. Passera. 1995. Primary and secondary sex ratios in monogyne colonies of the fire ant *Solenopsis invicta. Anim. Behav.* 49:749–757.

Atlan, A., H. Mercot, C. Landre, and C. Montchamp-Moreau. 1997. The sex-ratio trait in *Drosophila simulans*: Geographical distribution of distortion and resistance. *Evolution* 51:1886–1895.

Axelrod, R. 1984. *The evolution of cooperation.* New York: Basic Books.

Axelrod, R., and W. D. Hamilton. 1981. The evolution of cooperation. *Science (Wash.)* 211:1390–1396.

Balas, M. T., and E. S. Adams. 1996. The dissolution of cooperative groups—Mechanisms of queen mortality in incipient fire ant colonies. *Behav. Ecol. Sociobiol.* 38:391–399.

Barnes, J. 1984. *The complete works of Aristotle.* Princeton, N.J.: Princeton University Press.

Barton, R. A. 1996. Neocortex size and behavioural ecology in primates. *Proc. R. Soc. Lond. B* 263:173–177.

Bateman, A. J. 1948. Intra-sexual selection in *Drosophila. Heredity* 2:349–368.

Bateson, P. 1994. The dynamics of parent offspring relationships in mammals. *Trends Ecol. Evol.* 9:399–403.

Bechnel, J. J., and A. W. Sweeney. 1990. *Amblyospora trinus* (Microsporida: Amblyosporidae) in the Australian mosquito *Culex halifaxi* (Diptera: Culicidae). *J. Protozool.* 37:584–592.

Bednekoff, P. A. 1997. Mutualism among safe, selfish sentinels: A dynamic model. *Am. Nat.* 150:373–392.

Bednekoff, P. A., and S. L. Lima. 1998. Re-examining safety in numbers: Interactions between risk dilution and collective detection depend upon predator targeting behavior. *Proc. R. Soc. Lond. B* 265:2021–2026.

Beeman, R. W., K. S. Friesen, and R. E. Denell. 1992. Maternal-effect selfish genes in flour beetles. *Science (Wash.)* 256:89–92.

Bell, G. 1982. *The masterpiece of nature: The evolution and genetics of sexuality.* Berkeley: University of California Press.

———. 1985. The origin and early evolution of germ cells as illustrated by the Volvocales. In *The origin and evolution of sex*, ed. H. O. Halvorson and A. Monroy, 221–256. New York: Alan R. Liss, Inc.

Bell, G., and A. Burt. 1990. *B*-chromosomes: Germ-line parasites which induce changes in host recombination. *Parasitology* 100:S19-S26.

Benner, S. A., R. K. Allemann, A. D. Ellinton, L. Ge, A. Glasfeld, G. G. Leanz, T.

Krauch, L. J. MacPherson, S. Moroney, J. A. Piccirilli, and E. Weinhold. 1987. Natural selection, protein engineering, and the last ribo-organism: Rational model building in biochemistry. *Cold Spring Harbor Symp. Quant. Biol.* 52:56–63.

Benner, S. A., A. D. Ellington, and A. Tauer. 1989. Modern metabolism as a palimpsest of the RNA world. *Proc. Natl Acad. Sci. USA* 86:7054–7058.

Bercovitch, F. B. 1988. Coalitions, cooperation and reproductive tactics among adult male baboons. *Anim. Behav.* 36:1198–1209.

Berg, C. C. 1989. Classification and distribution of *Ficus. Experientia* 45:605–612.

Bergstrom, C. T., and M. Lachmann. 1997. Signalling among relatives. I. Is signalling *too* costly? *J. Theor. Biol.* 352:609–617.

Bernasconi, G., and L. Keller. 1996. Reproductive conflicts in cooperative associations of fire ant queens (*Solenopsis invicta*). *Proc. R. Soc. Lond. B* 263:509–513.

Bernasconi, G., M. J. B. Krieger, and L. Keller. 1997. Unequal partitioning of reproduction and investment between cooperating queens in the fire ant, *Solenopsis invicta*, as revealed by microsatellites. *Proc. R. Soc. Lond. B* 264:1331–1336.

Bernstein, H., H. C. Byerly, F. A. Hopf, R. A. Michod, and G. K. Vemulapalli. 1983. The Darwinian dynamic. *Q. Rev. Biol.* 58:185–207.

Bernstein, H., H. C. Byerly, F. A. Hopf, and R. A. Michod. 1984. Origin of sex. *J. Theor. Biol.* 110:323–351.

Bestor, T. H., and B. Tycko. 1996. Creation of genomic methylation patterns. *Nat. Genet.* 12:363–367.

Bickerton, D. 1990. *Language and species.* Chicago: University of Chicago Press.

Birkhead, T. R., A. P. Møller, and W. J. Sutherland. 1993. Why do females make it so difficult for males to fertilize their eggs? *J. Theor. Biol.* 161:51–60.

Bishop, J. D. D. 1996. Female control of paternity in the internally fertilizing compound ascidian *Diplosoma listerianum.* I. Autoradiographic investigation of sperm movements in the female reproductive tract. *Proc. R. Soc. Lond. B* 263:369–376.

Bishop, J. D. D., C. S. Jones, and L. R. Noble. 1996. Female control of paternity in the internally fertilizing compound ascidian *Diplosoma listerianum.* II. Investigation of male mating success using RAPD markers. *Proc. R. Soc. Lond. B* 263:401–407.

Black, J. M., and M. Owen. 1989. Agonistic behaviour in goose flocks: Assessment, investment and reproductive success. *Anim. Behav.* 37:187–198.

Blackstone, N. W. 1995. A units-of-evolution perspective on the endosymbiont theory of the origin of the mitochondrion. *Evolution* 49:785–796.

Boehm, C. 1992. Segmentary "warfare" and the management of conflict: Comparison of East African chimpanzees and patrilineal-patrilocal humans. In *Coalitions and alliances in humans and other animals,* ed. A. H. Harcourt and F. B. M. deWaal, 137–173. Oxford: Oxford University Press.

———. 1997. Impact of the human egalitarian syndrome on Darwinian selection mechanics. *Am. Nat.* 150:S100-S121.

Boerlijst, M. C., and P. Hogeweg. 1991. Spiral wave structure in pre-biotic evolution—hypercycles stable against parasites. *Physica D* 48:17–28.

Boesch, C. 1994a. Chimpanzees–red colobus monkeys: A predator-prey system. *Anim. Behav.* 47:1135–1148.

———. 1994b. Cooperative hunting in wild chimpanzees. *Anim. Behav.* 48:653–667.

———. 1996. The emergence of culture among wild chimpanzees. In *Evolution of social behaviour in primates and man*, ed. W. G. Runciman, J. Maynard Smith, and R. I. M Dunbar, 251–268. Oxford: Oxford University Press.

Boesch, C., and H. Boesch. 1989. Hunting behavior of wild chimpanzees in the Tai National Park. *Am. J. Phys. Anthropol.* 78:547–573.

Bonner, J. T. 1995. The evolution of life's complexity. *Nature (Lond.)* 374:508–509.

Boomsma, J. J. 1989. Sex-investment ratios in ants: Has female bias been systematically overestimated? *Am. Nat.* 133:517–532.

———. 1991. Adaptive colony sex ratios in primitively eusocial bees. *Trends Ecol. Evol.* 6:92–95.

———. 1993. Sex ratio variation in polygynous ants. In *Queen number and sociality in insects*, ed. L. Keller, 86–109. Oxford: Oxford University Press.

———. 1996. Split sex ratios and queen-male conflict over sperm allocation. *Proc. R. Soc. Lond. B* 263:697–704.

Boomsma, J. J., and A. Grafen. 1990. Intraspecific variation in ant sex ratios and the Trivers-Hare hypothesis. *Evolution* 44:1026–1034.

———. 1991. Colony-level sex ratio selection in the eusocial Hymenoptera. *J. Evol. Biol.* 3:383–407.

Boomsma, J. J., and F. L. W. Ratnieks. 1996. Paternity in eusocial Hymenoptera. *Philos. Trans. R. Soc. Lond. B* 351:947–975.

Boomsma, J. J., L. Keller, and M. G. Nielsen. 1995. A comparative analysis of sex ratio investment parameters in ants. *Funct. Ecol.* 9:743–753.

Boucek, Z. 1988. Family Agaonidae. In *Australasian Chalcidoidea (Hymenoptera) A biosystematic revision of genera of fourteen families, with a reclassification of species*, 156–209. Wallington, N. Z.: CABI.

———. 1993. The genera of chalcidoid wasps from *Ficus* fruit in the New World. *J. Nat. Hist.* 27:173–217.

Boucher, D. H. 1977. On wasting parental investment. *Am. Nat.* 111:786–788.

Bourke, A. F. G. 1988. Worker reproduction in the higher eusocial Hymenoptera. *Q. Rev. Biol.* 63:291–311.

———. 1994a. Indiscriminate egg cannibalism and reproductive skew in a multiple-queen ant. *Proc. R. Soc. Lond. B* 255:55–59.

———. 1994b. Worker matricide in social bees and wasps. *J. Theor. Biol.* 167:283–292.

———. 1997. Sociality and kin selection in insects. In *Behavioural ecology. An evolutionary approach*, 4th ed., ed. J. R. Krebs and N. B. Davies, 203–227. Oxford: Blackwell.

———. 1999. Colony size, social complexity and reproductive conflict in social insects. *J. Evol. Biol.* 12:245–257.

Bourke, A. F. G., and N. R. Franks. 1995. *Social evolution in ants*. Princeton, N.J.: Princeton University Press.

Bourke, A. F. G., and J. Heinze. 1994. The ecology of communal breeding: The case of multiple-queen leptothoracine ants. *Philos. Trans. R. Soc. Lond. B* 345:359–372.

Bourke, A. F. G., H. A. A. Green, and M. W. Bruford. 1997. Parentage, reproductive skew and queen turnover in a multiple-queen ant analysed with microsatellites. *Proc. R. Soc. Lond. B* 264:277–283.

Bourtzis, K., S. L. Dobson, H. R. Braig, and S. L. O'Neill. 1998. Rescuing *Wolbachia* have been overlooked. *Nature (Lond.)* 391:852–853.

Boyd, R., and P. J. Richerson. 1985. *Culture and the evolutionary process*. Chicago: University of Chicago Press.

———. 1992. Punishment allows the evolution of cooperation (or anything else) in sizable groups. *Ethol. Sociobiol.* 13:171–195.

———. 1996. Why culture is common, but cultural evolution is rare. In *Evolution of social behaviour in primates and man*, ed. W. G. Runciman, J. Maynard Smith, and R. I. M Dunbar, 77–93. Oxford: Oxford University Press.

Braidotti, G., and D. P. Barlow. 1997. Identification of a male meiosis-specific gene *Tcte2*, which is differentially in species that form sterile hybrids with laboratory mice and deleted in *t* chromosomes showing meiotic drive. *Dev. Biol.* 186:85–99.

Breed, M. D., C. K. Welch, and R. Cruz. 1994. Kin discrimination within honey bee (*Apis mellifera*) colonies: An analysis of the evidence. *Behav. Process.* 33:25–39.

Breeuwer, J. A., and J. H. Werren. 1990. Microorganisms associated with chromosome destruction and reproductive isolation between two insect species. *Nature (Lond.)* 346:558–560.

Bremmerman, H. J., and J. Pickering. 1983. A game-theoretical model of parasite virulence. *J. Theor. Biol.* 100:411–426.

Britten, R. J. 1996. DNA sequence insertion and evolutionary variation in gene regulation. *Proc. Natl. Acad. Sci. USA* 93:9374–9377.

Bronstein, J. L. 1991. The nonpollinating wasp fauna of *Ficus pertusa*: Exploitation of a mutualism? *Oikos* 61:175–186.

———. 1992. *Seed predators as mutualists: Ecology and evolution of the fig/pollinator interaction*. In *Insect-plant interactions*, vol. 4, ed. E. Bernays, 1–44. Boca Raton, Fla.: CRC Press.

Bronstein, J. L., and M. Hossaert-McKey. 1996. Variation in reproductive success within a subtropical fig/pollinator mutualism. *J. Biogeogr.* 23:433–446.

Brookfield, J. F. Y. 1991. Models of repression of transposition in P-M hybrid dysgenesis by P cytotype and by zygotically encoded repressor proteins. *Genetics* 128:471–486.

Brown, J. L. 1987. *Helping and communal breeding in birds: Ecology and evolution*. Princeton, N.J.: Princeton University Press.

Buckle, T. A., and D. H. Krüger. 1993. Biology of DNA restriction. *Microbiol. Rev.* 57:434–450.

Bull, J. J. 1994. The evolution of virulence. *Evolution* 48:1423–1437.

Bull, J. J., and W. R. Rice. 1991. Distinguishing mechanisms for the evolution of cooperation. *J. Theor. Biol.* 149:63–74.

Bull, J. J., I. J. Molineaux, and W. R. Rice. 1991. Selection of benevolence in a host-parasite system. *Evolution* 45:875–882.

Bulmer, M. G. 1986. Genetic models of endosperm evolution in higher plants. In *Evolutionary processes and theory*, eds. Karlin and E. Noveo, 245–267. New York: Academic Press.

Burke, T., N. B. Davies, M. W. Bruford, and B. J. Hatchwell. 1989. Parental care and mating behaviour of polyandrous dunnocks *Prunella modularis*. *Nature (Lond.)* 338:249–251.

Buss, L. W. 1987. *The evolution of individuality*. Princeton, N.J.: Princeton University Press.

Busse, C. D. 1978. Do chimpanzees hunt cooperatively? *Am. Nat.* 112:767–770.

Byle, P. A. F. 1990. Brood division and parental care in the period between fledging and independence in the dunnock (*Prunella modularis*). *Behaviour* 113:1–20.

Byrne, R., and A. Whitten. 1988. *Machiavellian intelligence*. Oxford: Clarendon.

Caballero, A. 1994. Developments in the prediction of effective population size. *Heredity* 73:657–679.

Cairns-Smith, A. G., and G. L. Walker. 1974. Primitive metabolism. *BioSystems* 5:173–186.

Cameron, I. L. 1971. Cell proliferation and renewal in the mammalian body. In *Cellular and molecular renewal in the mammalian body*, ed. I. L. Cameron and J. D. Thrasher, 45–85. New York: Academic Press.

Carlin, N. F., H. K. Reeve, and S. P. Cover. 1993. Kin discrimination and division of labour among matrilines in the polygynous carpenter ant *Camponotus planatus*. In *Queen number and sociality in insects*, ed. L. Keller, 362–401. Oxford: Oxford University Press.

Caro, T. M., L. Lombardo, A. W. Goldizen, and M. Kelly. 1995. Tail-flagging and other antipredator signals in white-tailed deer: New data and synthesis. *Behav. Ecol.* 6:442–450.

Carson, D. A., and J. M. Ribeiro. 1993. Apoptosis and disease. *Lancet* 341:1251–1254.

Cavalli-Sforza, L. L., and M. W. Feldman. 1981. *Cultural transmission and evolution: A quantitative approach*. Princeton, N.J.: Princeton University Press.

Cech, T. R. 1990. Self-splicing of group I introns. *Annu. Rev. Biochem.* 59:543–568.

Chan, G. L., and A. F. G. Bourke. 1994. Split sex ratios in a multiple-queen ant population. *Proc. R. Soc. Lond. B* 258:261–266.

Chapais, B. 1992. The role of alliances in social inheritance of rank among female primates. In *Coalitions and alliances in humans and other animals*, ed. A. H. Harcourt and F. B. M. deWaal, 29–59. Oxford: Oxford University Press.

Chapais, B., M. Girard, and G. Primi. 1991. Non-kin alliances, and the stability of matrilineal dominance relations in Japanese macaques. *Anim. Behav.* 41:481–491.

Chapman, T., L. F. Liddle, J. M. Kalb, M. F. Wolfner, and L. Partridge. 1995. Cost of mating in *Drosophila melanogaster* females is mediated by male accessory gland products. *Nature (Lond.)* 373:241–244.

Chapuisat, M., and L. Keller. 1999. Testing kin selection with sex ratio data in eusocial Hymenoptera. Heredity (in press).

Chapuisat, M., L. Sundström, and L. Keller. 1997. Sex ratio regulation: The economics of fraticide in ants. *Proc. R. Soc. Lond. B* 264:1255–1260.

Charlesworth, B., and D. L. Hartl. 1978. Population dynamics of the segregation distorter polymorphism of *Drosophila melanogaster*. *Genetics* 89:171–192.

Charlesworth, B., A. Lapid, and D. Canada. 1992. The distribution of transposable elements within and between chromosomes in a population of *Drosophila melanogaster*. II. Inferences on the nature of selection against elements. *Genet. Res.* 60:103–114.

Charlesworth, B., P. D. Sniegowski, and W. Stephan. 1994. The evolutionary dynamics of repetitive DNA in eukaryotes. *Nature (Lond.)* 371:215–220.

Charlesworth, D., and B. Charlesworth. 1995. Transposable elements in inbreeding and outbreeding populations. *Genetics* 140:415–417.

Charnov, E. L. 1978. Evolution of eusocial behavior: Offspring choice or parental parasitism. *J. Theor. Biol.* 75:451–465.

———. 1979 Simultaneous hermaphroditism and sexual selection. *Proc. Natl. Acad. Sci. USA* 76:2480–2484.

Charnov, E. L., and J. R. Krebs. 1975. The evolution of alarm calls: Altruism or manipulation? *Am. Nat.* 109:107–112.

Chase, I. D. 1980. Cooperative and noncooperative behavior in animals. *Am. Nat.* 115:827–857.

Chen, P. S., E. Stumm-Zollinger, T. Aigaki, J. Balmer, M. Bienz, and P. Bohjlen. 1988. A male accessory gland peptide that regulates reproductive behavior of female *D. melanogaster. Cell* 54:291–298.

Cheney, D. L. 1987. Interactions and relationships between groups. In *Primate societies*, ed. B. B. Smuts, D. L. Cheney, R. M. Seyfarth, R. W. Wrangham, and T. T. Struhsaker, 267–281. Chicago: University of Chicago Press.

Cheney, D. L., and R. M. Seyfarth. 1990. *How monkeys see the world.* Chicago: University of Chicago Press.

———. 1996. Function and intention in the calls of non-human primates. In *Evolution of social behaviour in primates and man*, ed. W. G. Runciman, J. Maynard Smith, and R. I. M Dunbar, 59–76. Oxford: Oxford University Press.

Chomsky, N. 1975. *Reflections on language.* New York: Pantheon Books.

Clancy, D. J., and A. A. Hoffman. 1996. Cytoplasmic incompatibility in *Drosophila simulans*: Evolving complexity. *Trends Ecol. Evol.* 11:145–146.

Clark, A. G., M. Augadé, T. Prout, L. G. Harshman, and C. H. Langley. 1995. Variation in sperm displacement and its association with accessory gland protein in *Drosophila melanogaster. Genetics* 139:189–201.

Clark, C. W., and R. C. Ydenberg. 1990. The risks of parenthood II. Parent-offspring conflict. *Evol. Ecol.* 4:312–325.

Clayton, D. H., and D. M. Tompkins. 1994. Ectoparasite virulence is linked to mode of transmission. *Proc. R. Soc. Lond. B* 256:211–217.

Clifton, K. E. 1989. Territory sharing by the Caribbean striped parrotfish, *Scarus iserti*: Patterns of resource abundance, group size, and behavior. *Anim. Behav.* 37:90–103.

Clotfelter, E. D. 1996. Mechanisms of facultative sex-ratio variation in zebra finches (*Taeniopygia guttata*). *Auk* 113:441–449.

Clutton-Brock, T. H. 1984. Reproductive effort and terminal investment in iteroparous animals. *Am. Nat.* 123:212–229.

———. 1991. *The evolution of parental care.* Princeton, N.J.: Princeton University Press.

Clutton-Brock, T. H., and H. C. J. Godfray. 1991. Parental investment. In *Behavioural ecology: An evolutionary approach*, 3d ed., ed. J. R. Krebs and N. B. Davies, 234–262. Oxford: Blackwell Scientific Publications.

Clutton-Brock, T. H., and G. A. Parker. 1995a. Sexual coercion in animal societies. *Anim. Behav.* 49:1345–1365.

———. 1995b. Punishment in animal societies. *Nature (Lond.)* 373:209–216.

Collinson, M. E. 1989. The fossil history of the Moraceae. In *Evolution, systematics, and fossil history of the Hamamelidae*, ed. P. R. Crane and S. Blackmore. Oxford: Clarendon Press.

Colwell, R. K. 1981. Group selection is implicated in the evolution of female-biased sex ratios. *Nature (Lond.)* 263:401–404.

Compton, S. G., and B. A. Hawkins. 1992. Determinants of species richness in South African fig wasp assemblages. *Oecologia* 91:68–74.

Compton, S. G., K. C. Holton, V. K. Rashbrook, S. van Noort, S. L. Vincent, and A. B. Ware. 1991. Studies of *Ceratosolen galili*, a nonpollinating Agaonid fig wasp. *Biotropica* 23:188–194.

Compton, S. G. 1990. A collapse of host specificity in some African fig wasps. *S. Afr. J. Sci.* 86:39–40.

———. 1993. One way to be a fig. *Afr. Entomol.* 1:151–158.

Compton, S. G., and H. G. Robertson. 1988. Complex interactions between mutualisms: Ants tending homopterans protect fig seeds and pollinators. *Ecology* 69:1302–1305.

Compton, S. G., and S. van Noort. 1992. Southern African fig wasps (Hymenoptera: Chalcidoidea): Resource utilization and host relationships. *Proc. K. Ned. Akad. Wet.* 95:423–435.

Compton, S. G., J. Y. Rasplus, and A. B. Ware. 1994. African fig wasp parasitoid communities. In *Parasitoid community ecology*, ed. B. A. Hawkins and W. Sheehan. Oxford: Oxford University Press.

Connor, R. C., and R. A. Smolker. 1995. Seasonal changes in the stability of male-male bonds in Indian Ocean bottlenose dolphins (*Tursiops* sp.). *Aquat. Mamm.* 21:213–216.

Connor, R. C., R. A. Smolker, and A. F. Richards. 1992. Two levels of alliance formation among male bottlenose dolphins (*Tursiops* sp.). *Proc. Natl. Acad. Sci. USA* 89:987–990.

Cook, J. M., and S. A. Power. 1996. Effects of within-tree flowering asynchrony on the dynamics of seed and wasp production in an Australian fig species. *J. Biogeogr.* 23:487–494.

Cook, S. E., J. G. Vernon, M. Bateson, and T. Guilford. 1994. Mate choice in the polymorphic African swallowtail butterfly, *Papilio dardanus*: Male-like females may avoid sexual harassment. *Anim. Behav.* 47:389–397.

Corlett, R. T., V. Boudville, and K. Seet. 1990. Seed and wasp production in five fig species (*Ficus*, Moraceae). *Malay. Nat. J.* 44:97–102.

Corner, E. J. H. 1940. Wayside trees of Malaya. Singapore: Government Printing Office.

———. 1964. *The life of plants*. Cleveland, Ohio: World Press.

———. 1985. *Ficus* (Moraceae) and Hymenoptera (Chalcidoidea): Figs and their pollinators. *Biol. J. Linn. Soc.* 25:187–195.

Cosmides, L. M., and J. Tooby. 1981. Cytoplasmic inheritance and intragenomic conflict. *J. Theor. Biol.* 89:83–129.

Cotton, P. A., A. Kacelnik, and J. Wright. 1996. Chick begging as a signal: Are nestlings honest? *Behav. Ecol.* 7:178–182.

Coulson, J. C. 1966. The influence of the pair-bond and age on the breeding biology of the kittiwake gull *Rissa tridactyla*. *J. Anim. Ecol.* 35:269–279.

Creel, S. R., and P. M. Waser 1991. Failure of reproductive suppression in dwarf mongooses (*Helogale parula*): Accident or adaptation? *Behav. Ecol.* 2:7–15.

Crepet, W. L. 1984. Advanced (constant) insect pollination mechanisms: Pattern of evolution and implications vis-à-vis angiosperm diversity. *Ann. Mo. Bot. Gard.* 71:607–630.

Crick, F. H. C. 1968. The origin of the genetic code. *J. Mol. Biol.* 38:367–379.

Crow, J. F. 1979. Genes that violate Mendel's laws. *Sci. Am.* 240:134–146.

Crow, J. F., and K. Aoki. 1982. Group selection for a polygenic behavioral trait: A differential proliferation model. *Proc. Natl. Acad. Sci. USA* 79:2628–2631.

Crozier, R. H. 1988. Kin recognition using innate labels: A central role for piggybacking. In *Invertebrate historecognition*, ed. R. K. Grosberg, D. Hegecock, and K. Nelson, 143–156. New York: Plenum Press.

Crozier, R. H., and R. E. Page. 1985. On being the right size: Male contributions and multiple mating in social Hymenoptera. *Behav. Ecol. Sociobiol.* 18:105–115.

Crozier, R. H., and P. Pamilo. 1996. *Evolution of social insect colonies: Sex allocation and kin selection.* Oxford: Oxford University Press.

Curtis, C. F. 1992. Selfish genes in mosquitoes. *Nature (Lond.)* 357:450–458.

Czárán, T., and E. Szathmáry. 1999. Coexistence of competitive-mutualist replicators in prebiotic evolution. In *The geometry of ecological interactions: Simplifying spatial complexity*, ed. U. Dieckmann, R. Law, and J. A. J. Metz. Cambridge: Cambridge University Press.

Dale, S., and T. Slagsvold. 1994. Polygyny and deception in the pied flycatcher: Can females determine male mating status? *Anim. Behav.* 48:1207–1217.

Dale, S., H. Rinden, and T. Slagsvold. 1992. Competition for a mate restricts mate search of female pied flycatchers. *Behav. Ecol. Sociobiol.* 30:165–176.

Daly, M., and M. Wilson. 1988. *Homicide.* New York: Aldine de Gruyter.

Darwin, C. R. 1859. *On the origin of species.* London: John Murray.

———. 1871. *The descent of man, and selection in relation to sex.* London: John Murray.

Davies, N. B. 1992. *Dunnock behaviour and social evolution.* Oxford: Oxford University Press.

Davies, N. B., and A. I. Houston. 1981. Owners and satellites: The economics of territory defense in the pied wagtail, *Motacilla alba. J. Anim. Ecol.* 50:157–180.

Dawkins, M. S., and T. Guilford. 1995. An exaggerated preference for simple neural network models of signal evolution. *Proc. R. Soc. Lond. B* 261:357–360.

———. 1996. Sensory bias and the adaptiveness of female choice. *Am. Nat.* 148: 937–942.

Dawkins, R. 1976. *The selfish gene.* Oxford: Oxford University Press.

———. 1982. *The extended phenotype.* Oxford: W. H. Freeman.

———. 1989. *The selfish gene,* new ed. Oxford: Oxford University Press.

Dawkins, R., and T. R. Carlisle. 1976. Parental investment, mate desertion and a fallacy. *Nature (Lond.)* 262:131–133.

Dawkins, R., and J. R. Krebs. 1979. Arms races within and between species. *Proc. R. Soc. Lond. B* 205:489–511.

de Carvalho, A. B., S. C. Vaz, and L. B. Klaczko. 1997. Polymorphism of Y-linked suppressors of sex-ratio in two natural populations of *Drosophila mediopunctata*. *Genetics* 146:891–902.

DeHeer, C. J., and K. G. Ross. 1997. Lack of detectable nepotism in multiple-queen colonies of the fire ant *Solenopsis invicta* (Hymenoptera, Formicidae). *Behav. Ecol. Sociobiol.* 40:27–33.

de Waal, F. B. M. 1996. *Good natured: The origins of right and wrong in humans and other animals.* Cambridge, Mass.: Harvard University Press.

Dickens, D. W., and R. A. Clark. 1987. Games theory and siblicide in the kittiwake gull, *Rissa tridactyla. J. Theor. Biol.* 125:301–305.

Dixon, A., D. Ross, S. L. C. O'Malley, and T. Burke. 1994. Paternal investment inversely related to degree of extra-pair paternity in the reed bunting. *Nature (Lond.)* 371:698–700.

Doolittle, W. F., and C. Sapienza. 1980. Selfish genes, the phenotype paradigm and genome evolution. *Nature (Lond.)* 284:601–603.

Douglas, A. E. 1994. *Symbiotic interactions.* Oxford: Oxford University Press.

Dugatkin, L. A., and H. K. Reeve. 1994. Behavioral ecology and levels of selection: Dissolving the group selection controversy. *Adv. Stud. Behav.* 23:101–133.

East, E. M. 1918. The role of reproduction in evolution. *Am. Nat.* 52:273–289.

East, R. 1984. Rainfall, soil nutrient status and biomass of large Africa savanna mammals. *Afr. J. Ecol.* 22:245–270.

Eberhard, W. G. 1980. Evolutionary consequences of intracellular organelle competition. *Q. Rev. Biol.* 55:231–249.

———. 1996. *Female control: Sexual selection by cryptic female choice.* Princeton, N.J.: Princeton University Press.

Ebert, D. 1994. Virulence and local adpatation of a horizontally transmitted parasite. *Science (Wash.)* 265:1084–1086.

Ebert, D., and E. A. Herre. 1996. The evolution of parasitic diseases. *Parasitol. Today* 12:96–101.

Eibel, H., and P. Philippsen. 1984. Preferential integration of yeast transposable element Ty into a promoter region. *Nature (Lond.)* 307:386–388.

Eickbush, D. G., and T. H. Eickbush. 1995. Vertical transmission of the retrotransposable elements R1 and R2 during the evolution of the *Drosophila melanogaster* species subgroup. *Genetics* 139:671–684.

Eigen, M. 1971. Self-organization of matter and the evolution of biological macromolecules. *Naturwissenschaften* 58:465–523.

———. 1992. *Steps towards life.* Oxford: Oxford University Press.

Eigen, M., and P. Schuster. 1977. The hypercycle: A principle of natural self-organization. Part A: Emergence of the hypercycle. *Naturwissenschaften* 64:541–565.

———. 1978. The hypercycle: A principle of natural self-organization. Part C: The realistic hypercycle. *Naturwissenschaften* 65:341–369.

———. 1979. *The hypercycle, a principle of natural self-organization.* Berlin: Springer-Verlag.

Eigen, M., P. Schuster, W. Gardiner, and R. Winkler-Oswatitsch. 1981. The origin of genetic information. *Sci. Am.* 244:78–94.

Eklöv, P. 1992. Group foraging versus solitary foraging efficiency in piscivorous predators: The perch, *Perca fluviatilis*, and the pike, *Esox lucius*, patterns. *Anim. Behav.* 44:313–326.

Ekman, P. 1973. Cross-cultural studies of facial expression. In *Darwin and facial expression, a century of research in review*, ed. P. Ekman, 169–222. New York: Academic Press.

Elfström, S. T. 1997. Fighting behaviour and strategy of rock pipit, *Anthus petrosus*, neighbors: Cooperative defence. *Anim. Behav.* 54:535–542.

Elgar, M. A. 1992. Sexual cannibalism in spiders and other invertebrates. In *Cannibalism: ecology and evolution among diverse taxa*, ed. M. A. Elgar and B. J. Crespi, 128–155. Oxford: Oxford University Press.

Emlen, S. T. 1982. The evolution of helping. II. The role of behavioural conflict. *Am. Nat.* 119:40–53.

———. 1995. An evolutionary theory of the family. *Proc. Natl. Acad. Sci. USA* 92:8092–8099.

———. 1996. Reproductive sharing in different types of kin associations. *Am. Nat.* 148:756–763.

———. 1997. Predicting family dynamics in social vertebrates. In *Behavioural ecology: An evolutionary approach*, 4th ed., ed. J. R. Krebs and N. B. Davies, 228–253. Oxford: Blackwell Scientific.

Emlen, S. T., J. M. Emlen, and S. A. Levin. 1986. Sex-ratio selection in species with helpers-at-the-nest. *Am. Nat.* 127:1–8.

Emslie, S. D., W. J. Sydeman, and P. Pyle. 1992. The importance of mate retention and experience on breeding success in Cassin's auklet (*Ptychoramphus aleuticus*). *Behav. Ecol.* 3:189–195.

Engels, W. R. 1989. *P* elements in *Drosophila*. In *Mobile DNA*, ed. D. Berg and M. Howe, 437–484. Washington D. C.: American Society of Microbiology.

———. 1992. The origin of *P* elements in *Drosophila melanogaster*. *Bioessays* 14:681–686.

Ens, B. J., U. N. Safriel, and M. P. Harris. 1993. Divorce in the long-lived and monogamous oystercatcher, *Haematopus ostralegus*: Incompatability or choosing the better option? *Anim. Behav.* 45:1199–1217.

Eshel, I., and M. W. Feldman. 1991. The handicap principle in parent-offspring conflict: Comparison of optimality and population genetic analyses. *Am. Nat.* 137:167–185.

Estoup, A., M. Solignac, and J. -M. Cornuet. 1994. Precise assessment of the number of patrilines and of genetic relatedness in honeybee colonies. *Proc. R. Soc. Lond. B* 258:1–7.

Evans, J. D. 1995. Relatedness threshold for the production of female. *Proc. Natl. Acad. Sci. USA* 92:6514–6517.

———. 1996. Competition and relatedness between queens of the facultatively polygynous ant *Myrmica tahoensis*. *Anim. Behav.* 51:831–840.

Evans, R. M., M. O. Wiebe, S. C. Lee, and S. C. Budgen. 1995. Embryonic and parental preferences for incubation temperature in herring gulls—implications for parent-offspring conflict. *Behav. Ecol. Sociobiol.* 36:17–23.

Ewald, P. W. 1983. Host-parasite relations, vectors and the evolution of parasite severity. *Annu. Rev. Ecol. Syst.* 14:465–485.

———. 1987. Transmission modes and the evolution of the parasitism-mutualism continuum. *Ann. N. Y. Acad. Sci.* 503:295–306.

Fabre, J.-H. 1989. *Souvenirs entomologiques*, 2 vols. Paris: Robert Laffont.

Fanshawe, J. H., and C. D. FitzGibbon. 1993. Factors influencing the hunting success of an African wild dog pack. *Anim. Behav.* 45:479–490.

Ferguson, J. W. H. 1987. Vigilance behaviour in white-browed sparrow-weavers *Plocepasser mahali*. *Ethology* 27:223–235.

Fischer, E. A. 1981. Sexual allocation in a simultaneously hermaphroditic coral reef fish. *Am. Nat.* 117:64–82.

———. 1988. Simultaneous hermaphroditism, tit-for-tat, and the evolutionary stability of social systems. *Ethol. Sociobiol.* 9:119–136.

Fisher, R. A. 1930. *The genetical theory of natural selection.* Oxford: Oxford University Press.

———. 1958. *The genetical theory of natural selection.* New York: Dover Press.

FitzGibbon, C. D. 1989. A cost to individuals with reduced vigilance in groups of Thomson's gazelles hunted by cheetahs. *Anim. Behav.* 37:508–510.

———. 1994. The costs and benefits of predator inspection behaviour in Thomson's gazelles. *Behav. Ecol. Sociobiol.* 34:139–148.

Fletcher, D. J. C., and K. G. Ross. 1985. Regulation of reproduction in eusocial Hymenoptera. *Annu. Rev. Entomol.* 30:319–343.

Foley, R. A. 1996. An evolutionary approach and chronological framework for human social behaviour. In *Evolution of social behaviour in primates and man,* ed. W. G. Runciman, J. Maynard Smith, and R. I. M. Dunbar, 95–117. Oxford: Oxford University Press.

Fontana, W., T. Griesmacher, W. Schnabl, P. F. Stadler, and P. Schuster. 1991. Statistics of landscapes based on free energies, replication and degradation rate constants of RNA secondary structures. *Monatsh. Chemie* 122:795–819.

Forbes, L. S. 1990. Insurance offspring and the evolution of avian clutch size. *J. Theor. Biol.* 147:345–359.

———. 1993. Avian brood reduction and parent-offspring conflict. *Am. Nat.* 142:82–117.

Forget, P.-M. 1991. Comparative recruitment patterns of two non-pioneer canopy tree species in French Guiana. *Oecologia* 85:434–439.

———. 1994. Recruitment pattern of *Vouacapoua americana* (Caesalpiniaceae), a rodent-dispersed tree species in French Guiana. *Biotropica* 26:408–419.

Forster, L. M. 1992. The stereotyped behaviour of sexual cannibalism in *Latrodectus hasselti* Thorell (Araneae: Theridiidae), the Australian redback spider. *Aust. J. Zool.* 40:1–11.

Fowler, K., and L. Partridge. 1989. A cost of mating in female fruitflies. *Nature (Lond.)* 338:760–761.

Frank, L. G., K. E. Holekamp, and L. Smale. 1995. Dominance, demography, and reproductive success of female spotted hyenas. In *Serengeti II: Dynamics, management, and conservation of an ecosystem,* ed. A. R. E. Sinclair and P. Arcese, 364–384. Chicago: University of Chicago Press.

Frank, S. A. 1984. The behavior and morphology of the fig wasps *Pegoscapus aseutus* and *P. jiminezi*: Descriptions and suggested behaviors for phylogenetic studies. *Psyche* 91:289–307.

———. 1985. Hierarchical selection theory and sex ratios. II. On applying the theory, and testing with fig wasps. *Evolution* 39:949–964.

———. 1987. Variable sex ratio among colonies of ants. *Behav. Ecol. Sociobiol.* 20:195–201.

———. 1992. A kin selection model for the evolution of virulence. *Proc. R. Soc. B* 250:195–197.

———. 1994. Recognition and polymorphism in host-parasite genetics. *Philos. Trans. R. Soc. Lond. B* 346:283–293.

———. 1995. Mutual policing and repression of competition in the evolution of cooperative groups. *Nature (Lond.)* 377:520–522.

———. 1996. The design of natural and artificial adaptive systems. In *Adaptation*, ed. M. R. Rose and G. V. Lauder, 451–505. New York: Academic Press.

———. 1997. Cytoplasmic incompatibility and population structure. *J. Theor. Biol.* 184:327–330.

———. 1996. Models of parasite virulence. *Q. Rev. Biol.* 71:37-78.

Franks, N. R. 1989. Army ants: A collective intelligence. *Am. Sci.* 77:138–145.

Frumhoff, P. C. 1991. The effect of the cordovan marker on apparent kin discrimination among nestmate honey bees. *Anim. Behav.* 42:854–856.

Gagneux, P., D. S. Woodruff, and C. Boesch. 1997. Furtive mating in female chimpanzees. *Nature (Lond.)* 387:358–359.

Galdikas, B. 1985. Orangutan sociality at Tunjung Puting A. *J. Primatol.* 9:101–119.

Galef, B. G. 1988. Imitation in animals. In *Social learning, psychological and biological perspectives*, ed. T. Zentall and B. G. Galef, 141–165. Hillsdale, N. J.: Lawrence Erlbaum.

Galil, J., and D. Eisikowitch. 1968. On the pollination ecology of *Ficus sycomorus* L. *Ecology* 49:259–269.

———. 1969. Further studies of the pollination ecology of *Ficus sycomorus* L. *Tidschr. Entomol.* 112:1–13.

———. 1971. Studies on mutualistic symbiosis between syconia and sycophilous wasps in monoecious figs. *New Phytol.* 70:773–787.

Gamboa, G. J., H. K. Reeve, and D. W. Pfennig. 1986. The evolution and ontogeny of nestmate recognition in social wasps. *Annu. Rev. Entomol.* 31:431–454.

Gánti, T. 1971. *The principle of life* [in Hungarian]. Budapest: Gondolat.

———. 1975. Organisation of chemical reactions into dividing and metabolizing units: The chemotons. *BioSystems* 7:189–195.

———. 1978. Chemical systems and supersystems III. Models of self-reproducing chemical supersystems: The chemotons. *Acta Chim. Acad. Sci. Hung.* 98:265–283.

———. 1979. *A theory of biochemical supersystems and its application to problems of natural and artificial biogenesis.* Budapest: Akadémiai Kiadó and Baltimore, Md.: University Park Press.

———. 1987. *The principle of life.* Budapest: OMIKK.

———. 1997. Biogenesis itself. *J. Theor. Biol.* 187:583–593.

Gaston, A. J. 1977. Social behaviour within groups of jungle babblers (*Turdoides striatus*). *Anim. Behav.* 25:828–848.

Gellner, E. 1988. Origins of society. In *Origins*, ed. A. C. Fabian, 128–140. Cambridge: Cambridge University Press.

Gems, D., and D. L. Riddle. 1996. Longevity in *Caenorhabditis elegans* reduced by mating but not gamete production. *Nature (Lond.)* 379:723–725.

Gerdes, K., P. B. Rasmussen, and S. Molin. 1986. Unique type of plasmid maintenance function: Postsegregational killing of plasmid-free cells. *Proc. Natl. Acad.
Sci. USA* 83:3116–3120.

Gerhart, J., and M. Kirschner. 1997. *Cells, embryos and evolution*. Oxford: Blackwell
Scientific.

Getty, T. 1996. Mate selection by repeated inspection: More on pied flycatchers.
Anim. Behav. 51:739–745.

Getz, W. M. 1981. Genetically based kin recognition systems. *J. Theor. Biol.* 92:209–
226.

Getz, W. M., and K. B. Smith. 1983. Genetic kin recognition: Honey bees discriminate between full and half sisters. *Nature (Lond.)* 302:147–148.

Gibernau, M., M. Hossaert-McKey, M.-C. Anstett, and F. Kjellberg. 1996. Consequences of protecting flowers in a fig: A one way trip for pollinators? *J. Biogeor.*
23:425–432.

Giblin-Davis, R. M., B. J. Center, H. Nadel, J. H. Frank, and W. Ramirez-B. 1995.
Nematodes associated with fig wasps, *Pegoscapus spp.* (Agaonidae), and syconia
of native Floridian figs (*Ficus* spp.). *J. Nematol.* 27:1–14.

Gilbert, W. 1986. The RNA world. *Nature (Lond.)* 319:818.

Gilson, É. 1971. *D'Aristote à Darwin et retour*. Paris: J. Vrin.

Gloor, G. B., N. A. Nassif, D. M. Johnson-Schiltz, C. R. Preston, and W. R. Engels.
1991. Targeted gene replacement in *Drosophila* via P-element-induced gap repair.
Science (Wash.) 253:1110–1117.

Godelle, B., and X. Reboud. 1995. Why are organelles uniparentally inherited. *Proc.
R. Soc. Lond. B* 259:27–33.

Godfray, H. C. J. 1987. The evolution of clutch size in parasitic wasps. *Am. Nat.*
129:221–233.

———. 1991. The signalling of need by offspring to their parents. *Nature (Lond.)*
352:328–330.

———. 1994. *Parasitoids, behavioral and evolutionary ecology*. Princeton, N.J.:
Princeton University Press.

———. 1995a. Evolutionary theory of parent-offspring conflict. *Nature (Lond.)*
376:133–138.

———. 1995b. Signalling of need between parents and young: Parent-offspring conflict and sibling rivalry. *Am. Nat.* 146:1–24.

Godfray, H. C. J., and A. B. Harper. 1990. The evolution of brood reduction by
siblicide in birds. *J. Theor. Biol.* 145:163–175.

Godfray, H. C. J., and G. A. Parker. 1991. Clutch size, fecundity and parent-offspring
conflict. *Philos. Trans. R. Soc. Lond. B* 332:67–79.

———. 1992. Sibling competition, parent-offspring conflict and clutch size. *Anim.
Behav.* 43:473–490.

Godfray, H. C. J., P. H. Harvey, and L. Partridge. 1991. Clutch size. *Annu. Rev. Ecol.
Syst.* 22:409–429.

Godin, J.-G. J., and S. A. Davis. 1995. Who dares, benefits: Predator approach behaviour in the guppy (*Poecilia reticulata*) deters predator pursuit. *Proc. R. Soc. Lond.
B* 259:193–200.

Goldschmidt, R. 1940. *The material basis of evolution*. New Haven, Conn.: Yale
University Press.

Goodall, J. 1986. *The chimpanzees of Gombe: Patterns of behavior.* Cambridge, Mass.: Belknap Press of Harvard University Press.

Goodall, J., A. Bandora, E. Bergmann, C. Busse, H. Matama, E. Mpongo, A. Pierce, and D. Riss. 1979. Intercommunity interactions in the chimpanzee population of the Gombe National Park. In *The great apes: Perspectives on human evolution,* vol. 5, ed. D. A. Hamburg and E. R. McCown, 13–53. London: Benjamin/Cummings.

Gopnik, M. 1990. Feature-blind grammar and dysphasia. *Nature (Lond.)* 344:715.

Gordh, G. 1975. The comparative external morphology and systematics of the neotropical parasitic fig wasp genus *Idarnes* (Hymenoptera: Torymidae). *Univ. Kans. Sci. Bull.* 50:389–455.

Gottlander, K. 1987. Parental feeding behaviour and sibling competition in the pied flycatcher *Ficedula hypoleuca. Ornis Scand.* 18:269–276.

Gould, S. J. 1984. Caring groups and selfish genes. In *Conceptual issues in evolutionary biology: An anthology,* ed. E. Sober, 119–124. Cambridge, Mass.: MIT Press.

Gould, S. J., and R. C. Lewontin. 1979. The spandrels of San Marco and the Panglossian paradigm: A critique of the adaptationist programme. *Proc. R. Soc. B* 205:281–288.

Gowaty, P. A., and D. L. Droge. 1991. Sex ratio conflict and the evolution of sex-biased provisioning in birds. *Proc. Int. Ornithol. Cong.* 20:932–945.

Gradwohl, J., and R. Greenberg. 1980. The formation of antwren flocks on Barro Colorado Island, Panamá. *Auk* 97: 385–395.

Grafen, A. 1982. How not to measure inclusive fitness. *Nature (Lond.)* 298:425.

———. 1984. Natural selection, kin selection and group selection. In *Behavioural ecology. An evolutionary approach,* 2d ed., ed. J. R. Krebs and N. B. Davies, 62–84. Oxford: Blackwell.

———. 1985. A geometric view of relatedness. *Oxford Surv. Evol. Biol.* 2:28–89.

———. 1990a. Biological signals as handicaps. *J. Theor. Biol.* 144:517–546.

———. 1990b. Sexual selection unhandicapped by the Fisher process. *J. Theor. Biol.* 144:473–516.

Grafen, A., and R. Sibly. 1978. A model of mate desertion. *Anim. Behav.* 26:645–652.

Gray, J. A. B., and E. J. Denton. 1991. Fast pressure pulses and communication between fish. *J. Mar. Biol. U.K.* 71:83–106.

Grey, D., V. Hutson, and E. Szathmáry. 1995. A re-examination of the stochastic corrector model. *Proc. R. Soc. Lond. B* 262:29–35.

Grinnell, J., C. Packer, and A. E. Pusey. 1995. Cooperation in male lions: Kinship, reciprocity or mutualism? *Anim. Behav.* 49:95–105.

Grosberg, R. K., and J. F. Quinn. 1989. The evolution of selective aggression based on allorecognition specificity. *Evolution* 43:504–515.

Haartman, L. von. 1969. Nest-site and the evolution of polygamy in European passerine birds. *Ornis Fenn.* 46:1–12.

Hahn, D. C. 1981. Asynchronous hatching in the laughing gull: Cutting losses and reducing sibling rivalry. *Anim. Behav.* 29:421–427.

Haig, D. 1992a. Brood reduction in gymnosperms. In *Cannibalism: Ecology and evolution among diverse taxa,* ed. M. A. Elgar and B. J. Crespi, 63–84. Oxford: Oxford University Press.

———. 1992b. Genomic imprinting and the theory of parent-offspring conflict. *Semin. Dev. Biol.* 3:153–160.

———. 1993a. Alternatives to meiosis. *J. Theor. Biol.* 163:15–31

———. 1993b. Genetic conflicts in human pregnancy. *Q. Rev. Biol.* 68:495–532.

———. 1996. Placental hormones, genomic imprinting, and maternal-fetal communication. *J. Evol. Biol.* 9:357–380.

———. 1997. Parental antagonism, relatedness asymmetries, and genomic imprinting. *Proc. R. Soc. Lond. B* 264:1657–1662.

Haig, D., and A. Grafen. 1991. Genetic scrambling as a defence against meiotic drive. *J. Theor. Biol.* 153:531–558.

Haig, D., and C. Graham. 1995. Genomic imprinting and the strange case of the insulin-like growth factor. *Cell* 64:1045–1046.

Haig, D., and M. Westoby. 1989. Parent-specific gene expression and the triploid endosperm. *Am. Nat.* 134:147–155.

Haigh, J. 1978. The accumulation of deleterious genes in a population—Muller's ratchet. *Theor. Pop. Biol.* 14:251–267.

Hailman, J. P., K. J. McGowan, and G. E. Woolfenden. 1994. Role of helpers in the sentinel behaviour of the Florida scrub jay (*Aphelocoma c. coerulescens*). *Ethology* 97:119–140.

Haldane, J. B. S. 1937. The effect of variation on fitness. *Am. Nat.* 71:337–349.

Haldane, J. B. S., and S. D. Jayakar. 1963. Polymorphism due to selection of varying direction. *J. Genet.* 58:318–323.

Hall, M., P. G. Norris, and R. T. Johnson. 1995. Human repair deficiencies and predisposition to cancer. In *The genetics of cancer*, ed. B. A. J. Ponder and M. J. Waring, 123–157. Dordrecht: Kluwer.

Hamilton, W. D. 1964a. The genetical evolution of social behavior. 1. *J. Theor. Biol.* 7:1–16.

———. 1964b. The genetical evolution of social behavior. 2. *J. Theor. Biol.* 7:17–52.

———. 1967. Extraordinary sex ratios. *Science (Wash.)* 156:477–488.

———. 1971. Geometry for the selfish herd. *J. Theor. Biol.* 31:295–311.

———. 1972. Altruism and related phenomena, mainly in social insects. *Annu. Rev. Ecol. Syst.* 3:193–232.

———. 1979. Wingless and fighting males in fig wasps and other insects. In *Sexual selection and reproductive competition in insects*, ed. M. S. Blum and N. A. Blum, 167–220. New York: Academic Press.

Hamilton, W. D., and M. Zuk. 1982. Heritable true fitness and bright birds. A role for parasites? *Science (Wash.)* 218:384–387.

Hanson, M. R. 1992. Plant mitochondrial mutations and plant sterility. *Annu. Rev. Genet.* 25:461–486.

Hansson, B., S. Bensch, and D. Hasselquist. 1997. Infanticide in great reed warblers: Secondary females destroy eggs of primary females. *Anim. Behav.* 54:297–304.

Harada, Y., and Y. Iwasa. 1996. Female mate preference to receive maximum paternal care: A two-step game. *Am. Nat.* 147:996–1027.

Hardin, G. 1968. The tragedy of the commons. *Science (Wash.)* 162:1243–1248.

Harper, A. B. 1986. The evolution of begging: Sibling competition and parent-offspring conflict. *Am. Nat* 128:99–114.

Harper, D. G. C. 1985. Brood division in robins. *Anim. Behav.* 33:466–480.

Harshman, L. G., and T. Prout. 1994. Sperm displacement without sperm transfer in *Drosophila melanogaster. Evolution* 48:758–766.

Harvey, P. H., and P. M. Bennett. 1985. Sexual dimorphism and reproductive strategies. In *Human sexual dimorphism*, ed. J. Ghesquire, R. D. Martin, and F. Newcombe, 43–59. London: Taylor and Francis.

Haskell, D. 1994. Experimental evidence that nestling begging behaviour incurs a cost due to predation. *Proc. R. Soc. Lond. B* 257:161–164.

Hastings, I. M. 1992. Population genetic aspects of deleterious cytoplasmic genomes and their effect on the evolution of sexual reproduction. *Genet. Res.* 59:215–225.

Hausfater, G. 1984. Infanticide in non-human primates: An introduction and perspective. In *Infanticide: Comparative and evolutionary perspectives*, ed. G. Hausfater and S. Blaffer Hrdy, 145–150. Hawthorne, N.Y.: Aldine.

Hawkins, B. A., and S. G. Compton. 1992. African fig wasp communities: Undersaturation and latitudinal gradients in species richness. *J. Anim. Ecol.* 61:361–372.

Heinsohn, R., and C. Packer. 1995. Complex cooperative strategies in group-territorial African lions. *Science (Wash.)* 269:1260–1262.

Heinze, J. 1993. Queen-queen interactions in polygynous ants. In *Queen number and sociality in insects*, ed. L. Keller, 308–333. Oxford: Oxford University Press.

———. 1995. Reproductive skew and genetic relatedness in *Leptothorax* ants. *Proc. R. Soc. Lond. B* 261:375–379.

Heinze, J., B. Hölldobler, and C. Peeters. 1994. Conflict and cooperation in ant societies. *Naturwissenschaften* 81:489–497.

Heppner, F. 1997. Three-dimensional structure and dynamics of bird flocks. In *Animal groups in three dimensions*, ed. J. K. Parrish and W. M. Hamner, 68–89. Cambridge: Cambridge University Press.

Herbers, J. M. 1990. Reproductive investment and allocation ratios for the ant *Leptothorax longispinosus*: Sorting out the variation. *Am. Nat.* 136:178–208.

Herre, E. A. 1985. Sex ratio adjustment in fig wasps. *Science (Wash.)* 228:896–898.

———. 1987. Optimality, plasticity, and selective regime in fig wasp sex ratios. *Nature (Lond.)* 329:627–629.

———. 1989. Coevolution of reproductive characteristics in 12 species of New World figs and their pollinator wasps. *Experientia* 45:637–647.

———. 1993. Population structure and the evolution of virulence in nematode parasites of fig wasps. *Science (Wash.)* 259:1442–1445.

———. 1995. Factors affecting the evolution of virulence: Nematode parasites of fig wasps as a case study. *Parasitology* 111 (suppl.):179–191.

———. 1996. An overview of studies on a community of Panamanian figs. *J. Biogeogr.* 23:593–607.

Herre, E. A., and S. A. West. 1997. Conflict of interest in a mutualism: Documenting the elusive fig wasp-seed tradeoff. *Proc. R. Soc. Lond. B* 264: 1501–1507.

Herre, E. A., C. A. Machado, E. Bermingham, J. D. Nason, D. M. Windsor, S. S. McCafferty, W. Van Houten, and K. Bachmann. 1996. Molecular phylogenies of figs and their pollinating wasps. *J. Biogeogr.* 23:531–542.

Herre, E. A., S. A. West, J. M. Cook, S. G. Compton, and F. Kjellberg. 1997. Fig wasp mating systems: Pollinators and parasites, sex ratio adjustment and male polymorphism, population structure and its consequences. In *Social competition and cooperation in insects and arachnids. 1. Evolution of mating systems*, ed. J. C. Choe and B. Crespi, 226–239. Cambridge: Cambridge University Press.

Herrera, J. A., D. M. López-León, J. Cabrero, M. W. Shaw, and J. P. M. Camacho. 1996. Evidence for B chromosome drive suppression in the grasshopper *Eyprepocnemis plorans. Heredity* 76:633–639.

Hethcote, H. W., and A. G. Knudson. 1978. Model for the incidence of embryonal cancers: Application to retinoblastoma. *Proc. Natl. Acad. Sci. USA* 75:2453–2457.

Hill, G. E., and R. Montgomerie. 1994. Plumage colour signals nutritional condition in the house finch. *Proc. R. Soc. Lond. B* 258:47–52.

Hinde, R. A. 1976. Interactions, relationships and social structure. *Man* 11:1–17.

Hiraiwa Hasegawa, M., and T. Hasegawa. 1994. Infanticide in non-human primates: Sexual selection and local resource competition. In *Infanticide and parental care*, ed. S. Parmigiani and F. S. vom Saal, 137–154. Chur, Switzerland: Harwood.

Hochstenbach, R., H. Harhangi, K. Shouren, and W. Hennig. 1994. Degenerating gypsy retrotransposons in a male-fertility gene in the Y-chromosome of *D. hydei. J. Mol. Evol.* 39:452–465.

Hoekstra, R. F. 1990. Evolution of uniparental inheritance of cytoplasmic DNA. In *Organizational constraints on the dynamics of evolution*, ed. J. Maynard Smith and G. Vida, 269–278. New York: Manchester University Press.

Hoffman, A. A., and M. Turelli. 1997. Cytoplasmic incompatibility. In *Insects*, ed. F. L. O'Neill, A. A. Hoffman, and J. H. Werren, 42–80. Oxford: Oxford University Press.

Hoffman, A. A., D. J. Clancy, and E. Merton. 1994. Cytoplasmic incompatibility in Australian populations of *Drosophila melanogaster. Genetics* 136:993–999.

Hoffman, A. A., M. Turelli, and G. M. Simmons. 1990. Factors affecting the distribution of cytoplasmic incompatibility in *Drosophila simulans. Genetics* 126:933–948.

Hoffman, A. A., D. J. Clancy, and J. Duncan. 1996. Naturally occurring *Wolbachia* infection in *Drosophila simulans* that does not cause cytoplasmic incompatibility. *Heredity* 76:1–8.

Holekamp, K. E., E. E. Boydston, and L. Smale. In press. Group movements in social carnivores. In *Group movement: Patterns, Processes, and cognitive implications in primates and other animals*, ed. S. Boinski and P. A. Garber. Chicago: University of Chicago Press.

Hölldobler, B., and E. O. Wilson. 1990. *The ants*. Berlin: Springer-Verlag.

Hoogland, C., and C. Biémont. 1996. Chromosomal distribution of transposable elements in *Drosophila melanogaster*: Test of the ectopic recombination model for maintenance of insertion site number. *Genetics* 144:197–204.

Hoogland, C., Vieira, C., and Biémont, C. 1997. Chromosomal distribution of the 412 retrotransposon in natural populations of *Drosophila simulans. Heredity* 79:128–134.

Hopf, F., R. E. Michod, and M. J. Sanderson. 1988. The effect of reproductive system on mutation load. *Theor. Pop. Biol.* 33:243–265.

Hossaert-McKey, M., M. Gibernau, and J. E. Frey. 1994. Chemosensory attraction of fig wasps to substances produced by receptive figs. *Entomol. Exp. Appl.* 70:185–191.

Houston, A. I., and N. B. Davies. 1985. The evolution of cooperation and life history in the dunnock. In *Behavioural ecology*, ed. R. M. Sibly and R. H. Smith, 471–487. Oxford: Blackwell Scientific.

Houston, A. I., C. E. Gasson, and J. M. McNamara. 1997. Female choice of matings to maximize parental care. *Proc. R. Soc. Lond. B* 264:173–179.

Hull, D. L. 1980. Individuality and selection. *Annu. Rev. Ecol. Syst.* 11:311–332.

Hurd, P. L. 1995. Communication in discrete action-response games. *J. Theor. Biol.* 174:217–222.

Hurst, G. D. D., and M. E. N. Majerus. 1993. Why do maternally inherited microorganisms kill males? *Heredity* 71:81–95.

Hurst, G. D. D., L. D. Hurst, and M. E. N. Majerus. 1997. Cytoplasmic sex-ratio distorters. In *Influential passengers: Inherited microorganisms and arthropod reproduction*, ed. F. L. O'Neill, A. A. Hoffman, and J. H. Werren, 125–154. Oxford: Oxford University Press.

Hurst, G. D. D., E. L. Purvis, J. J. Sloggett, and M. E. N. Majerus. 1994. The effect of infection with male-killing *Rickettsia* on the demography of female *Adalia bipunctata* L. (two-spot ladybird). *Heredity* 73:309–316.

Hurst, L. D. 1990. Parasite diversity and the evolution of diploidy, multicellularity and anisogamy. *J. Theor. Biol.* 144:429–443.

———. 1991a. The evolution of cytoplasmic incompatibility or when spite can be successful. *J. Theor. Biol.* 148:269–277.

———. 1991b. The incidences and evolution of cytoplasmic male killers. *Proc. R. Soc. Lond. B* 244:91–99.

———. 1992a. Is *Stellate* a relict meiotic driver? *Genetics* 130:229–230.

———. 1992b. Intragenomic conflict as an evolutionary force. *Proc. R. Soc. Lond. B* 248:135–140.

———. 1993a. The incidences, mechanisms and evolution of cytoplasmic sex ratio distorters in animals. *Biol. Rev.* 68:121–193.

———. 1993b. *Scat*[+] is a selfish gene analogous to *Medea* in *Tribolium castaneum*. *Cell* 75:407–408.

———. 1995. Selfish genetic elements and their role in evolution: The evolution of sex and some of what that entails. *Philos. Trans. R. Soc. B* 349:321–332.

———. 1996a. Adaptation and selection of genomic parasites. In *Adaptation*, ed. M. R. Rose and G. V. Lauder, 407–449. San Diego: Academic Press.

———. 1996b. Further evidence consistent with Stellate's involvement in meiotic drive. *Genetics* 142:641–643.

Hurst, L. D., and G. T. McVean. 1996. Clade selection, reversible evolution and the persistence of selfish elements: The evolutionary dynamics of cytoplasmic incompatibility. *Proc. R. Soc. Lond. B* 263:97–104.

Hurst, L. D., A. Atlan, and B. O. Bengtsson. 1996. Genetic conflicts. *Q. Rev. Biol.* 71:317–364.

Hurst, L., and A. Pomiankowski. 1991. Causes of sex ratio bias may account for unisexual sterility in hybrids: A new explanation of Haldane's rule and related phenomena. *Genetics* 128:841–858.

Hutchinson, G. E. 1959. Homage to Santa Rosalia, or Why are there so many kinds of animals. *Am. Nat.* 93:145–159.

Huxley, T. H. 1894. *Evolution and ethics*. London: Macmillan.

Jablonka, E., and M. J. Lamb. 1995. *Epigenetic inheritance and evolution, the Lamarckian dimension*. Oxford: Oxford University Press.

Jackendoff, R. 1993. *Patterns in the mind*. New York: Harvester Wheatsheaf.

Jaenike, J. 1996. Sex-ratio meiotic drive in the *Drosophila quinaria* group. *Am. Nat.* 148:237–254.

Jaki, S. L. 1983. *Angels, apes and men*. Peru, Ill.: Sherwood Sugden.

Jamieson, I. G. 1997. Testing reproductive skew models in a communally breeding bird, the pukeko, *Porphyrio porphyrio*. *Proc. R. Soc. Lond. B* 264:335–340.

Janzen, D. H. 1979. How to be a fig. *Annu. Rev. Ecol. Syst.* 10:13–51.

Jarvis, J. U. M., and N. C. Bennett. 1991. Ecology and behavior of the family Bathyergidae. In *The biology of the naked mole-rat*, ed. P. W. Sherman, J. U. M. Jarvis, and R. D. Alexander, 66–96. Princeton, N.J.: Princeton University Press.

Johns, P. M., and M. R. Maxwell. 1997. Sexual cannibalism: Who benefits? *Trends Ecol. Evol.* 12:127–128.

Johnson, L. R., S. H. Pilder, J. L. Bailey, and P. Olds-Clarke. 1995. Sperm from mice carrying one or two *t*-haplotypes are deficient in investment and oocyte penetration. *Dev. Biol.* 168:138–149.

Johnstone, R. A. 1995a. Honest advertisement of multiple qualities using multiple signals. *J. Theor. Biol.* 177:87–94.

———. 1995b. Sexual selection, honest advertisement and the handicap principle: Reviewing the evidence. *Biol. Rev.* 70:1–65.

———. 1996. Multiple displays in animal communication: 'Backup signals' and 'multiple messages.' *Phil. Trans. R. Soc. B* 351:329–338.

———. 1997. The tactics of mutual mate choice and competitive search. *Behav. Ecol. Sociobiol.* 40:51–59.

Johnstone, R. A., and A. Grafen. 1992. Error-prone signalling. *Proc. R. Soc. Lond. B* 248:229–233.

———. 1993. Dishonesty and the handicap principle. *Anim. Behav.* 46:759–764.

Johnstone, R. A., J. D. Reynolds, and J. C. Deutsch. 1996. Mutual mate choice and sex differences in choosiness. *Evolution* 50:1382–1391.

Jolly, A. 1991. Conscious chimpanzees? A review of recent literature. In *Cognitive ethology*, ed. C. Ristau, 231–252. Hillsdale, N.J.: Lawrence Erlbaum.

Jones, I. L., and F. M. Hunter. 1993. Mutual sexual selection in a monogamous seabird. *Nature (Lond.)* 362:238–239.

Jones, R. N. 1991. B-chromosome drive. *Am. Nat.* 137:430–442.

Juchault, P., T. Rigaud, and J.-P. Mocquard. 1992. Evolution of sex-determining mechanisms in a wild population of *Armadillidium vulgare*: Competition between two feminizing parasitic sex factors. *Heredity* 69:382–390.

Kalko, E. K. V., E. A. Herre, and C. O. Handley, Jr. 1996. Relation of fig fruit characteristics to fruit eating bats in the New and Old World tropics. *J. Biogeogr.* 23:565–576.

Kathuria, P., K. N. Ganeshaiah, R. Uma Shaanker, and R. Vasudeva. 1995. Is there dimorphism for style lengths in monoecious figs? *Curr. Sci.* 68:1047–1049.

Keane, B., P. M. Waser, S. R. Creel, N. M. Creel, L. F. Elliott, and D. J. Minchella. 1994. Subordinate reproduction in dwarf mongooses. *Anim. Behav.* 47:65–75.

Keller, L. 1991. Queen number, mode of colony founding and queen reproductive success in ants (Hymenoptera, Formicidae). *Ethol. Ecol. Evol.* 4:307–316.

———. 1993a. The assessment of reproductive success of queens in ants and other social insects. *Oikos* 67:177–180.

———. 1993b. *Queen number and sociality in insects.* Oxford: Oxford University Press.

———. 1995. Social life: The paradox of multiple-queen colonies. *Trends Ecol. Evol.* 10:355–360.

———. 1997. Indiscriminate altruism: Unduly nice parents and siblings. *Trends Ecol. Evol.* 12:99–103.

Keller, L., and P. Nonacs. 1993. The role of queen pheromones in colonies of social insects: Queen control or queen signal? *Anim. Behav.* 45:787–794.

Keller, L., and L. Passera. 1989. Size and fat content of gynes in relation to the mode of colony founding in ants (Hymenoptera; Formicidae). *Oecologia* 80:236–240.

———. 1993. Incest avoidance, fluctuating asymmetry, and the consequences of inbreeding in *Iridomyrmex humilis*, an ant with multiple queen colonies. *Behav. Ecol. Sociobiol.* 33:191–199.

Keller, L., and H. K. Reeve. 1994a. Genetic variability, queen number, and polyandry in social Hymenoptera. *Evolution* 38:694–704.

———. 1994b. Partitioning of reproduction in animal societies. *Trends Ecol. Evol.* 9:98–102.

———. 1995. Why do females mate with multiple males? The sexually selected sperm hypothesis. *Adv. Study Behav.* 24:291–315.

Keller, L., and K. G. Ross. 1993. Phenotypic basis of reproductive success in a social insect: Genetic and social determinants. *Science (Wash.)* 260:1107–1110.

———. 1998. Selfish genes: A green beard in the red fire ant. *Nature* 394:573–575.

Keller, L., and E. L. Vargo. 1993. Reproductive structure and reproductive roles in colonies of eusocial insects. In *Queen number and sociality in insects*, ed. L. Keller, 16–44. Oxford: Oxford University Press.

Keller, L., S. Aron, and L. Passera. 1996a. Internest sex-ratio variation and male brood survival in the ant *Pheidole pallidula. Behav. Ecol.* 7:292–298.

Keller, L., G. L'Hoste, F. Balloux, and O. Plumey. 1996b. Queen number influences the primary sex ratio in the Argentine ant, *Linepithema humile* (= *Iridomyrmex humilis*). *Anim. Behav.* 51:445–449.

Keller, L., L. Passera, and J. P. Suzzoni. 1989. Queen execution in the Argentine ant *Iridomyrmex humilis* (Mayr). *Physiol. Entomol.* 14:157–163.

Kempenaers, B. 1995. Polygyny in the blue tit: Intra- and inter-sexual conflicts. *Anim. Behav.* 49:1047–1064.

Kempenaers, N., and B. C. Sheldon. 1996. Why do male birds not discriminate between their own and extra-pair offspring? *Anim. Behav.* 51:1165–1173.

Kerdelhue, C., and J.-Y. Rasplus. 1996a. Nonpollinating Afrotropical fig wasps affect the fig-pollinator mutualism in *Ficus* within the subgenus *Sycomorus. Oikos* 75:3–14.

———. 1996b. The evolution of dioecy among *Ficus* (Moraceae): An alternative hypothesis involving non-pollinating fig wasp pressure on the fig-pollinator mutualism. *Oikos* 77:163–166.

———. 1997. Active pollination of *Ficus sur* by two sympatric fig wasp species in West Africa. *Biotropica* 29:69–74.

Kilner, R. 1997. Mouth colour is a reliable signal of need in begging canary nestlings. *Proc. R. Soc. Lond. B* 264:963–968.

Kinsey, P. T., and S. B. Sandmeyer. 1991. Adjacent pol II and pol III promoters: Transcription of the yeast retrotransposon Ty3 and a target tRNA gene. *Nucleic Acid Res.* 19:1317–1324.

Kirkpatrick, M. 1987. Sexual selection by female choice in polygynous animals. *Annu. Rev. Ecol. Syst.* 18:43–70.

Kjellberg, F., and S. Maurice. 1989. Seasonality in the reproductive phenology of *Ficus*: Its evolution and its consequences. *Experientia* 45:653–660.

Kjellberg, F., M. C. Anstett, and E. A. Herre. 1994. Yucca sex. *Nature (Lond.)* 370:604.

Kjellberg, F., P.-H. Gouyon, M. Ibrahim, M. Raymond, and G. Valdeyron. 1987. The stability of the symbiosis between dioecious figs and and their pollinators: A study of *Ficus carica* and *Blastophaga pneses. Evolution* 41:693–704.

Kleckner, N. 1990. Regulating Tn10 and IS10 transposition. *Genetics* 124:449–454.

Knudson, A. G. 1971. Mutation and cancer: Statistical study of retinoblastoma. *Proc. Natl. Acad. Sci. USA* 68:820–823.

Kölliker, M., H. Richner, I. Werner, and P. Heeb. 1998. Begging signals and biparental care: Nestling great tits discriminate between parents. *Anim. Behav.* 55:215–222.

Kondrashov, A. S. 1982. Selection against harmful mutations in large sexual and asexual populations. *Genet. Res.* 40:325–332.

———. 1993. Classification of hypotheses on the advantage of amphimixis. *J. Hered.* 84:372–387.

Korona, R., and B. R. Levin. 1993. Phage-mediated selection and the evolution and maintenance of restriction-modification. *Evolution* 47:556–575.

Kruuk, H. 1972. *The spotted hyena: A study of predation and social behavior.* Chicago: University of Chicago Press.

Kruuk, H., and D. MacDonald. 1985. Group territories of carnivores: Empires and enclaves. In *Behavioural ecology: Ecological consequences of adaptive behaviour,* ed. R. M. Sibly and R. H. Smith, 521–536. Oxford: Blackwell Scientific.

Kryger, P., and R. F. A. Moritz. 1997. Lack of kin recognition in swarming honeybees (*Apis mellifera*). *Behav. Ecol. Sociobiol.* 40:271–276.

Kummer, H. 1968. *Social organization of Hamadryas baboons.* Chicago: University of Chicago Press.

Lacey, E. A., and P. W. Sherman. 1991. Social organization of naked mole-rat colonies: Evidence for division of labor. In *The biology of the naked mole-rat,* ed. P. W. Sherman, J. U. M. Jarvis, and R. D. Alexander, 275–336. Princeton, N.J.: Princeton University Press.

Lachmann, M., and C. T. Bergstrom. 1998. Signalling among relatives—Beyond the Tower of Babel. *J. Theor. Biol.* 54:146–160.

Lack, D. 1947. The significance of clutch size. *Ibis* 89:302–352.

———. 1954. *The natural regulation of animal numbers.* Oxford: Clarendon Press.

Lacy, R. C., and P. W. Sherman. 1983. Kin recognition by phenotype matching. *Am. Nat.* 121:489–512.

Lande, R. 1980. Sexual dimorphism, sexual selection, and adaptation in polygenic characters. *Evolution* 34:292–305.

———. 1987. Genetic correlations between the sexes in the evolution of sexual dimorphism and mating preferences. In *Sexual selection: Testing the alternatives*, ed. J. Bradbury and M. B. Andersson, 83–94. Chichester: Wiley.

Langley, C. H., E. A. Montgomery, R. Hudson, N. Kaplan, and B. Charlesworth. 1988. On the role of unequal exchange in the containment of transposable element copy number. *Genet. Res.* 52:223–235.

Laser, K. D., and N. R. Lersten. 1972. Anatomy and cytology of microsporogenesis in cytoplasmic male sterile angiosperms. *Bot. Rev.* 38:425–454.

Law, R., and C. Cannings. 1984. Genetic analysis of conflicts arising during development of seeds in the Angiospermophyta. *Proc. R. Soc. Lond. B* 221:53–70.

Law, R., and V. Hutson. 1992. Intracellular symbionts and the evolution of uniparental cytoplasmic inheritance. *Proc. R. Soc. Lond. B* 248:69–77.

Lawrence, S. E. 1992. Sexual cannibalism in the praying mantid, *Mantis religiosa*: A field study. *Anim. Behav.* 43:569–583.

Lazarus, J. 1990. The logic of mate desertion. *Anim. Behav.* 39:672–684.

Lazarus, J., and B. Inglis. 1986. Shared and unshared parental investment, parent-offspring conflict, and brood size. *Anim. Behav.* 34:1791–1804.

Le Masurier, A. D. 1987. A comparative study of the relationship between host size and brood size in *Apanteles* spp. (Hymenoptera: Braconidae). *Ecol. Entomol.* 12: 383–393.

Lee, D. 1980. *The sinking ark: Environmental problems in Malaysia and Southeast Asia*. Kuala Lumpur: Heinemann.

Lee, D. H., J. R. Granja, J. A. Martinez, K. Severin, and M. R. Ghadiri. 1996. A self-replicating peptide. *Nature (Lond.)* 382:525–528.

Lee, D. H., K. Severin, Y. Yokobayashi, and M. R. Ghadiri. 1997. Emergence of symbiosis in peptide self-replication through a hypercyclic network. *Nature (Lond.)* 390:591–594.

Leigh, E. G., Jr. 1971. *Adaptation and diversity*. San Francisco: Freeman Cooper.

———. 1977. How does selection reconcile individual advantage with the good of the group? *Proc. Natl. Acad. Sci. USA* 74:4524–4646.

———. 1983. When does the good of the group override the advantage of the individual? *Proc. Natl. Acad. Sci. USA* 80:2985–2989.

———. 1987. Ronald Fisher and the development of evolutionary theory. 2. Influences of new variation on evolutionary process. *Oxf. Surv. Evol. Biol.* 4:212–263.

———. 1991. Genes, bees and ecosystems: The evolution of a common interest among individuals. *Trends Ecol. Evol.* 6:257–262.

———. 1995. The major transitions of evolution. *Evolution* 49:1302–1306.

Leigh, E. G., Jr., and T. E. Rowell. 1995. The evolution of mutualism and other forms of harmony at various levels of biological organization. *Ecology* 26:131–158.

Lemaitre, B., S. Ronsseray, and D. Coen. 1993. Maternal repression of the *P* element promoter in the germline of *Drosophila melanogaster. Genetics* 135:149–160.

Lessells, C. M. 1991. The evolution of life histories. In *Behavioural ecology*, 3d ed., ed. J. R. Krebs and N. B. Davies, 32–68. Oxford: Blackwell Scientific.

———. 1994. Baby bunting in paternity probe. *Nature (Lond.)* 371:655–656.

————. 1998. A theoretical framework for sex-biased parental care. *Anim. Behav.* 56:395–407.

Lessells, C. M., and M. I. Avery. 1987. Sex-ratio selection in species with helpers at the nest: Some extensions of the repayment model. *Am. Nat.* 129:610–620.

Levin, B. R. 1988. The evolution of sex in bacteria. In *The evolution of sex*, ed. R. E. Michod and B. R. Levin, 194–211. Sunderland, Mass.: Sinauer.

Levin, S. A., and D. Pimentel. 1981. Selection of intermediate rates of increase in parasite-host systems. *Am. Nat.* 117:308–315.

Lewontin, R. C. 1970. The units of selection. *Annu. Rev. Ecol. Syst.* 1:1–18.

————. 1978. Adaptation. *Sci. Am.* 239:212–230.

Liersch, S., and P. Schmid-Hempel. 1998. Genetic variation within social insect colonies reduces parasite load. *Proc. R. Soc. Lond. B* 265:221–225.

Lima, S. L. 1989. Iterated prisoner's dilemma: An approach to evolutionary stable cooperation. *Am. Nat.* 134:828–834.

————. 1995. Collective detection of predatory attack by social foragers: Fraught with ambiguity? *Anim. Behav.* 50:1097–1108.

Lima, S. L., and P. A. Zollner. 1996. Anti-predatory vigilance and limits to collective detection: Spatial and visual separation between foragers. *Behav. Ecol. Sociobiol.* 38:355–363

Lindsey, D. L., and G. C. Zimm. 1992. *The genome of Drosophila melanogaster.* San Diego: Academic Press.

Livak, K. J. 1990. Detailed structure of the *Drosophila melanogaster Stellate* genes and their copy number. *Genetics* 124:303–316.

López-León, D. M., J. Cabrero, M. C. Pardo, E. Viseras, J. P. M. Camacho, and J. L. Santos. 1993. Generating high variability of B chromosomes in the grasshopper *Eyprepocnemis plorans. Heredity* 71:352–362.

Lorenz, K. 1978. *Behind the mirror.* San Diego: Harcourt Brace Jovanovich.

Løvtrup, S. 1976. On the falsifiability of neo-Darwinism. *Evol. Theory* 1:267–283.

Lundberg, S., and H. Smith. 1994. Parent-offspring conflicts over reproductive efforts: Variations upon a theme by Charnov. *J. Theor. Biol.* 170:215–218.

Lynch, M., R. Burger, D. Butcher, and W. Gabriel. 1993. The mutational meltdown in asexual populations. *J. Hered.* 84:339–344.

Lyon, M. F. 1991. The genetic basis of transmission-ratio distortion and male sterility due to the *t*-complex. *Am. Nat.* 137:349–358.

————. 1992. Deletion of mouse *t*-complex distorter-1 produces an effect like that of the *t*-form of the distorter. *Genet. Res.* 59:27–33.

Lyttle, T. W. 1977. Experimental population genetics of meiotic drive systems. 1. Pseudo-Y chromosomal drive as a means of eliminating cage populations of *Drosophila melanogaster. Genetics* 86:413–445.

————. 1979. Experimental population genetics of meiotic drive systems. 2. Accumulation of genetic modifiers of segregation distorter (*SD*) in laboratory populations. *Genetics* 91:339–357.

————. 1991. Segregation distorters. *Annu. Rev. Genet.* 25:511–557.

Machado, C. A., E. A. Herre, S. McCafferty, and E. Bermingham. 1996. Molecular phylogenies of fig pollinating and non-pollinating wasps and the implications for the origin and evolution of the fig-fig wasp mutualism. *J. Biogeogr.* 23:531–542.

Macnair, M. R., and G. A. Parker. 1978. Models of parent-offspring conflict. 2. Promiscuity. *Anim. Behav.* 26:111–122.

———. 1979. Models of parent-offspring conflict. 3. Intra brood conflict. *Anim. Behav.* 27:1202–1209.

Magrath, R. D. 1990. Hatching asynchrony in altricial birds. *Biol. Rev.* 65:587–622.

Magurran, A. E., and B. H. Seghers. 1994. Sexual conflict as a consequence of ecology: Evidence from guppy, *Poecilia reticulata*, populations in Trinidad. *Proc. R. Soc. Lond. B* 255:31–36.

Malagolowkin, C., and C. G. Carvalho. 1961. Direct and indirect transfer of maternally inherited abnormal sex ratio in different species of *Drosophila*. *Genetics* 46:1009–1013.

Mangel, M., and C. W. Clark. 1988. *Dynamic modeling in behavioral ecology.* Princeton, N.J.: Princeton University Press.

Margulis, L. 1981. *Symbiosis in cell evolution.* San Francisco: W. H. Freeman.

———. 1993. *Symbiosis in cell evolution, microbial communities in the Archean and Proterozoic eons*, 2d ed. New York: W. H. Freeman.

Markman, S., Y. Yom-Tov, and J. Wright. 1995. Male parental care in the orange-tufted sunbird: behavioural adjustments in provisioning and nest guarding effort. *Anim. Behav.* 50:655–669.

Markman, S., Y. Yom-Tov, and J. Wright. 1996. The effect of male removal on female parental care in the orange-tufted sunbird. *Anim. Behav.* 52:437–444.

May, R. M., and R. M. Anderson. 1983. Epidemiology and genetics in the coevolution of parasites and hosts. *Proc. R. Soc. Lond. B* 219:281–313.

May, R. M., and M. A. Nowak. 1994. Superinfection, metapopulation dynamics, and the evolution of diversity. *J. Theor. Biol.* 170:95–114.

Maynard Smith, J. 1958. *The theory of evolution.* Harmondsworth, U.K.: Penguin.

———. 1964. Group selection and kin selection. *Nature (Lond.)* 201:1145–1147.

———. 1971. The origin and maintenance of sex. In *Group selection*, ed. G. C. Williams, 163–175. Chicago: Aldine-Atherton.

———. 1976. Group selection. *Q. Rev. Biol.* 51:277–283.

———. 1977. Parental investment: A prospective analysis. *Anim. Behav.* 25:1–9.

———. 1978. *The evolution of sex.* Cambridge: Cambridge University Press.

———. 1979. Hypercycles and the origin of life. *Nature (Lond.)* 280:445–446.

———. 1982. *Evolution and the theory of games.* Cambridge: Cambridge University Press.

———. 1983a. Game theory and the evolution of cooperation. In *Evolution from molecules to man*, ed. D. S. Bendall, 445–456. Cambridge: Cambridge University Press.

———. 1983b. Models of evolution. *Proc. R. Soc. Lond. B* 219:315–325.

———. 1987. How to model evolution. In *The latest on the best: Essays on evolution and optimality*, ed. J. Dupré, 119–131. Cambridge, Mass.: MIT Press.

———. 1988. Evolutionary progress and levels of selection. In *Evolutionary Progress*, ed. M. H. Nitecki, 219–230. Chicago: University of Chicago Press.

———. 1990. Models of a dual inheritance system. *J. Theor. Biol.* 143:41–53.

———. 1991a. A Darwinian view of symbiosis. In *Symbiosis as a source of evolutionary innovation*, ed. L. Margulis and R. Fester, 26–39. Cambridge, Mass.: MIT Press.

————. 1991b. Honest signalling—the Philip Sidney Game. *Anim. Behav.* 42:1034–1035.

Maynard Smith, J., and D. G. C. Harper. 1995. Animal signals: Models and terminology. *J. Theor. Biol.* 177:305–312.

Maynard Smith, J., and E. Szathmáry. 1993. The origin of chromosomes 1. Selection for linkage. *J. Theor. Biol.* 164:437–446.

————. 1995. *The major transitions in evolution.* San Francisco: W. H. Freeman.

McComb, K., C. Packer, and A. Pusey. 1994. Roaring and numerical assessment in contests between groups of female lions, *Panthera leo. Anim. Behav.* 47:379–387.

McGowan, K. J., and G. E. Woolfenden. 1989. A sentinel system in the Florida scrub jay. *Anim. Behav.* 37:1000–1006.

McKean, K. A., and M. Zuk. 1995. An evolutionary perspective on signaling in behavior and immunology. *Naturwissenschaften* 82:509–516.

McKey, D. 1989. Population biology of figs: Applications for conservation. *Experientia* 45:661–673.

McLaughlin, R. L., and R. D. Montgomerie. 1985. Brood division by lapland longspurs. *Auk* 102:687–695.

McLean, J. R., C. J. Merrill, P. A. Powers, and B. Ganetsky. 1994. Functional identification of the *Segregation distorter* locus of *Drosophila melanogaster* by germline transformation. *Genetics* 137:201–209.

McNamara, J. M., and A. I. Houston. 1992. Evolutionarily stable levels of vigilance as a function of group size. *Anim. Behav.* 43:641–658.

McRae, S. B., P. J. Weatherhead, and R. Montgomerie. 1993. American robin nestlings compete by jockeying for position. *Behav. Ecol. Sociobiol.* 33:102–106.

Mercot, H., A. Atlan, M. Jacques, and C. Montchamp-Moreau. 1995. Sex-ratio distortion in *Drosophila simulans*: Co-occurrence of a meiotic drive and a suppressor of drive. *J. Evol. Biol.* 8:283–300.

Merilä, J., B. C. Sheldon, and H. Ellegren. 1997. Antagonistic natural selection revealed by molecular sex identification of nestling collared flycatchers. *Mol. Ecol.* 6:1167–1175.

————. 1998. Quantitative genetics of sexual size dimorphism in the collared flycatcher, *Ficedula albicollis. Evolution* 52:870–876.

Metcalf, J. S. , J. Stamps, and V. Krishnan. 1979. Parent-offspring conflict which is not limited by degree of kinship. *J. Theor. Biol.* 76:99–107.

Michaloud, G., S. Michaloud-Pelletier, J. T. Wiebes, and C. C. Berg. 1985. The co-occurrence of two pollinating species of fig wasp and one species of fig. *Proc. K. Ned. Akad. Wet. Ser. C* 88:93–119.

Michiels, N. K. 1998. Mating conflicts and sperm competition in simultaneous hermaphrodites. In *Sperm competition and sexual selection*, ed. T. R. Birkhead and A. P. Møller., 219–254. London: Academic Press.

Michod, R. E. 1981. Positive heuristics in evolutionary biology. *Br. J. Philos. Sci. (Wash.)* 32:1–36.

————. 1983. Population biology of the first replicators: On the origin of the genotype, phenotype and organism. *Am. Zool.* 23:5–14.

————. 1986. On fitness and adaptedness and their role of evolutionary explanation. *J. Hist. Biol.* 19:289–302.

———. 1991. Sex and evolution. In *1990 Lectures in Complex Systems, SFI Studiesin the Sciences of Complexity, Lect. Vol. 3*, ed. L. Nadel and D. Stein, 285–320. Reading, Mass.: Addison-Wesley.

———. 1995. *Eros and evolution: A natural philosophy of sex*. Reading, Mass.: Addison-Wesley.

———. 1996. Cooperation and conflict in the evolution of individuality. 2. Conflict mediation. *Proc. R. Soc. Lond. B* 263:813–822.

———. 1997a. Cooperation and conflict in the evolution of individuality. 1. Multilevel selection of the organism. *Am. Nat.* 149:607–645.

———. 1997b. Evolution of the individual. *Am. Nat.* 150:S5–S21

———. 1997c. What good is sex? *The Sciences* 37:42–46.

———. 1998. Origin of sex for error repair. 3. Selfish sex. *Theor. Pop. Biol.* 53:60–74.

———. 1999. *Darwinian dynamics, evolutionary transitions in fitness and individuality*. Princeton, N.J.: Princeton University Press

Michod, R. E., and B. R. Levin, eds. 1988. *Evolution of sex: An examination of current ideas*. Sunderland, Mass.: Sinauer.

Michod, R. E., and D. Roze. 1997. Transitions in individuality. *Proc. R. Soc. Lond. B* 264:853–857.

———. 1999. Cooperation and conflict in the evolution of individuality. 3. Transitions in the unit of fitness. In *Mathematical and computational biology; computational morphogenesis, hierarchical complexity, and digital evolution*, ed. C. L. Nehaniv, 26:47–92. Providence, R.I.: American Mathematical Society. Series title: Lectures on Mathematics in the Life Sciences. No series editors.

Miklos, G. L. G., M. T. Yamamoto, J. Davies, and V. Pirrotta. 1988. Microcloning reveals a high frequency of repetitive sequences characteristic of chromosome 4 and *b*-heterochromatin of *Drosophila melanogaster. Proc. Natl. Acad. Sci. USA* 85:2051–2055.

Milton, K. 1991. Leaf change and fruit production in six neotropical Moraceae species. *J. Ecol.* 79:1–26.

Milton, K., D. M. Windsor, D. W. Morrison, and M. A. Estribi. 1982. Fruiting phenologies of two neotropical *Ficus* species. *Ecology* 63:752–762.

Mock, D. W. 1987. Siblicide, parent-offspring conflict, and unequal parental investment by egrets and herons. *Behav. Ecol. Sociobiol.* 20:247–256.

Mock, D. W., and L. S. Forbes. 1992. Parent-offspring conflict: A case of arrested development. *Trends Ecol. Evol.* 7:409–413.

Mock, D. W., and G. A. Parker. 1986. Advantages and disadvantages of egret and heron brood reduction. *Evolution* 40:459–470.

———. 1997. *The evolution of sibling rivalry*. Oxford: Oxford University Press.

Møller, A. P., and Birkhead, T. R. 1993. Certainty of paternity covaries with paternal care in birds. *Behav. Ecol. Sociobiol.* 33, 261–268.

Montgomery, E. A., and C. H. Langley. 1983. Transposable elements in Mendelian populations. 2. Distribution of three *copia*-like elements in a natural population of *Drosophila melanogaster. Genetics* 104:473–483.

Montgomery, E. A., S. M. Huang, C. H. Langley, and B. H. Judd. 1991. Chromosome rearrangement by ectopic recombination in *Drosophila melanogaster*: Genome structure and evolution. *Genetics* 129:1085–1098.

Moran, G. 1984. Vigilance behaviour and alarm calls in a captive group of meerkats, *Suricata suricatta. Z. Tierpsychol.* 65:228–240.

Morita, T., H. Kubota, K. Murata, M. Nozaki, C. Delarbre, K. Willison, Y. Satta, M. Sakaizumi, N. Takahata, G. Gachelin, and A. Matcuchiro. 1992. Evolution of the mouse *t* haplotype: Recent and worldwide introgression to *Mus musculus. Proc. Natl. Acad. Sci. USA* 89:6851–6855.

Moschetti, R., R. Caizzi, and S. Pimpinelli. 1996. Segregation distortion in *Drosophila melanogaster*: Genomic organization of responder sequences. *Genetics* 144: 1665–1671.

Moss, C. J. 1988. *Elephant memories: Thirteen years in the life of an elephant family.* New York: William Morrow.

Moss, C. J., and J. H. Poole. 1983. Relationships and social structure in African elephants. In *Primate social relationships: An integrated approach*, ed. R. A. Hinde, 315–325. Oxford: Blackwell Scientific.

Motro, U. 1994. Evolutionary and continuous stability in asymmetric games with continuous strategy sets: The parental investment conflict as an example. *Am. Nat.* 144:229–241.

Moynihan, M. 1962. The organization and probable evolution of some mixed species flocks of Neotropical birds. *Smithson. Misc. Publ.* 143:1–140.

———. 1968. Social mimicry: Character convergence versus character displacement. *Evolution* 22:315–331.

———. 1979. *Geographic variation in social behavior and in adaptations to competition among Andean birds.* Cambridge, Mass.: Nuttall Ornithological Club.

———. 1998. *The social regulation of competition and aggression in animals.* Washington, D. C.: Smithsonian Institution Press.

Mueller, U. G. 1991. Haplodiploidy and the evolution of facultative sex ratios in a primitively eusocial bee. *Science (Wash.)* 254:442–444.

Mueller, U. G., G. C. Eickwort, and C. F. Aquadro. 1994. DNA fingerprinting analysis of parent-offspring conflict in a bee. *Proc. Natl. Acad. Sci. USA* 91:5143–5147.

Mulder, R. A., and N. E. Langmore. 1993. Dominant males punish helpers for temporary defection in superb fairy-wrens. *Anim. Behav.* 45:830–833.

Muller, H. J. 1932. Some genetic aspects of sex. *Am. Nat.* 66:118–138.

———. 1964. The relation of recombination to mutational advance. *Mutat. Res.* 1:2–9.

Munn, C. A. 1985. Permanent canopy and understory flocks in Amazonia: Species composition and population density. In *Neotropical ornithology*, ed. P. A. Buckley, M. S. Foster, E. S. Morton, R. S. Ridgely and F. G. Buckley, 683–712. Washington, D.C.: American Ornithologists' Union.

Naito, T., K. Kusano, and I. Kobayashi. 1995. Selfish behavior of restriction-modification systems. *Science (Wash.)* 267:897–899.

Nason, J. D., E. A. Herre, and J. L. Hamrick. 1996. Paternity analysis of the breeding structure of strangler fig populations: Evidence for substantial long-distance wasp dispersal. *J. Biogeogr.* 23:501–512.

———. 1998. The breeding structure of a tropical keystone plant resource. *Nature (Lond.)* 391:685–687.

Natsoulis, G., W. Thomas, M. Roughman, F. Winston, and J. E. Boynton. 1989. Ty1 transposition in *Saccharomyces cerevisiae* is nonrandom. *Genetics* 123:269–279.

Nee, S., and J. Maynard Smith. 1990. The evolutionary biology of molecular parasites. *Parasitology* 100:S5–S18.

Nefdt, R. J. C., and S. G. Compton. 1996. Regulation of seed and pollinator production in the fig-fig wasp mutualism. *J. Anim. Ecol.* 65:170–182.

Newton, L. E., and A. Lomo. 1979. The pollination of *Ficus vogelii* in Ghana. *Bot. J. Linn. Soc.* 78:21–30.

Nishiumi, I., S. Yamagishi, H. Maekawa, and C. Shimoda. 1996. Paternal expenditure is related to brood sex ratio in polygynous great reed warblers. *Behav. Ecol. Sociobiol.* 39:211–217.

Nöe, R. 1990. A veto game played by baboons: A challenge to use of the prisoner's dilemma as a paradigm for reciprocity and cooperation. *Anim. Behav.* 39:78–90.

Nonacs, P. 1986a. Ant reproductive strategies and sex allocation theory. *Q. Rev. Biol.* 61:1–21.

———. 1986b. Sex ratio determination within colonies of ants. *Evolution* 40:199–204.

———. 1993a. The effects of polygyny and colony life history on optimal sex investment. In *Queen number and sociality in insects*, ed. L. Keller, 110–131. Oxford: Oxford University Press.

———. 1993b. Male parentage and sexual deception in the social Hymenoptera. In *Evolution and diversity of sex ratio in insects and mites*, ed. D. L. Wrensch and M. Ebberts, 384–401. New York: Chapman and Hall.

Nonacs, P., and N. F. Carlin. 1990. When can ants discriminate the sex of brood ? A new aspect of queen-worker conflict. *Proc. Natl. Acad. Sci. USA* 87:9670–9673.

Nowak, M. A., and R. M. May. 1994. Superinfection and the evolution of virulence. *Proc. R. Soc. B.* 255:81–89.

Nowak, M., and K. Sigmund. 1993. A strategy of win-stay, lose-shift that outperforms tit-for-tat in the Prisoner's Dilemma game. *Nature (Lond.)* 364:56–58.

Nunney, L. 1985. Group selection, altruism, and structured-deme models. *Am. Nat.* 126:212–230.

———. 1989. The maintenance of sex by group selection. *Evolution* 43:245–257.

———. 1993. Review of "Natural selection: Domains, levels, and challenges" by G. C. Williams, Oxford University Press, Oxford. *J. Evol. Biol.* 6:773–75.

———. 1999. Lineage selection and the evolution of multistage carcinogenesis. *Proc. R. Soc. Lond. B* 266:493–498.

Nunney, L., and D. R. Elam. 1994. Estimating the effective size of conserved populations. *Conserv. Biol.* 8:175–184.

Nur, U., J. H. Werren, D. G. Eickbush, D. Burke, and T. H. Eickbush. 1988. A "selfish" B chromosome that enhances its transmission by eliminating the paternal genome. *Science (Wash.)* 240:512–515.

O'Brien, T. G., E. S. Kinnaird, N. L. Dierenfeld, R. W. Conklin-Brittain, R. W. Wrangham, and S. C. Silver. 1998. What's so special about figs? *Nature (Lond.)* 392:668.

O'Connor, R. J. 1978. Brood reduction in birds: Selection for fratricide, infanticide and suicide? *Anim. Behav.* 26:79–96.

Oldroyd, B. P., T. E. Rinderer, and S. M. Buco. 1990. Nepotism in honey bees. *Nature (Lond.)* 346:707–708.

Olds-Clarke, P., and L. R. Johnson. 1993. *t* haplotypes in the mouse compromise sperm flagellar function. *Dev. Biol.* 155:14–25.

Orell, M., S. Rytkönen, and K. Koivula. 1994. Causes of divorce in the monogamous willow tit, *Parus montanus*, and consequences for reproductive success. *Anim. Behav.* 48:1143–1154.

Orgel, L. E. 1968. Evolution of the genetic apparatus. *J. Mol. Biol.* 38:381–393.

———. 1990. Adding to the genetic alphabet. *Nature (Lond.)* 343:18–20.

———. 1992. Molecular replication. *Nature (Lond.)* 358:203–209.

Orgel, L. E., and F. H. C. Crick. 1980. Selfish DNA: The ultimate parasite. *Nature (Lond.)* 284:604–607.

Orians, G. H. 1969. On the evolution of mating systems in birds and mammals. *Am. Nat.* 103:589–603.

Oster, G. F., and E. O. Wilson. 1978. *Caste and ecology in the social insects*. Princeton, N.J.: Princeton University Press.

Otter, K., and L. Ratcliffe. 1996. Female initiated divorce in a monogamous songbird: Abandoning mates for males of higher quality. *Proc. R. Soc. Lond. B* 263:351–354.

Ottosson, U., J. Bäckman, and H. G. Smith. 1997. Begging affects parental effort in the pied flycatcher, *Ficedula hypoleuca. Behav. Ecol. Sociobiol.* 41:381–384.

Pacala, S. W., D. M. Gordon, and H. C. J. Godfray. 1996. Effects of social group size on information transfer and task allocation. *Evol. Ecol.* 10: 127–165.

Packer, C., and P. Abrams. 1990. Should co-operative groups be more vigilant than selfish groups? *J. Theor. Biol.* 142:341–357.

Packer, C., and A. E. Pusey. 1984. Infanticide in carnivores. In *Infanticide: Comparative and evolutionary perspectives*, ed. G. Hausfater and S. Blaffer Hrdy, 31–42. Hawthorne, N. Y.: Aldine.

Packer, C., and L. Ruttan. 1988. The evolution of cooperative hunting. *Am. Nat.* 132:159–198.

Packer, C., D. Scheel, and A. E. Pusey. 1990. Why lions form groups: Food is not enough. *Am. Nat.* 136:1–19.

Packer, C., D. A. Gilbert, A. E. Pusey, and S. J. O'Brian. 1991. A molecular genetic analysis of kinship and cooperation in African lions. *Nature (Lond.)* 351:562–565.

Packer, L., and R. E. Owen. 1994. Relatedness and sex ratio in a primitively eusocial halictine bee. *Behav. Ecol. Sociobiol.* 34:1–10.

Page, R. E. 1986. Sperm utilization in social insects. *Annu. Rev. Entomol.* 31:297–320.

Page, R. E. J., G. E. Robinson, and M. K. Fondrk. 1989. Genetic specialists, kin recognition and nepotism in honey-bee colonies. *Nature (Lond.)* 338:576–579.

Paine, R. T. 1966. Food web complexity and species diversity. *Am. Nat.* 100:65–75.

Palopoli, M. F., and C.-I. Wu. 1996. Rapid evolution of a co-adapted gene-complex: Evidence from the *Segregation Distorter* (*SD*) system of meiotic drive in *Drosophila melanogaster. Genetics* 143:1675–1688.

Pamilo, P. 1990. Sex allocation and queen-worker conflict in polygynous ants. *Behav. Ecol. Sociobiol.* 27:31–36.

———. 1991a. Evolution of colony characteristics in social insects. 1. Sex allocation. *Am. Nat.* 137:83–107.

————. 1991b. Evolution of colony characteristics in social insects. 2. Number of reproductive individuals. *Am. Nat.* 138:412–433.

Pamilo, P., and R. Rosengren. 1983. Sex ratio strategies in *Formica* ants. *Oikos* 40:24–35.

Pamilo, P., and P. Seppä. 1994. Reproductive competition and conflicts in colonies of the ant *Formica sanguinea. Anim. Behav.* 48:1201–1206.

Parker, G. A. 1970. Sperm competition and its evolutionary consequences in the insects. *Biol. Rev.* 45:525–567.

————. 1978. Selection on non-random fusion of gametes during the evolution of anisogamy. *J. Theor. Biol.* 73:1–28.

————. 1979. Sexual selection and sexual conflict. In *Sexual selection and reproductive competition in insects*, ed. M. S. Blum and N. B. Blum, 123–166. New York: Academic Press.

————. 1982. Why are there so many tiny sperm? Sperm competition and the maintenance of two sexes. *J. Theor. Biol.* 96:281–294.

————. 1983. Mate quality and mating decisions. In *Mate choice*, ed. P. P. G. Bateson, 141–166. Cambridge: Cambridge University Press.

————. 1985. Models of parent-offspring conflict. 5. Effects of the behaviour of the two parents. *Anim. Behav.* 33:519–533.

————. 1989. Hamilton's rule and conditionality. *Ethol. Ecol. Evol.* 1:195–211.

Parker, G. A., and P. Hammerstein. 1985. Game theory and animal behaviour. In *Evolution: Essays in honour of John Maynard Smith*, ed. P. J. Greenwood and P. H. Harvey, 73–94. Cambridge: Cambridge University Press.

Parker, G. A., and M. R. Macnair. 1978. Models of parent-offspring conflict. 1. Monogamy. *Anim. Behav.* 26:97–110.

————. 1979. Models of parent-offspring conflict. 4. Suppression: Evolutionary retaliation by the parent. *Anim. Behav.* 27:1210–1235.

Parker, G. A., and D. W. Mock. 1987. Parent-offspring conflict over clutch size. *Evol. Ecol.* 1:161–174.

Parker, G. A., and L. Partridge. 1998. Sexual conflict and speciation. *Philos. Trans. R. Soc. Lond. B* 353:261–274.

Parker, G. A., R. R. Baker, and V. G. F. Smith. 1972. The origin and evolution of gamete dimorphism and the male-female phenomenon. *J. Theor. Biol.* 36:529–553.

Parrish, J. K., and P. Turchin. 1997. Individual decisions, traffic rule, and emergent pattern in schooling fish. In *Animal groups in three dimensions*, ed. J. K. Parrish and W. M. Hamner, 126–142. Cambridge: Cambridge University Press.

Partridge, L., A. Green, and K. Fowler. 1987. Effects of egg-production and of exposure to males on female survival in *Drosophila melanogaster. J. Insect Physiol.*, 33:745–749.

Passera, L. 1994. Characteristics of tramp species. In *Exotic ants, biology, impact, and control of introduced species*, ed. D. F. Williams, 23–43. Boulder, Colo.: Westview Press.

Passera, L., E. Roncin, B. Kaufmann, and L. Keller. 1996. Increased soldier production in ant colonies exposed to intraspecific competition. *Nature (Lond.)* 379:630–631.

Patel, A. 1996. Variation in a mutualism: Phenology and the maintenance of gynodioecy in two Indian fig species. *J. Ecol.* 84:667–680.

Patiño, S., E. A. Herre, and M. T. Tyree. 1994. Physiological determinants of *Ficus* fruit temperature and the implications for survival of pollinator wasp species: Comparative physiology through an energy budget approach. *Oecologia* 100: 13–20.

Patiño, S., M. T. Tyree, and E. A. Herre. 1995. A comparison of hydraulic architecture of woody plants of differing phylogeny and growth form with special reference to free-standing and hemiepiphytic *Ficus* species from Panama. *New Phytol.* 129:125–134.

Peeters, C. 1993. Monogyny and polygyny in ponerine ants with and without queens. In *Queen number and sociality in insects*, ed. L. Keller, 234–261. Oxford: Oxford University Press.

Pellmyr, O. 1989. The cost of mutualism: Interactions between *Trollius europaeus* and its pollinating parasites. *Oecologia* 78:53–59.

Pellmyr, O., and C. J. Huth. 1994. Evolutionary stability of mutualism between yuccas and yucca moths. *Nature (Lond.)* 372:257–260.

Pellmyr, O., J. Leebens-Mack, C. J. Huth. 1996. Non-mutualistic yucca moths and their evolutionary consequences. *Nature (Lond.)* 380:155–156.

Perrins, C. M. 1970. The timing of birds' breeding seasons. *Ibis* 112:242–253.

———. 1996. Eggs, egg formation and the timing of breeding. *Ibis* 138:2–15.

Persson, O., and P. Öhrström. 1989. A new avian mating system: Ambisexual polygamy in the penduline tit *Remiz pendulinus*. *Ornis Scand.* 20:105–111.

Peters, L. L., and J. E. Barker. 1993. Novel inheritance of the murine severe combined anemia and thrombocytopenia (*Scat*) phenotype. *Cell* 74:135–142.

Petes, T. D., and C. W. Hill. 1988. Recombination between repeated genes in microorganisms. *Annu. Rev. Genet.* 22:147–168.

Petrie, M., and A. P. Møller. 1991. Laying eggs in others' nests: Intraspecific brood parasitism in birds. *Trends Ecol. Evol.* 6:315–320.

Piccirilli, J. A., T. Krauch, S. E. Moroney, and S. A. Benner. 1990. Enzymatic incorporation of a new base pair into DNA and RNA extends the genetic alphabet. *Nature (Lond.)* 343:33–37.

Pinker, S. 1994. *The language instinct.* New York: Morrow.

———. 1997. Words and rules in the human brain. *Nature (Lond.)* 387:547–548.

Pitcher, T. J., and J. K. Parrish. 1993. Functions of shoaling behaviour in teleosts. In *Behaviour of teleost fishes*, ed. T. J. Pitcher, 363–439. London: Chapman and Hall.

Poinar, G. O., Jr. 1979. *Parasitodiplogaster sycophilon* gen. n., sp. n. (Diplogasteridae: Nematoda), a parasite of *Elizabethiella stuckenbergi* Grandi (Agaonidae: Hymenoptera) in Rhodesia. *Proc. K. Ned. Akad. Wet* 82:375–381.

Poinar, G. O., Jr., and E. A. Herre 1991. Speciation and adaptive radiation in the fig wasp nematode, *Parasitodiplogaster* (Diplogasteridae: Rhabditida), in Panama. *Rev. Nematol.* 14:361–374.

Poinsot, D., and H. Mercot. 1997. *Wolbachia* infection in *Drosophila simulans*: Does the female host bear a physiological cost? *Evolution* 51:180–186.

Polanyi, M. 1958. *Personal knowledge.* Chicago: University of Chicago Press.

Pomiankowski, A. 1988. The evolution of female mate preferences for male genetic quality. *Oxf. Surv. Evol. Biol.* 5:136–184.

Pomiankowski, A., Y. Iwasa, and S. Nee. 1991. The evolution of costly mate preferences. 1. Fisher and biased mutation. *Evolution* 45:1422–1430.

Poole, J. H., K. Payne, W. R. Langbauer, Jr., and C. Moss. 1988. The social contexts of some very low frequency calls of African elephants. *Behav. Ecol. Sociobol.* 22:385–392.

Popper, K. R. 1991. *The poverty of historicism.* London: Routledge.

Potts, W. K. 1984. The chorus-line hypothesis of manoeuvre coordination in avian flocks. *Nature (Lond.)* 309:344–345.

Potts, W. K., and E. K. Wakeland. 1993. Evolution of MHC genetic diversity: A tale of incest, pestilence and sexual preference. *Genetics* 9:408–412.

Presgraves, D., E. Severance, and G. S. Wilkinson. 1997. Sex chromosome meiotic drive in stalk-eyed flies. *Genetics* 147:1169–1180.

Price, D. K. 1996. Sexual selection, selection load and quantitative genetics of zebra finch bill colour. *Proc. R. Soc. Lond. B* 263:217–221.

Price, D. K., and N. T. Burley. 1993. Constraints on the evolution of attractive traits: Genetic (co)variation for zebra finch bill color. *Heredity* 71:405–412.

———. 1994. Constraints on the evolution of attractive traits. *Am. Nat.* 144:908–934.

Price, G. R. 1970. Selection and covariance. *Nature (Lond.)* 227: 529–531.

———. 1972. Extension of covariance selection mathematics. *Ann. Hum. Genet.* 35:485–490.

———. 1995. The nature of selection. *J. Theor. Biol.* 175:389–396.

Price, K., H. Harvey, and R. C. Ydenberg. 1996. Begging tactics of nestling yellow-headed blackbirds, *Xanthocephalus xanthocephalus*, in relation to need. *Anim. Behav.* 51:421–435.

Price, T. D., and H. L. Gibbs. 1987. Brood division in Darwin's ground finches. *Anim. Behav.* 35:299–301.

Prins, H. H. T. 1996. *Ecology and behaviour of the African buffalo.* London: Chapman and Hall.

Proctor, H. C. 1991. Courtship in the water mite *Neumania papillator*: Males capitalize on female adaptations for predation. *Anim. Behav.* 42:589–598.

———. 1992. Sensory exploitation and the evolution of male mating behaviour: A cladistic test using water mites (Acari: Parasitengona). *Anim. Behav.* 44:745–752.

Prout, T., J. Bundgaard, and S. Bryant. 1973. Population genetics of modifiers of meiotic drive. 1. The solution of a special case and some general implications. *Theor. Pop. Biol.* 4:446–465.

Provine, W. B. 1971. *The origins of theoretical population genetics.* Chicago: University of Chicago Press.

———. 1977. Role of mathematical population geneticists in the evolutionary synthesis of the 1930's and 40's. In *Mathematical models in biological discovery*, ed. D. L. Solomon and C. Walter, 2–30. Berlin: Springer.

———. 1986. *Sewall Wright and evolutionary biology.* Chicago: University of Chicago Press.

Prud'homme, N., M. Gans, M. Mason, C. Terzian, and A. Bucherton. 1995. *Flamenco*, a gene controlling the *gypsy* retrovirus of *Drosophila melanogaster. Genetics* 139:697–711.

Pulliam, H. R. 1973. On the advantages of flocking. *J. Theor. Biol.* 38:419–422.

Pulliam, H. R., G. H. Pyke, and T. Caraco. 1982. The scanning behaviour of juncos: A game-theoretical approach. *J. Theor. Biol.* 95:89–103.

Queller, D. C. 1984. Kin selection and frequency dependence: A game theoretic approach. *Biol. J. Linn. Soc.* 23:143.

———. 1989. Inclusive fitness in a nutshell. *Oxf. Surv. Evol. Biol.* 6:73–109.

———. 1993. Worker control of sex ratios and selection for extreme multiple mating by queens. *Am. Nat.* 142:346–351.

———. 1994. Male-female conflict and parent-offspring conflict. *Am. Nat.* 144:S84–S99.

Queller, D. C., and J. E. Strassmann. 1998. Kin selection and social insects. *Bioscience* 48:165–175.

Queller, D. C., C. R. Hughes, and J. E. Strassmann. 1990. Wasps fail to make distinction. *Nature (Lond.)* 344:388.

Queller, D. C., J. E. Strassmann, C. R. Solis, C. R. Hughes, and D. M. Deloack. 1993. A selfish strategy of social insect workers that promotes social cohesion. *Nature (Lond.)* 365: 639–641.

Ramirez, B. W. 1969. Fig wasps: Mechanism of pollen transfer. *Science (Wash.)* 163:580–581.

———. 1974. Coevolution of *Ficus* and Agaonidae. *Ann. Mo. Bot. Gard.* 61:770–780.

———. 1980. Evolution of the monoecious and dioecious habit in *Ficus* (Moraceae). *Brenesia* 18:207–216.

Ramirez B. W., and J. Malavasi. 1997. Fig wasps: Mechanisms of pollen transfer in *Malvanthera* and *Pharmacosycea* figs (Moraceae). *Rev. Biol. Trop.* 46:1635–1640.

Rasa, O. A. E. 1977. The ethology and sociology of the dwarf mongoose (*Helogale undulata rufula*). *Z. Tierpsychol.* 43:337–406.

———. 1983. Dwarf mongoose and hornbill mutualism in the Taru Desert, Kenya. *Behav. Ecol. Sociobiol.* 12:181–190.

———. 1986. Coordinated vigilance in dwarf mongoose family groups: The "watchman's song" hypothesis and the cost of guarding. *Ethology* 7:1340–344.

———. 1987. Vigilance behavior in individual mongooses: Selfish or altruistic? *S. Afr. J. Sci.* 83:587–590.

———. 1989a. Behavioural parameters of vigilance in the dwarf mongoose: Social acquisition of a sex-biased role. *Behaviour* 110:125–145.

———. 1989b. The costs and effectiveness of vigilance behaviour in the dwarf mongoose: Implications for fitness and optimal group size. *Ethol. Ecol. Evol.* 1:265–282.

Rasplus, J.-Y., C. Kerdelhue, I. Le Chainche, and G. Mondor. 1998. Molecular phylogeny of fig wasps. Agaonidae are not monophyletic. *C. R. Acad. Sci. (Paris)* 321:517–527.

Ratnieks, F. L. W. 1988. Reproductive harmony via mutual policing by workers in eusocial Hymenoptera. *Am. Nat.* 132:217–236.

———. 1991. The evolution of genetic odor-clue diversity in social Hymenoptera. *Am. Nat.* 137:202–226.

———. 1996. Evolution of unstable and stable biparental care. *Behav. Ecol.* 7:490–493.

Ratnieks, F. L. W., and J. J. Boomsma. 1995. Facultative sex allocation by workers and the evolution of polyandry by queens in social Hymenoptera. *Am. Nat.* 145: 969–993.

Ratnieks, F. L. W., and H. K. Reeve. 1992. Conflict in single-queen hymenopteran societies: The structure of conflict and processes that reduce conflict in advanced eusocial species. *J. Theor. Biol.* 158:33–65.

Ratnieks, F. L. W., and P. K. Visscher. 1989. Worker policing in the honeybee. *Nature (Lond.)* 342:796–797.

Rätti, O., and R. V. Alatalo. 1993. Determinants of the mating success of polyterritorial pied flycatcher males. *Ethology* 94:137–146.

Read, A. F., and P. H. Harvey. 1993. The evolution of virulence. *Nature (Lond.)* 362:500–501.

Rebek, J. 1994. Synthetic self-replicating molecules. *Sci. Am.* 271:34–40.

Reeve, H. K. 1989. The evolution of conspecific acceptance thresholds. *Am. Nat.* 133:407–435.

———. 1991. *Polistes*. In *The social biology of wasps*, ed. K. G. Ross and R. W. Matthews, 99–148. Ithaca, N.Y.: Cornell University Press.

———. 1997. Evolutionarily stable communication between kin—a general model. *Proc. R. Soc. Lond. B* 264:1037–1040.

———. 1998a. Acting for the good of of others: Kinship and reciprocity, with some new twists. In *Handbook of evolutionary psychology*, ed. C. Crawford and D. Krebs, 43–85. Hillsdale, N.J.: Lawrence Erlbaum.

———. 1998b. Game theory, reproductive, skew, and nepotism. In *Game theory and animal behavior*, ed. L. Dugatkin and H. K. Reeve, 118–145. New York: Oxford University Press.

Reeve, H. K., and L. Keller. 1995. Partitioning of reproduction in mother-daughter versus sibling associations: A test of optimal skew theory. *Am. Nat.* 145:119–132.

———. 1996. Relatedness asymmetry and reproductive sharing in animal societies. *Am. Nat.* 148:764–769.

———. 1997. Reproductive bribing and policing evolutionary mechanisms for the suppression of within-group selfishness. *Am. Nat.* 150:S42–S58.

Reeve, H. K., and P. Nonacs. 1992. Social contracts in wasp societies. *Nature (Lond.)* 359:823–825.

———. 1993. Weak queen or social contract? *Nature (Lond.)* 363:503.

———. 1997. Within-group aggression and the value of group members—theory and a field test with social wasps. *Behav. Ecol.* 8:75–82.

Reeve, H. K., and F. L. W. Ratnieks. 1993. Queen-queen conflicts in polygynous societies: mutual tolerance and reproductive skew. In *Queen number and sociality in insects*, ed. L. Keller, 45–85. Oxford: Oxford University Press.

Reeve, H. K., S. T. Emlen, and L. Keller. 1998. Reproductive sharing in animal societies: Reproductive incentives or incomplete control by dominant breeders? *Behav. Ecol.* 9:267–278.

Regal, P. J. 1977. Ecology and evolution of flowering plant dominance. *Science (Wash.)* 196:622–629.

Reyer, H.-U. 1980. Flexible helper structure as an ecological adaptation in the pied kingfisher (*Ceryle rudis*). *Behav. Ecol. Sociobiol.* 6:219–227.

———. 1984. Investment and relatedness: A cost/benefit analysis of breeding and helping in the pied kingfisher (*Ceryle rudis*). *Anim. Behav.* 32:1163–1178.

———. 1986. Breeder-helper-interactions in the pied kingfisher reflect the costs and benefits of cooperative breeding. *Behaviour* 96:277–303.

Ricard, J., and G. Noat. 1986. Electrostatic effects and the dynamics of enzyme reactions at the surface of plant cells. 1. A theory of the ionic control of a complex multi-enzyme system. *Eur. J. Biochem.* 155:183–190.

Rice, W. R. 1984. Sex chromosomes and the evolution of sexual dimorphism. *Evolution* 38:735–742.

———. 1987. The accumulation of sexually antagonistic genes as a selective agent promoting the evolution of reduced recombination between primitive sex chromosomes. *Evolution* 41:911–914.

———. 1992. Sexually antagonistic genes: experimental evidence. *Science (Wash.)* 256:1436–1439.

———. 1996. Sexually antagonistic male adaptation triggered by experimental arrest of female evolution. *Nature (Lond.)* 381:232–234.

Richards, P. W. 1996. *The tropical rain forest.* Cambridge: Cambridge University Press.

Rigaud, T. 1997. Inherited microorganisms and sex determination of arthropod hosts. In *Influential passengers: Inherited microorganisms and arthropod reproduction*, ed. F. L. O'Neill, A. A. Hoffman and J. H. Werren, 81–101. Oxford: Oxford University Press.

Rigaud, T., and P. Juchault. 1993. Conflict between feminizing sex ratio distorters and an autosomal masculinizing gene in the terrestrial isopod *Armadillidium vulgare*. *Genetics* 133:247–252.

Robertson, H. M. 1985. Female dimorphism and mating behaviour in a damselfly, *Ischnura ramburi*: Females mimicking males. *Anim. Behav.* 33:805–809.

———. 1993. The *mariner* transposable element is widespread in insects. *Nature (Lond.)* 362:241–245.

Robinson, M. H. 1985. Predator-prey interactions, informational complexity, and the origins of intelligence. *J. Wash. Acad. Sci.* 75:91–104.

Rodriguez-Girones, M. A., P. A. Cotton, and A. Kacelnik. 1996. The evolution of begging: Signalling and sibling competition. *Proc. Natl. Acad. Sci. USA* 93:14637–14641.

Roitberg, B. D., and M. Mangel. 1993. Parent-offspring conflict and life-history consequences in herbivorous insects. *Am. Nat.* 142:443–456.

Romey, W. L. 1997. Inside or outside? Testing evolutionary predictions of positional effects. In *Animal groups in three dimensions*, ed. J. K. Parrish and W. M. Hamner, 172–188. Cambridge: Cambridge University Press.

Rosengren, R., and P. Pamilo. 1983. The evolution of polygyny and polydomy in mound-building *Formica* ants. *Acta Entomol. Fenn.* 42:65–77.

Ross, K. G., and L. Keller. 1995. Ecology and evolution of social organization—insights from fire ants and other highly eusocial insects. *Annu. Rev. Ecol. Syst.* 26:631–656.

Rousset, F., and M. Raymond. 1991. Cytoplasmic incompatibility in insects: Why sterilize females? *Trends Ecol. Evol.* 6:54–57.

Rousset, F., D. Bouchon, B. Pintreau, P. Juchault, and M. Solignac. 1992. *Wolbachia* endosymbionts responsible for various alterations of sexuality in arthropods. *Proc. R. Soc. Lond. B* 250:91–98.

Runciman, W. G., J. Maynard Smith, and R. I. M. Dunbar. 1996. *Evolution of social behaviour in primates and man.* Oxford: Oxford University Press.

Ruse, M. 1973. *The philosophy of biology.* London: Hutchinson University Library.

Ryan, M. J. 1983. Sexual selection and communication in a neotropical frog, *Physalaemus pustulosus. Evolution* 37:261–272.

Ryan, M. J., and A. Keddy-Hector. 1992. Directional patterns of female mate choice and the role of sensory biases. *Am. Nat.* 139:S4–S35.

Ryan, M. J., and A. S. Rand. 1993. Sexual selection and signal evolution: The ghost of biases past. *Philos. Trans. R. Soc. Lond.* B 340:187–195.

Ryan, M. J., J. H. Fox, W. Wilczynski, and A. S. Rand. 1990. Sexual selection for sensory exploitation in the frog *Physalaemus pustulosus. Nature (Lond.)* 343:66–67.

Rydén, O., and H. Bengtsson. 1980. Differential begging and locomotory behavior by early and late hatched nestlings affecting distribution of food in asynchronously hatched broods of altricial birds. *Z. Tierpsychol.* 53:209–224.

Saino, N., A. M. Bolzern, and A. P. Møller. 1997. Immunocompetence, ornamentation, and viability of male barn swallows (*Hirundo rustica*). *Proc. Natl. Acad. Sci. USA* 94:549–552.

Sakaluk, S. K., P. J. Bangert, A.-K. Eggert, C. Gack, and L. V. Swanson. 1995. The gin trap as a device facilitating coercive mating in sagebrush crickets. *Proc. R. Soc. Lond.* B 261:65–71.

Sandell, M. I., and Smith, H. G. 1996. Already mated females constrain male mating success in the European starling. *Proc. R. Soc. Lond.* B 263:743–747.

Sano, Y. 1990. The genic nature of gamete eliminator in rice. *Genetics* 125:183–191.

Saumitou-Laprade, P., J. Cuguen, and P. Vernet. 1994. Cytoplasmic male sterility in plants: Molecular evidence and the nucleocytoplasmic conflict. *Trends Ecol. Evol.* 9:431–435.

Schaller, G. 1972. *The Serengeti lion: A study of predator-prey relations.* Chicago: University of Chicago Press.

Scheel, D., and C. Packer. 1991. Group hunting behaviour of lions: A search for cooperation. *Anim. Behav.* 41:697–709.

Schilt, C. R., and K. S. Norris. 1997. Perspectives on sensory integration systems: Problems, opportunities, and predictions. In *Animal groups in three dimensions*, ed. J. K. Parrish and W. M. Hamner, 225–244. Cambridge: Cambridge University Press.

Schilthuizen, M. and R. Stouthamer. 1997. Horizontal transmission of parthenogenesis-inducing microbes in *Trichogamma* wasps. *Proc. R. Soc. Lond.* B 264:361–366.

Schmid-Hempel, P. 1998. *Parasites in social insects.* Princeton, N.J.: Princeton University Press.

Schneider, J. M., and Y. Lubin. 1996. Infanticidal male eresid spiders. *Nature (Lond.)* 381: 655–656.

———. 1997. Infanticide by males in a spider with suicidal maternal care, *Stegodyphus lineatus* (Eresidae). *Anim. Behav.* 54:305–312.

Schuster, P. 1993. RNA based evolutionary optimization. *Orig. Life Evol. Biosphere* 23:373–391.

Schwartz, A. W. 1997. Speculation on the RNA precursor problem. *J. Theor. Biol.* 187:523–527.

Scott, D. K. 1988. Breeding success in Bewick's swans. In *Reproductive success*, ed. T. H. Clutton-Brock, 220–236. Chicago: University of Chicago Press.

Scott, M. P. 1994. Competition with flies promotes communal breeding in the burying beetle, *Nicrophorus tomentosus*. *Behav. Ecol. Sociobiol.* 34:367–373.

Searcy, W. A. 1992. Song repertoire and mate choice in birds. *Am. Zool.* 32:71–80.

Searcy, W. A., D. Eriksson, and A. Lundberg. 1991. Deceptive behavior in pied flycatchers. *Behav. Ecol. Sociobiol.* 29:167–175.

Seeley, T. D. 1985. *Honeybee ecology: A study of adaptation in social life.* Princeton, N.J.: Princeton University Press.

———. 1995. *The wisdom of the hive: The social physiology of honey bee colonies.* Cambridge, Mass.: Harvard University Press.

———. 1997. Honey bee colonies are group-level adaptive units. *Am. Nat.* 150:S22–S41.

Selker, E. U. 1990. Premeiotic instability of repeated sequences in *Neurospora crassa*. *Annu. Rev. Genet.* 24:579–613.

Selous, E. 1931. *Thought-transference (or what?) in birds.* New York: Richard R. Smith.

Shackney, S. E., and T. V. Shankey. 1997. Common patterns of genetic evolution in human solid tumors. *Cytometry* 29:1–27.

Shellman-Reeve, J. S. 1997. The spectrum of eusociality in termites. In *Social behavior in insects and arachnids*, ed. J. C. Choe and B. J. Crespi, 52–93. Cambridge: Cambridge University Press.

Sherman, P. W. 1977. Nepotism and the evolution of alarm calls. *Science (Wash.)* 197:1246–1253.

Sherman, P. W., T. D. Seeley, and H. K. Reeve. 1988. Parasites, pathogens, and polyandry in social Hymenoptera. *Am. Nat.* 131:602–610.

Sherman, P. W., E. A. Lacey, H. K. Reeve, and L. Keller. 1995. The eusociality continuum. *Behav. Ecol.* 6:102–108.

Sherman, P. W., H. K. Reeve, and D. W. Pfennig. 1997. Recognition systems. In *Behavioural ecology: An evolutionary approach*, 4th ed., ed. J. R. Krebs and N. B. Davies, pp. 69–96. Oxford: Blackwell.

Sievers, D., and G. von Kiedrowski. 1995. Self-replication of complementary nucleotide-based oligomers. *Nature (Lond.)* 369:221–224.

Sigg, H., A. Stolba, J. J. Abegglen, and V. Dasser. 1982. Life history of hamadryas baboons: Physical development, infant mortality, reproductive parameters, and family relationships. *Primates* 23:473–487.

Silver, L. M., K. Artzt, D. Barlow, K. Fischer-Lindahl, M. F. Lyon, J. Klein, and L. Snyder. 1992. Mouse chromosome 17. *Mamm. Genome* 3:S241–260.

Sinervo, B., and C. M. Lively. 1996. The rock-paper-scissors game and the evolution of alternative male strategies. *Nature (Lond.)* 380:240–243.

Sjerps, M., and P. Haccou. 1994. A war of attrition between larvae on the same host plant—stay and starve or leave and be eaten. *Evol. Ecol.* 8:269–287.

Skinner, S. W. 1992. Maternally inherited sex ratio in the parasitoid wasp *Nasonia vitripennis*. *Science (Wash.)* 215:1133–1134.

Slagsvold, T., and S. Dale. 1994. Why do female pied flycatchers mate with already mated males: Deception or restricted mate sampling? *Behav. Ecol. Sociobiol.* 34:239–250.

Slagsvold, T., and J. T. Lifjeld. 1989. Hatching asynchrony in birds: The hypothesis of sexual conflict over parental investment. *Am. Nat.* 134:239–253.

———. 1994. Polygyny in birds: The role of competition between females for male parental care. *Am. Nat.* 143:59–94.

Slagsvold, T., T. Amundsen, S. Dale, and H. Lampe. 1992. Female-female aggression explains polyterritoriality in male pied flycatchers. *Anim. Behav.* 43:397–407.

Slagsvold, T., T. Amundsen, and S. Dale. 1994. Selection by sexual conflict for evenly spaced offspring in blue tits. *Nature (Lond.)* 370:136–138.

Slatkin, M. 1984. Ecological causes of sexual dimorphism. *Evolution* 38:622–630.

Smith, H. G., and R. Montgomerie. 1991. Nestling American robins compete with siblings by begging. *Behav. Ecol. Sociobiol.* 29:307–312.

Smythe, N. 1989. Seed survival in the palm *Astrocaryum standleyanum*: Evidence for dependence on its seed dispersers. *Biotropica* 21:50–56.

———. 1991. Steps toward domesticating the paca (*Agouti = Cuniculus paca*) and prospects for the future. In *Neotropical wildlife use and conservation*, ed. J. G. Robinson and K. H. Redford, 202–216. Chicago: University of Chicago Press.

Sniegowski, P. D., and B. Charlesworth. 1994. Transposable element numbers in cosmopolitan inversions from a natural population of *Drosophila melanogaster. Genetics* 137:815–827.

Snyder, L. E. 1993. Non-random behavioural interactions among genetic subgroups in a polygynous ant. *Anim. Behav.* 46:431–439.

Sober, E. 1984. *The nature of selection.* Cambridge, Mass.: MIT Press.

Sober, E., and R. C. Lewontin. 1984. Artifact, cause and genic selection. In *Conceptual issues in evolutionary biology: An anthology*, ed. E. Sober, 210–231. Cambridge, Mass.: MIT Press.

Sober, E., and D. S. Wilson. 1998. *Unto others: The evolution and psychology of unselfish behavior.* Cambridge, Mass.: Harvard University Press.

Solomon, N. G., and J. A. French, eds. 1997. *Cooperative breeding in mammals.* Cambridge: Cambridge University Press.

Spencer, H., G. Weiblen, and B. Flick. 1996. Phenology of *Ficus variegata* in a seasonal wet tropical forest at Cape Tribulation, Australia. *J. Biogeogr.* 23:467–476.

Stacey, P. B., and W. D. Koenig, eds. 1990. *Cooperative breeding in birds: Long-term studies of ecology and behavior.* Cambridge: Cambridge University Press.

Stammbach, E. 1987. Desert, forest and montane baboons: Multilevel societies. In *Primate societies*, ed. B. B. Smuts, D. L. Cheney, R. M. Seyfarth, R. W. Wrangham, and T. T. Struhsaker, 112–120. Chicago: University of Chicago Press.

Stamps, J., R. Metcalf, and V. Krishnan. 1978. A genetic analysis of parent-offspring conflict. *Behav. Ecol. Sociobiol.* 3:367–392.

Stamps, J., A. Clark, P. Arrowood, and B. Kus. 1985. Parent-offspring conflict in budgerigars. *Behaviour* 94:1–40.

Stamps, J., A. Clark, B. Kus, and P. Arrowood. 1987. The effects of parent and offspring gender on food allocation in budgerigars. *Behaviour* 101:177–199.

Stander, P. E. 1992a. Cooperative hunting in lions: The role of the individual. *Behav. Ecol. Sociobiol.* 29:445–454.

———. 1992b. Foraging dynamics of lions in a semi-arid environment. *Can. J. Zool.* 70:8–21.

Stanley, S. M. 1975. Clades versus clones in evolution: Why we have sex. *Science (Wash.)* 190:382–383.

———. 1979. *Macroevolution: Pattern and process*. San Francisco: W. H. Freeman.

Starks, P. T., and E. S. Poe. 1997. Male-stuffing in wasp societies. *Nature (Lond.)* 389:450.

Starr, C. K. 1984. Sperm competition, kinship, and sociality in the aculeate Hymenoptera. In *Sperm competition and the evolution of animal mating systems*, ed. S. L. Smith, 428–464. London: Academic Press.

Stavenhagen, J. B., and D. M. Robbins. 1988. An ancient provirus has imposed androgen regulation on the adjacent mouse sex-limited protein gene. *Cell* 55:247–254.

Stearns, S. C. 1986. Natural selection and fitness, adaptation and constraint. In *Patterns and processes in the history of life*, ed. D. Jablonski and D. M. Raup, 23–44. Berlin: Springer.

———, ed. 1987. *The evolution of sex and its consequences*. Basel: Birkhauser Verlag.

Steinemann M., and S. Steinemann. 1992. Degenerating Y chromosome of *Drosophila miranda*: A trap for retrotransposons. *Proc. Natl. Acad. Sci. USA* 89:7591–7595.

Stenmark, G., T. Slagsvold, and J. T. Lifjeld. 1988. Polygyny in the pied flycatcher, *Ficedula hypoleuca*: A test of the deception hypothesis. *Anim. Behav.* 36:1646–1657.

Stille, M. 1996. Queen/worker thorax volume ratios and nest-founding strategies in ants. *Oecologia* 105:87–93.

Stockley, P. 1997. Sexual conflict resulting from adaptations to sperm competition. *Trends Ecol. Evol.* 12:154–159.

Stouthamer, R., and D. J. Kazmer. 1994. Cytogenetics of microbe-associated parthenogenesis and its consequences for gene flow in *Trichogamma* wasps. *Heredity* 73:317–327.

Stouthamer, R., and R. F. Luck. 1993. Influence of microbe associated parthenogenesis on the fecundity of *Trichogamma deion* and *T. pretiosum. Entomol. Exp. Appl.* 67:183–192.

Stouthamer, R., R. F. Luck, and W. D. Hamilton. 1990. Antibiotics cause parthenogenetic *Trichogramma* (Hymenoptera; Trichogrammatidae) to revert to sex. *Proc. Natl. Acad. Sci. USA* 87:2424–2427.

Stouthamer, R., J. A. J. Breeuwer, R. F. Luck, and J. H. Werren. 1993. Molecular identification of the microorganisms associated with parthenogenesis. *Nature (Lond.)* 361:66–68.

Strassmann, J. E. 1993. Weak queen or social contract ? *Nature (Lond.)* 363:502–503.

Summers, D. K. 1996. *The biology of plasmids*. Oxford: Blackwell Science.

Sundström, L. 1994. Sex ratio bias, relatedness asymmetry and queen mating frequency in ants. *Nature (Lond.)* 367:266–268.

Sundström, L., M. Chapuisat, and L. Keller. 1996. Conditional manipulation of sex ratios by ant workers—a test of kin selection theory. *Science (Wash.)* 274: 993–995.

Sweeney, A. W., M. F. Graham, and E. I. Hazard. 1988. Life cycle of *Amblyospora dyxenoides* in mosquito *Culex annulirostris* and the copepod *Mesocyclops albicans*. *J. Invert. Pathol.* 51:46–57.

Szathmáry, E. 1984. The roots of individual organisation. In *Evolution 4. Frontiers of Evolution Research* [in Hungarian], ed. G. Vida, 37–157. Budapest: Natura.

———. 1988. A hypercyclic illusion. *J. Theor. Biol.* 134:561–563.

———. 1989a. The emergence, maintenance, and transitions of the earliest evolutionary units. *Oxf. Surv. Evol. Biol.* 6:169–205.

———. 1989b. The integration of the earliest genetic information. *Trends Ecol. Evol.* 4:200–204.

———. 1990a. Towards the evolution of ribozymes. *Nature (Lond.)* 344:115.

———. 1990b. RNA worlds in test tubes and protocells. In *Organizational constraints on the dynamics of evolution*, ed. J. Maynard Smith and G. Vida, 3–14. Manchester: Manchester University Press.

———. 1991a. Simple growth laws and selection consequences. *Trends Ecol. Evol.* 6:366–370.

———. 1991b. Four letters in the genetic alphabet: A frozen evolutionary optimum? *Proc. R. Soc. Lond. B* 245:91–99.

———. 1992a. Natural selection and the dynamical coexistence of defective and complementing virus segments. *J. Theor. Biol.* 157:383–406.

———. 1992b. What determines the size of the genetic alphabet? *Proc. Natl. Acad. Sci. USA* 89:2614–2618.

———. 1994. Self-replication and reproduction: From molecules to protocells. In *Self-production of supramolecular structures: From synthetic structures to models of minimal living systems*, ed. G. R. Fleischaker, S. Colonna, and P. L. Luisi, 65–73. NATO ASI Series. Dordrecht: Kluwer.

———. 1995. A classification of replicators and lambda-calculus models of biological organization. *Proc. R. Soc. Lond. B* 260:279–286.

Szathmáry, E., and L. Demeter. 1987. Group selection of early replicators and the origin of life. *J. Theor Biol.* 128:463–486.

Szathmáry, E., and I. Gladkih. 1989. Sub-exponential growth and coexistence of non-enzymatically replicating templates. *J. Theor. Biol.* 138:55–58.

Szathmáry, E., and J. Maynard Smith. 1993a. The origin of chromosomes 2. Molecular mechanisms. *J. Theor. Biol.* 164:447–454.

———. 1993b. The origin of genetic systems. *Abstr. Bot.* 17:197–206.

———. 1995. The major evolutionary transitions. *Nature (Lond.)* 374:227–232.

———. 1997. From replicators to reproducers: The first major transitions leading to life. *J. Theor. Biol.* 187:555–571.

Szostak, J. W. 1992. *In vitro* genetics. *Trends Biochem. Sci.* 17:89–93.

Taylor, P. D., and M. G. Bulmer. 1980. Local mate competition and the sex ratio. *J. Theor. Biol.* 86:409–419.

Te Boekhorst, I. J. A., and P. Hogeweg. 1994. Self-structuring in artificial "chimps" offers new hypotheses for male grouping in chimpanzees. *Behaviour* 130:229–252.

Thisted, T., N. S. Sørensen, and K. Gerdes. 1994. Mechanism of post-segregational killing I. *EMBO J.* 13:1960–1968.

Thompson, D. W. 1942. *On growth and form*. Cambridge: Cambridge University Press.

Thompson, J. N. 1994. *The coevolutionary process*. Chicago: University of Chicago Press.

Thompson, J. N., and O. Pellmyr. 1992. Mutualism with pollinating seed parasites amid co-pollinators: Constraints on specializations. *Ecology* 73:1780–1791.

Ting, C. N., M. P. Rosenberg, C. M. Snow, L. C. Samuelson, and M. H. Meisler. 1992. Endogenous retroviral sequences are required for tissue-specific expression of a human salivary amylase gene. *Genes Dev.* 6:1457–1465.

Treisman, M. 1975. Predation and the evolution of gregariousness. *Anim. Behav.* 23:779–800.

Trench, R. K. 1991. *Cyanophora paradoxa* Korschikoff and the origins of chloroplasts. In *Symbiosis as a source of evolutionary innovation*, ed. L. Margulis and R. Fester, 143–150. Cambridge, Mass.: MIT Press.

Trivers, R. L. 1971. The evolution of reciprocal altruism. *Q. Rev. Biol.* 46:35–57.

———. 1972. Parental investment and sexual selection. In *Sexual selection and the descent of man, 1871–1971*, ed. B. Campbell, 136–179. Chicago: Aldine.

———. 1974. Parent-offspring conflict. *Am. Zool.* 14:249–264.

Trivers, R. L., and H. Hare. 1976. Haplodiploidy and the evolution of the social insects. *Science (Wash.)* 191:249–263.

Tsaur, S. C., and C.-I. Wu. 1997. Positive selection and the molecular evolution of a gene of male reproduction, Acp26Aa of *Drosophila*. *Mol. Biol. Evol.* 14:544–549.

Tsuji, K., and K. Yamauchi. 1994. Colony level sex allocation in a polygynous and polydomous ant. *Behav. Ecol. Sociobiol.* 34:157–167.

Turner, B. C., and D. D. Perkins. 1991. Meiotic drive in *Neurospora* and other fungi. *Am. Nat.* 137:416–429.

Ulenberg, S. A. 1985. The phylogeny of the genus *Apocrypta* Coquerl in relation to its hosts, *Ceratosolen* Mayr (Agaonidae) and *Ficus* L. In *The systematics of the fig wasp parasites of the genus Apocrypta coquerl*, ed. S. A. Ulenberg. Amsterdam: North Holland.

Uma Shaanker, R., K. N. Ganeshaiah, and K. S. Bawa. 1988. Parent-offspring conflict, sibling rivalry, and brood size patterns in plants. *Annu. Rev. Ecol. Syst.* 19: 177–205.

Valera, F., H. Hoi, and B. Schleicher. 1997. Egg burial in penduline tits, *Remiz pendulinus*: Its role in mate desertion and female polyandry. *Behav. Ecol.* 8:20–27.

van Damme, J. M. M., and W. van Delden. 1984. Gynodioecy in *Plantago lanceolata*. 4. Fitness components of sex types in different life cycle stages. *Evolution.* 38:1326–1336.

van Gent, D. C., K. Mizuuchi, and M. Gellert. 1996. Similarities between initiation of V(D)J recombination and retroviral integration. *Science (Wash.)* 271:1592–1594.

van Noort, S., and S. G. Compton. 1996. Convergent evolution of agaonine and sycoecine (Agaonidae, Chalcidoidea) head shape in response to the constraints of host fig morphology. *J. Biogeogr.* 23:415–424.

van Noort, S., A. B. Ware, and S. G. Compton. 1989. Pollinator specific volitile attractants released from the the figs of *Ficus bert-davyi*. *S. Afr. J. Sci.* 85:323–324.

Van Orsdol, K. G. 1981. Lion predation in Rwenzori National Park, Uganda. Ph.D. thesis, Cambridge University.

Van Valen, L. 1973. A new evolutionary law. *Evol. Theory* 1:1–30.

———. 1975. Group selection, sex, and fossils. *Evolution* 29:87–94.

Varandas, F. R., M. C. Sampaio, and A. B. de Carvalho. 1996. Heritability of sexual proportion in experimental sex-ratio populations of *Drosophila mediopunctata*. *Heredity* 79:104–112.

Varela, F. J. 1979. *Principles of biological autonomy*. New York: North Holland.

Varga, Z., and E. Szathmáry. 1997. An extremum principle for parabolic competition. *Bull. Math. Biol.* 59:1145–1154.

Vargo, E. L. 1996. Sex investment ratios in monogyne and polygyne colonies of the fire ant *Solenopsis invicta*. *J. Evol. Biol.* 9:783–802.

Vargo, E. L., and L. Passera. 1991. Pheromonal and behavioral queen control over the production of gynes in the Argentine ant *Iridomyrmex humilis* (Mayr). *Behav. Ecol. Sociobiol.* 28:161–169.

Vehrencamp, S. L. 1983. A model for the evolution of despotic versus egalitarian societies. *Anim. Behav.* 31:667–682.

Veiga, J. P. 1990. Sexual conflict in the house sparrow: Interference between polygynously mated females versus asymmetric male investment. *Behav. Ecol. Sociobiol.* 27:345–350.

Verkerke, W. 1989. Structure and function of the fig. *Experientia* 45:612–622.

Verner, J. 1964. Evolution of polygamy in the long-billed marsh wren. *Evolution* 18:252–261.

Verner, J., and M. F. Willson. 1966. The influence of habitats on mating systems of North American passerine birds. *Ecology* 47:143–147.

Vieira, C., and C. Biémont. 1996. Selection against transposable elements insertions in *Drosophila simulans* and *D. melanogaster*. *Genet. Res.* 68:9–16.

Visscher, P. K. 1989. A quantitative study of worker reproduction in honey bee colonies. *Behav. Ecol. Sociobiol.* 25:247–254.

Visscher, P. K., and R. Dukas. 1995. Honey bees recognize development of nestmates' ovaries. *Anim. Behav.* 49:542–544.

Voelker, R. A. 1972. Preliminary characterization of "sex ratio" and rediscovery and reinterpretation of "male sex ratio" in *Drosophila affinis*. *Genetics* 71:597–606.

von Kiedrowski, G. 1986. A self-replicating hexadeoxy nucleotide. *Angew. Chem. Int. Ed. Engl.* 25: 932–935.

———. 1993. Minimal replicator theory. 1. Parabolic versus exponential growth. *Bioorg. Chem. Front.* 3:113–146.

Vrba, E. S. 1989. Levels of selection and sorting with special reference to the species level. *Oxf. Surv. Evol. Biol.* 6:111–168.

Wächtershäuser, G. 1988. Before enzymes and templates: Theory of surface metabolism. *Microbiol. Rev.* 52:452–484.

———. 1992. Groundworks for an evolutionary biochemistry: The iron-sulphur world. *Prog. Biophys. Mol. Biol.* 58:85–201.

———. 1994. Life in a ligand sphere. *Proc. Natl. Acad. Sci. USA* 91:4283–4287.

Wade, M. J. 1985. Soft selection, hard selection, kin selection, and group selection. *Am. Nat.* 125:61–73.

Wade, M. J., and F. Breden. 1980. The evolution of cheating and selfish behavior. *Behav. Ecol. Sociobiol.* 7:167–172.

Wade, M. J., and N. W. Chang. 1995. Increased male fertility in *Tribolium confusum* beetles after infection with the intracellular parasite *Wolbachia*. *Nature (Lond.)* 373:72–74.

Wagner, G. P. 1995. The biological role of homologues: A building block hypothesis. *Neues. Jahrb. Geol. Palaeont. Abh. B* 195:279–288.

———. 1996. Homologues, natural kinds, and the evolution of modularity. *Am. Zool.* 36:36–43.

Wagner, G. P., and L. Altenberg. 1996. Complex adaptations and the evolution of evolvability. *Evolution* 50:967–976.

Waldman, B. 1987. Mechanisms of kin recognition. *J. Theor. Biol.* 128:159–185.

Ware, A. B., and S. G. Compton. 1992. Breakdown of pollinator specificity in an African fig tree. *Biotropica* 24:544–549.

Ware, A. B., T. K. Perry, S. G. Compton, and S. van Noort. 1993. Fig volatiles: Their role in attracting pollinators and maintaining pollinator specificity. *Plant Syst. Evol.* 186:147–156.

Warner, R. R., D. Y. Shapiro, A. Marcanato, and C. W. Petersen. 1995. Sexual conflict: Males with the highest mating success convey the lowest fertilization benefits to females. *Proc. R. Soc. Lond. B* 262:135–139.

Watson, J. D., N. H. Hopkins, J. W. Roberts, J. A. Steitz, and A. M. Weiner. 1987. *Molecular biology of the gene*, vol. 1. Menlo Park, Calif.: Benjamin/Cummings.

Weichenhan, D., W. Traut, B. Kunze, and H. Winking. 1996. Distortion of Mendelian recovery ratio for a mouse *HSR* is caused by maternal and zygotic effects. *Genet. Res.* 68:125–129.

Weismann, A. 1890. Prof. Weismann's theory of heredity. *Nature (Lond.)* 41:317–323.

Werren, J. H. 1987. The coevolution of autosomal and cytoplasmic sex ratio factors. *J. Theor. Biol.* 124:317–334.

———. 1997. Biology of *Wolbachia. Annu. Rev. Entomol.* 42:587–609.

Werren, J. H., L. Guo, and D. M. Windsor. 1995. Distribution of *Wolbachia* in Neotropical arthropods. *Proc. R. Soc. Lond. B* 262:197–204.

Werren, J. H., S. W. Skinner, and A. M. Huger. 1986. Male-killing bacteria in a parasite wasp. *Science (Wash.)* 231:990–992.

West, S. A, and E. A. Herre. 1994. The ecology of the New World fig-parasitizing wasps *Idarnes* and implications for the evolution of the fig-pollinator mutualism. *Proc. R. Soc. Lond. B* 258:67–72.

West, S. A., E. A. Herre, D. M. Windsor, and P. R. S. Green. 1996. The ecology and evolution of the New World non-pollinating fig wasp communities. *J. Biogeogr.* 23:447–458.

West-Eberhard, M. J. 1975. The evolution of social behavior by kin selection. *Q. Rev. Biol.* 50:1–33.

———. 1978. Polygyny and the evolution of social behavior in wasps. *J. Kan. Entomol. Soc.* 51:832–856.

———. 1987. Flexible strategy and social evolution. In *Animal societies: Theories and facts*, ed. Y. Itô, J. L. Brown, and J. Kikkawa, 33–51. Tokyo: Japan Scientific Society Press.

Westneat, D. F., and P. W. Sherman. 1993. Parentage and the evolution of parental behavior. *Behav. Ecol.* 4:66–77.

Wheeler, D. E. 1986. Developmental and physiological determinants of caste in social Hymenoptera: Evolutionary implications. *Am. Nat.* 128:13–34.

————. 1991. The developmental basis of worker caste polymorphism in ants. *Am. Nat.* 138:1218–1238.

Wheeler, W. M. 1928. *The social insects: Their origin and evolution*. London: Kegan Paul.

White, S. E., L. F. Habera, and S. R. Wessler. 1994. Retrotransposons in the flanking regions of normal plant genes. *Proc. Natl. Acad. Sci. USA* 91:11792–11796.

Wickler, W. 1985. Coordination of vigilance in bird groups: The "watchman's song" hypothesis. *Z. Tierpsychol.* 69:250–253.

Wiebes, J. T. 1979. Co-evolution of figs and their insect pollinators. *Annu. Rev. Ecol. Syst.* 10:1–12.

————. 1982. The phylogeny of the agaonidae (Hymenoptera, Chalcidoidea). *Neth. J. Zool.* 32:395–411.

————. 1995. Agaonidae (Hymenoptera Chalcidoidea) and *Ficus* (Moraceae): Fig wasps and their figs (Meso-American Pegoscapus). *Proc. K. Ned. Akad. Wet.* 98: 167–183.

Wiley, R. H., and J. Poston. 1996. Indirect mate choice, competition for mates, and coevolution of the sexes. *Evolution* 50:1371–1381.

Williams, G. C. 1966a. *Adaptation and natural selection*. Princeton, N.J.: Princeton University Press.

————. 1966b. Natural selection, the costs of reproduction and a refinement of Lack's principle. *Am. Nat.* 100:687–690.

————, ed. 1971. *Group selection*. Chicago: Aldine-Atherton.

————. 1975. *Sex and evolution*. Princeton, N.J.: Princeton University Press.

————. 1979. The question of adaptive sex ratio in outcrossed vertebrates. *Proc. R. Soc. Lond. B* 205:567–580.

————. 1988. Retrospect on sex and kindred topics. In *The evolution of sex: An examination of current ideas*, ed. R. E. Michod and B. R. Levin, 287–298. Sunderland, Mass.: Sinauer.

————. 1989. A sociobiological expansion of *Evolution and Ethics*. In *Evolution and ethics: T. H. Huxley's* Evolution and Ethics, *with new essays on its Victorian and sociobiological context*, ed. J. G. Paradis and G. C. Williams, 179–214. Princeton, N.J.: Princeton University Press

————. 1992. *Natural selection: Domains, levels, and challenges*. Oxford: Oxford University Press.

Willis, E. O. 1972. *The behavior of spotted antbirds*. Ornithological Monograph no. 10. Washington, D.C.: American Ornithologists' Union.

Wilson, D. S. 1997a. Incorporating group selection into the adaptationist program: A case study involving human decision making. In *Evolutionary social psychology*, ed. J. A. Simpson and D. T. Kenrick, 345–386. Mahwah, N.J.: Erlbaum Press.

————. 1997b. Altruism and organism: Disentangling the themes of multilevel selection theory. *Am. Nat.* 150:S122–S134.

Wilson, D. S. 1980. *The natural selection of populations and communities*. Menlo Park, Calif.: Benjamin/Cummings.

Wilson, D. S., and E. Sober. 1994. Reintroducing group selection to the human behavioral sciences. *Behav. Brain Sci.* 17:585–608.

————. 1989. Reviving the superorganism. *J. Theor. Biol.* 136:337–356.

Wilson, E. O. 1971. *The insect societies.* Cambridge, Mass.: Harvard University Press.

————. 1975. *Sociobiology: The new synthesis.* Cambridge, Mass.: Belknap Press of Harvard University Press.

————. 1980. Caste and division of labor in leaf-cutter ants (Hymenoptera: Formicidae: *Atta*) 1. The overall pattern in *A. sexdens. Behav. Ecol. Sociobiol.* 7:143–156.

Wimsatt, W. 1980. Reductionistic research strategies and their biases in the units of selection controversy. In *Scientific discovery.* Vol. 2: *Case Studies,* ed. T. Nichols, 213–259. Dordrecht: Reidel.

Windsor, D. M., D. W. Morrison, M. A. Estribi, and B. de Leon. 1989. Phenology of fruit and leaf production by "strangler" figs on Barro Colorado Island, Panama. *Experientia* 45:647–653.

Winkler, D. 1987. A general model for parental care. *Am. Nat.* 130:526–543.

Wirtz, P. 1997. Sperm selection by females. *Trends Ecol. Evol.* 12:172–173.

Woese, C. R. 1967. *The genetic code.* New York: Harper and Row.

Wong, J. T.-F. 1991. Origin of genetically encoded protein synthesis: A model based on selection for RNA peptidation. *Orig. Life Evol. Biosphere* 21:165–176.

Wood, R. J., and M. E. Newton. 1991. Sex-ratio distortion caused by meiotic drive in mosquitoes. *Am. Nat.* 137:379–391.

Woyciechowski, M., and A. Łomnicki. 1987. Multiple mating of queens and the sterility of workers among eusocial Hymenoptera. *J. Theor. Biol.* 128:317–327.

Wright, J., and I. Cuthill. 1989. Manipulation of sex differences in parental care. *Behav. Ecol. Sociobiol.* 25:171–181.

Wright, S. 1931. Evolution in Mendelian populations. *Genetics* 16:97–159.

Wu, C.-I. 1983a. Virility deficiency and the sex ratio trait in *Drosophila pseudoobscura.* 1. Sperm displacement and sexual selection. *Genetics* 105:651–662.

————. 1983b. Virility deficiency and the sex ratio trait in *Drosophila pseudoobscura.* 2. Multiple mating and overall virility selection. *Genetics* 105:663–679.

Wu, C.-I., T. W. Lyttle, M.-L. Wu, and G.-F. Lin. 1988. Association between a satellite DNA sequence and the responder of segregation distorter in *Drosophila melanogaster. Cell* 54:179–189.

Wu, C.-I., R. True, and N. Johnstone. 1989. Fitness reduction associated with a satellite DNA sequence and the responder of *Segregation distorter* in *Drosophila melanogaster. Nature (Lond.)* 341:248–251.

Wynne-Edwards, V. C. 1962. *Animal dispersion in relation to social behaviour.* Edinburgh: Oliver and Boyd.

————. 1964. A reply to Maynard Smith. *Nature (London),* 201:1147.

————. 1993. A rationale for group selection. *J. Theor. Biol.* 162:1–22.

Yamamura, N. 1993. Vertical transmission and the evolution of mutualism from parasitism *Theor. Pop. Biol.* 44:95–109.

Yamamura, N., and M. Higashi. 1992. An evolutionary theory of conflict resolution between relatives: Altruism, manipulation, compromise. *Evolution* 46:1236–1239.

Yoder, J. A., C. P. Walsh, and T. H. Bestor. 1997. Cytosine methylation and the ecolgy of intragenomic parasites. *Trends Genet.* 13:335–340.

Zahavi, A. 1975. Mate selection—a selection for a handicap. *J. Theor. Biol.* 53:205–214.

————. 1977a. Reliability in communication systems and the evolution of altruism. In *Evolutionary ecology*, ed. B. Stonehouse and C. Perrins, 253–259. Baltimore, Md.: University Park Press.

————. 1977b. The cost of honesty (further remarks on the handicap principle). *J. Theor. Biol.* 67:146–151.

Zeyl, C., and G. Bell. 1996. Symbiotic DNA in eukaryotic genomes. *Trends. Ecol. Evol.* 11:10–15.

Zielinski, W. S., and L. E. Orgel. 1987. Autocatalytic synthesis of a tetranucleotide analogue. *Nature (Lond.)* 327:346–347.

Author Index

Please note that authors embedded within "et al." text cites are also indexed.

Subject Index